# SILFS

Volume 3

# New Directions in Logic and the Philosophy of Science

Volume 1
New Essays in Logic and Philosophy of Science
Marcello D'Agostino, Guiio Giorello, Federico Laudisa, Telmo Pievani and Corrado Sinigaglia, eds.

Volume 2
Open Problems in Philosophy of Sciences
Pierluigi graziani, Luca Guzzardi and Massimo Sangoi, eds.

Volume 3
New Directions in Logic and the Philosophy of Science
Laura Felline, Antonio Ledda, Francesco Paoli and Emanuele Rossanese, eds.
Daniel Zingaro

**SILFS Series Editor**
Marcello D'Agostino						marcello.dagostino@lunimi.it

# New Directions in Logic and the Philosophy of Science

Edited by

Laura Felline

Antonio Ledda

Francesco Paoli

Emanuele Rossanese

© Individual author and College Publications 2016. All rights reserved.

ISBN 978-1-84890-160-5

College Publications
Scientific Director: Dov Gabbay
Managing Director: Jane Spurr
Department of Computer Science

http://www.collegepublications.co.uk

Original cover design by Laraine Welch
Printed by Lightning Source, Milton Keynes, UK

All rights reserved. No part of this publication may be reproduced, stored in a retrieval system or transmitted in any form, or by any means, electronic, mechanical, photocopying, recording or otherwise without prior permission, in writing, from the publisher.

# Table of contents

| | |
|---|---|
| Editors' preface | ix |
| List of contributors | xi |

**PART I  Epistemology and general philosophy of science**

| | |
|---|---|
| MARIO ALAI<br>Stars and Minds<br>*Empirical* Realism and *Metaphysical* Antirealism in Liberalized Neopositivism | 3 |
| KARIM BSCHIR<br>Realism, Empiricism, and Ontological Relativity: A Happy Ménage à Trois? | 17 |
| ALBERTO CORDERO<br>Content Reduction for Robust Realism | 31 |
| LUIGI SCORZATO<br>A Simple Model of Scientific Progress | 45 |
| BENJAMIN BEWERSDORF<br>Total Evidence, Uncertainty and A Priori Beliefs | 57 |
| NEVIA DOLCINI<br>The Pragmatics of Self-Deception | 67 |
| MARCO FENICI<br>Succeeding in the False Belief Test: Why Does Experience Matter? | 77 |
| ANDREAS BARTELS<br>How to Bite the Bullet of Quidditism – Why Bird's Argument against Categoricalism in Physics fails | 87 |
| MARCO GIUNTI<br>A Real World Semantics for Deterministic Dynamical Systems with Finitely Many Components | 97 |
| SIMONE PINNA<br>An Embodied-Extended Approach to the Acquisition of Numerical Skills | 111 |

**PART II  Logic and Philosophy of Logic**

| | |
|---|---|
| MASSIMILIANO CARRARA AND CIRO DE FLORIO<br>On an Account of Logicality | 127 |

CLAUDIA CASADIO AND MEHRNOOSH SADRZADEH
Cyclic Properties: from Linear Logic to Pregroups — 139

GUSTAVO CEVOLANI
Another Way Out of the Preface Paradox? — 155

ROBERTO CIUNI AND MASSIMILIANO CARRARA
Characterizing Logical Consequence in Paraconsistent Weak Kleene — 165

ALESSANDRO GIORDANI
Logic of Implicit and Explicit Justifiers — 177

SARA NEGRI AND GIORGIO SBARDOLINI
A System of Proof for Lewis Counterfactual — 189

PAOLO PISTONE
On the "No Deadlock Criterion": from Herbrand's Theorem to *Geometry of Interaction* — 205

JAN VON PLATO
Wittgenstein's Struggles With the Quantifiers — 219

## PART III  Philosophy of natural sciences

MASSIMILIANO BADINO
Typicality in Statistical Mechanics: An Epistemological Approach — 233

MARTA BERTOLASO
Disentangling Context Dependencies in Biological Sciences — 245

LAURA FELLINE
Mechanistic Causality and the Bottoming-Out Problematic — 257

MARIO HUBERT
Quantity of Matter or Intrinsic Property: Why Mass Cannot Be Both — 267

ROBERTO LALLI
"Geometry as a Branch of Physics": Philosophy at Work in Howard P. Robertson's Contributions to Relativity Theories — 279

J. BRIAN PITTS
Historical and Philosophical Insights about General Relativity and Space-time from Particle Physics — 291

DAVIDE ROMANO
Bohmian Classical Limit in Bounded Regions — 303

EMANUELE ROSSANESE
Structural Realism and Algebraic Quantum Field Theory — 319

FRANCO STROCCHI
Symmetries, Symmetry Breaking, Gauge Symmetries — 331

ANTONIO VASSALLO
A Metaphysical Reflection on the Notion of Background in
Modern Spacetime Physics                                              349

# Editors' preface

The Italian Society for Logic and the Philosophy of Science (SILFS) was founded in 1952 with the aim of promoting and encouraging research in logic and philosophy of science in Italy. To this aim, it awards study grants, supports relevant publications, organizes conferences, and collaborates with national and international bodies on projects of common interest.

SILFS is the Italian representative of the Division of Logic, Methodology and Philosophy of Science (DLMPS) in the Union of History and Philosophy of Science, (IUHPS) – affiliated to the International Council for Science (ICSU – formerly the International Council of Scientific Unions).

In 2014, from June 18 to June 20, the Society held in Rome its Triennial International Conference, SILFS 14, at the Department of Philosophy of the University of Rome "Roma TRE".

The conference included three plenary sessions held by John Norton (University of Pittsburgh), Hannes Leitgeb (Ludwig-Maximilians-Universität München) and Tarja Knuuttila (University of Helsinki) and a special lecture held by Stefano Giaimo, the winner of the SILFS prize for the best PhD thesis, awarded for a dissertation entitled 'The Pleiotropy Theory of Ageing: Conceptual, Methodological and Empirical Issues'.

The conference was divided into several sessions, each centred on one of the main current topics in logic and philosophy of science, with a special focus on interdisciplinary approaches to logical and epistemological issues in the foundations of special sciences (both natural, social and human). The topics included:

- Causation
- Epistemology
- General philosophy of science
- Historical considerations in mathematics and philosophy
- Logic
- Logic and philosophical paradoxes
- Mathematics and computation
- Metaphysics and science
- Philosophical and historical issues in logic
- Philosophy of biology
- Philosophy of mathematics and truth

- Philosophy of perception
- Philosophy of physics
- Philosophy of the social sciences
- Scientific Models and Realism
- Scientific Realism and Antirealism
- Quantum logic and computation
- Topics in the philosophy of science

The 100 contributed papers presented at the conference underwent a further selection, resulting into the 28 double-blind peer reviewed paper that are here published. The 28 selected papers are here organized into three macro-areas: *Epistemology and General Philosophy of Science, Logic and Philosophy of Logic, Philosophy of Natural Sciences*.

Part I of the volume is devoted to the Epistemology and General Philosophy of Science. The first four papers (written by Mario Alai, Karim Bshir, Alberto Cordero and Luigi Scorzato) discuss topics of scientific realism. The next three papers (whose authors are Benjamin Bewersdorf, Nevia Dolcini e Marco Fenici) are devoted to specific debates in philosophy of knowledge. Andreas Bartels analyses and objects to Alexander Bird's a priori argument against Categoricalism with respect to fundamental physics properties. Marco Giunti proposes a real world semantics for Deterministic Dynamical Systems with Finitely Many Components. Finally, Simone Pinna illustrates and discusses the virtues of an embodied-extended approach to the acquisition of numerical skills.

Part II of the volume centres on Logic and the Philosophy of Logic. Three papers (by Casadio and Sadrzadeh, by Negri and Sbardolini, and by Pistone) share a marked proof-theoretic flavour, although the former two papers also discuss, from different perspectives, issues in formal linguistics and the semantics of natural language. The problem of logicality, one of the key issues in the philosophy of logic, is at the heart of Carrara and De Florio's contribution, while the paper by Ciuni and Carrara deals with the 3-valued Kleene logic with weak tables and two designated values. Cevolani's and Giordani's papers tackle several questions of interest to the logician, to the epistemologist and to the philosopher of science alike, while Wittgenstein's troubled relationship with logical quantifiers is described by von Plato.

Part III of the volume collects ten papers that are devoted to the Philosophy of Natural Sciences. Massimiliano Badino investigates the notion of tipicality in the context of statistical mechanics. Marta Bertolaso proposes an analysis of the status of context dependencies in biological sciences. James Brian Pitts and Roberto Lalli discuss some historical and philosophical issues concerning Relativity Theory, by focusing respectively on the connection between General Relativity and particle physics, and on the work of Howard P. Robertson and the role of geometry as a branch of physics. Laura Felline analyses the bottoming-out problem for Stuart Glennan's mechanistic theory of causality. Mario Hubert proposes an analysis of the notion of mass in Newtonian Mechanics. Davide Romano discusses the classical limit in the context of Bohmian Mechanics applied to bounded regions. Emanuele Rossanese proposes a structuralist interpretation of Algebraic Quantum Field Theory and discusses some possible objections to this interpretation. Franco Strocchi analyses the role of symmetries and gauge symmetries in contemporary physics. Antonio Vassallo offers some philosophical considerations on the notion of background in modern space-time physics.

We would like to conclude this introduction by acknowledging our gratitude to all those who helped making this conference possible, including the speakers and all the people involved, on different levels, in its organization. In particular, we would like to to thank:

- The members of the Scientific Committee: Roberto Arpaia (University of Bergamo), Giovanni Boniolo (University of Milan and IFOM) Chair of the Programm Committee, Giovanna Corsi (University of Bologna) Chair of the Program Committee, Massimiliano Carrara (University of Padua), Mauro Ceruti (University of Bergamo), Mauro Dorato (University of Roma Tre) - SILFS President, Vincenzo Fano (University of Urbino), Laura Felline (University of Roma Tre), Roberto Giuntini (University of Cagliari), Federico Laudisa (University of Milan-Bicocca), Sabina Leonelli (University of Exeter), Massimo Marraffa (University of Roma Tre), Pierluigi Minari (University of Florence), Matteo Morganti (University of Roma Tre), Francesco Paoli (University of Cagliari), Federica Russo (University of Ferrara);

- The Organizing Commitee: Massimiliano Carrara (University of Padua), Angelo Cei (University of Roma Tre), Mauro Dorato (University of Roma Tre), Pierluigi Graziani (University of Urbino), Matteo Morganti (University of Roma Tre), Emanuele Rossanese (University of Rome Tre);

- The group of volunteering graduate students, whose role was essential in all the practicalities: Mariaflavia Castelli, Matteo Grasso and Mattia Sorgon.

- Finally, we would like to thank all the anonymous referees who cooperated in the peer review process.

Laura Felline, Antonio Ledda, Francesco Paoli, Emanuele Rossanese
March 9, 2016

# List of contributors

MARIO ALAI, Università Carlo Bo – Urbino
MASSIMILIANO BADINO, University of Barcelona
ANDREAS BARTELS, University of Bonn
MARTA BERTOLASO, University Campus Bio-Medico of Rome
BENJAMIN BEWERSDORF, University of Groningen
KARIM BSCHIR, ETH Zurich
MASSIMILIANO CARRARA, Università di Padova
CLAUDIA CASADIO, Università di Chieti-Pescara
GUSTAVO CEVOLANI, Università di Torino
ROBERTO CIUNI, University of Amsterdam
ALBERTO CORDERO, The City University of New York
CIRO DE FLORIO, Università Cattolica del Sacro Cuore
NEVIA DOLCINI, University of Macau
LAURA FELLINE, Università Roma Tre
MARCO FENICI, Università di Siena
ALESSANDRO GIORDANI, Università Cattolica del Sacro Cuore
MARCO GIUNTI, Università di Cagliari
MARIO HUBERT, University of Lausanne
ROBERTO LALLI, Max Planck Institute for the History of Science
SARA NEGRI, University of Helsinki
SIMONE PINNA, University of Cagliari
JAMES BRIAN PITTS, University of Cambridge
PAOLO PISTONE, University of Marseille
DAVIDE ROMANO, University of Lausanne
EMANUELE ROSSANESE, Università Roma Tre
MEHRNOOSH SADRZADEH, Queen Mary University of London
GIORGIO SBARDOLINI, Università di Milan
LUIGI SCORZATO, INFN
FRANCO STROCCHI, Università di Pisa
ANTONIO VASSALLO, University of Lausanne
JAN VON PLATO, University of Helsinki

# PART I

# EPISTEMOLOGY AND GENERAL PHILOSOPHY OF SCIENCE

# Stars and Minds. Empirical Realism and Metaphysical Antirealism in Liberalized Neopositivism

Mario Alai

ABSTRACT. In 1936, after the "liberalization of empiricism", Schlick and Carnap thought they could accept C.I. Lewis' claim "If all minds disappeared from the universe, the stars would still go on in their courses" as a scientific truth, without accepting the metaphysical thesis of the mind-independence of the external world. But this compromise failed, as it was based on mistaken antirealist semantic views: they accepted Lewis' *sentence*, not the *proposition* it expresses, or its consequences. This is not to say that science is enough to support metaphysical realism, since it takes philosophy to show where they went wrong.

Originally the neopositivists thought that the whole cognitive meaning, or content, of a statement consisted just in the set of possible experiences which would verify it. Meaning was thus the method of verification: any empirical proposition was only about the regular connection of certain experiences ([7], p. xi; Section 179; [8], Section 7; [29], 100, 107, 111; etc.).

Therefore unverifiable sentences[1] were meaningless. Among them were all the philosophical sentences, especially metaphysical sentences. In particular, they rejected as meaningless both metaphysical realism

(MR) the external world is mind-independent

and idealism

(I) the external world is mind-dependent ([25], 107).

However, they granted that some traditional philosophical claims could be reinterpreted so to become verifiable, hence meaningful ([32], 6). For instance, two geographers disagreeing on the real existence of a mountain in an unexplored area of Africa could solve the problem by travelling there and observing whether there was a mountain, measuring it, etc.: this would settle the *empirical* question of reality. But if they further disagreed whether the mountain was *metaphysically* real, in

---
[1] More or less like [30], here I use 'sentence' for an uninterpreted (possibly meaningless) expression, and 'proposition' for an interpreted expression.

the sense of being something over and above their perceptions, this question could not possibly be solved by any experience, so it would be cognitively meaningless ([8], Section 10).

They could even attach an empirical meaning to the claim that the external world is independent of the mind in the sense that, for instance,

($\alpha$) a castle in the park existed hundreds of years ago (well before my existence), as it can be recognized from the time-worn state of its walls;

($\beta$) it must have existed even last night, when nobody observed it, since experience tells us that it couldn't have been built in a few hours this morning. ([29], 103);

($\gamma$) The back of an apple is there even when nobody observes it ([7], Section 135).[2]

This was possible because according to them "the meaning of every proposition is exhaustively determined by its verification in the given" ([29], 110), so that "the claim that a thing is real is a statement about lawful connections of experiences" (ibid., 100), and "propositions about bodies can be transformed into propositions of like meaning about the regularity of occurrence of sensations (ibid., 111).[3]

Therefore ($\alpha$), ($\beta$), and ($\gamma$) said nothing about objective states of things, but only about various patterns of connections among past perceptions and perceptions one could inductively forecast for the future. In particular, they did not convey the realist and commonsense idea of the independence of physical objects from actual or possible perceptions. Hence, this "empirical" reformulation of the mind-independence of the external world was neither a form of realism nor a philosophical doctrine: Moritz Schlick explicitly said that he accepted it in the same sense of Berkeley and Kant ([29], 98-99).

Progressively, however, a closer attention to the factual procedures of science modified the neopositivists' stand. They realized that the empirical confirmation does not concern a single proposition, but a system of propositions ([9], Section 3). Besides, Neurath [23] [24] argued against Schlick and Carnap that the actual language of science is physicalist, rather than phenomenalist. Carnap replied that the choice between phenomenalism and physicalism was not a substantial question, but a pragmatic decision about language, and both languages could be used as a basis for the unification of science [10]. Eventually, however, he granted that the physicalist language was preferable, and most importantly, he acknowledged that commonsense and scientific statements about physical objects cannot be exhaustively translated into statements about perceptions: a physicalist proposition, like

(i) On May 6, 1935, at 4 P.M., there is a round black table in my room

should be translated by an infinite conjunction of propositions of the form:

---

[2] Just like the other face of the moon for Kant: [29], 88.
[3] See also ibid., 98,102, etc.

(ii) If on May 6, 1935, at 4 P.M., somebody is in my room and looks in such and such direction, he has a visual perception of such and such a kind.

But first, we cannot produce or understand infinite conjunctions; and second, (i) would not be equivalent to such a conjunction, anyway: for even if (i) were false (ii) would be (trivially) true if nobody were in my room on May 6, 1935, at 4 P.M. (since any material conditional with a false antecedent is true).[4]

Moreover, it became clear that we cannot completely define the non observational concepts of scientific theories by observational ones; we can only "reduce" them, i.e. show which difference they can make with respect to possible observations, which however do not exhaust their original content ([11] 52-53, [12] Section 9; [21], chs. I-III).

From all of this there followed that no proposition of science or commonsense could ever be verified (i.e., made definitely certain). Thus, during the first half of the Thirties, the neopositivists proceeded to a "liberalization of empiricism", by substituting

(I) the requirement of verifiability with that of confirmability (i.e. the possibility of raising or lowering the probability of a proposition by testing its empirical consequences);

(II) complete with incomplete definability of theoretical terms on the basis of observation terms ([11], Sections 11 ff.);

(III) the phenomenalist language with a physicalist one.

This is why in a number of papers[5] Gino Tarozzi has made two claims:

(1) *contra* the original claims of the Vienna Circle, there are genuinely *philosophical* doctrines supported by experience, hence non metaphysical;

(2) with the liberalization of empiricism the neopositivists came to accept some of them.

But I shall argue that (2) is not quite the case.

---

[4][11] 68-69; 80, passim. The latter problem arises because Carnap (who only 11 years later was to develop his intensional logic) interpreted (ii) as a material conditional; hence, this problem would disappear if (ii) were interpreted as an implication (i.e., the intensional conditional of ordinary language), for then it would be false whenever (i) is false, even if nobody is in the room. But unlike material conditionals, implications and counterfactuals require objective truth-makers: in the case at hand, for example, only the objective existence of the table could entitle us to claim that if, counterfactually, somebody were in the room, she would have such and such perceptions. When so understood, therefore, (ii) can be made true only by the existence of an objective table; hence, statements about physical objects are actually not dispensable in favour of statements about perceptions.

[5]See references in [5].

At first sight their departure from verificationism may seem radical: Schlick, for instance, accepted as meaningful assertions about:

(a) future events ([30], 345),

(b) the existence of a 10 dimensions universe (ibi, 355)

(c) the existence of perceptions altogether different from those of humans (ibid.);[6]

(d) the survival after death of myself or of others (ibi, 357).

But they never admitted any change of mind with respect to metaphysics, and even after 1936 they were adamant in rejecting metaphysical realism and idealism ([13], Section 4A, and [14], xi; [30], p. 368). So, we must assume that claims about (a)-(d) were accepted by them only in a non-philosophical interpretation, similar to that they gave to "realist" claims like ($\alpha$), ($\beta$), and ($\gamma$) in the Twenties.

Tarozzi's favourite example is this: in [22] C.I. Lewis criticized verificationism, arguing that it excluded even scientifically sound claims like

(1) If all minds disappeared from the universe, the stars would still go on in their courses.

But Carnap [11] (pp.87-88) and Schlick [30] 368 replied that (1) was empirically meaningful by their criteria. Yet, (1) implies

(2) stars are something over and above all possible perceptions,

and

(MR) the external world is mind-independent,

both of which they had earlier rejected as meaningless. So, how could they accept (1) without contradictions? Schlick claimed that the mistake of metaphysicians was rejecting the empirical interpretation of (1) and look for some *further* mysterious sense for it ([30], 368), but in *which* sense (1) would not entail (2) and (MR)?

To begin with, let's see how Schlick explains his acceptance of (1). First, he reformulates it as

(1') If all *living beings* disappeared from the universe, the stars would still go on in their courses,

which he takes to be equivalent; subsequently, he justifies it in a somewhat odd way:

---

[6]While in [29] (pp.93-95) he had claimed that a similar hypothesis, that of inverted colour spectra, was meaningless.

> The laws of motion of the celestial bodies are formulated entirely without reference to any human bodies, and this is the reason why we are justified in maintaining that they will go on in their courses after mankind has vanished from the Earth (ibid.).

But in speaking of living beings, and more precisely of human bodies, he is obviously changing subject: the question was whether stars are independent of minds, not of bodies.

In fact, a few pages earlier he had argued that (a) what we call 'my body' is constituted by (actual and possible: 345) sense data (360); and (b) there is nothing in sense data which qualifies them as belonging to me, to an "I", or a to a "mind": the only meaningful content of saying that my sense data are perceived by me, or by my mind, is that all the data about the "external world" have a special relation to the data which constitute "my body" (e.g., whenever I shut my eyes all visual data disappear, etc.). But this, he says, is a purely empirical fact: it is logically possible that this relation is broken: for instance, it might be possible to feel sensations of "other human bodies"; or to go on having data about the world, without having any more those data which constitute "my body". So, unlike Carnap, he is still holding a phenomenalist semantics; hence, what he means by (1') is actually that

(1") If all *sense data* about *human bodies* disappeared, *sense data about the stars* [not stars themselves] would still go on as usual.

He then adds:

> Experience shows no connention between the two kinds of events. We observe that the course of the stars is no more changed by the death of human beings than, say, by the eruption of a volcano, or by a change of government in China. Why should we suppose that there would be any difference if all living beings on our planet, or indeed everywhere in the universe, were extinguished?" (358).

Now, literally understood this would be mean that *the very laws of science* imply that

(2) stars are something over and above all possible perceptions,

and

(MR) the external world is mind-independent.

But in Schlick's phenomenalist semantics these actually mean, respectively,

(2') *sense data about the stars* are something over and above all possible perceptions,

and

(MR') sense data about the external world are independent of sense data about human bodies.

This explains why Schlick could consistently reject metaphysical realism while holding the *sentences* (not propositions) (1), (2), and (MR): because he interpreted them respectively as (1"), (2'), and (MR'), which obviously are not realist doctrines. But it must be noticed that the propositions (1"), (2') and (MR') actually held by Schlick are (a) false, (b) not what science tells us, and of course, (c) not what Lewis had in mind.

(1") is false because when we will no longer have any sense data about our bodies (i.e., when mankind will be extinguished) we will no longer have any sense data about the stars, either.[7]

Hence (MR') is also false, because when sense data about our bodies will cease, also sense data about the external world in general will. And of course, (2') is not just false but inconsistent, as it claims that sense data about the stars are something over and above all possible sense data.

Moreover, what science teaches, and what Lewis meant by (1) and (2), is not (1") and (2'), but rather that *stars themselves* will exist when no sense data will exist anymore, hence *stars* (and the external world in general) are independent of sense data.

So, *pace* Schlick, it is not the metaphysical realist which gives (1) some further mysterious sense, but Schlick himself: the realist understands (1) in the same sense of science and common sense, while Schlick reinterprets it phenomenistically (even if no longer verificationistically). So, he did not actually meet Lewis' challenge, and got science's deliverances wrong.

But, Schlick might object, how is it possible that science supports a metaphysical doctrine? Must it not stick to empirical data, merely describing possible experiences, without venturing to claim anything beyond them? in particular, should not science refrain from claiming either that physical bodies are something over and above sets of data (realism) or that they are nothing more than them (idealism)?

Before answering these questions, however, let's examine Carnap's argument for (1), for it is also somewhat puzzling:[8] he explains that for some **i** and some **j** it follows from our empirically confirmed astronomical laws $L_1 \ldots L_n$ that

(3) $Si_{now} \supset Sj_{1my}$, ( = if the stars are in state **i** now, then they will be in state **j** in 1 million years).

Moreover, we observe that

(4) $Si_{now}$ ( = the stars are in state **i** now).

---

[7] Perhaps there might be sense data of the stars perceived by other animals; but this is doubtful, and it is even more doubtful that our language might be given a semantics in terms of sense data of other animals.

[8] I am only slightly adapting the wording of his exposition in ([11], 86-88).

Therefore, *by modus ponens*, it follows from scientific laws and observation that

(5) $Sj_{1my}$ ( = the stars will be in state **j** in 1 million years).

But (5) implies

(6) $\sim M_{1my} \supset Sj_{1my}$ ( = if in 1 million years there will be no minds, the stars will be in state **j**)

(because any conditional with a true consequent is true). But (6) is equivalent to (1), so (1) can be established just by observation, induction and propositional logic.

It might be objected that (6) is neither equivalent to (1), nor what Lewis meant by it: for (6) is a logical consequence of (5), so it says *nothing* more than (5). But (1) means *something* more than (5): not only that there will be stars in one million years, but besides, that they *would* be there *even if* there were no minds at that time: namely

(7) $Sj_{1my}$ & $\Diamond(\sim M_{1my}$ & $Sj_{1my})$
( = the stars will be in state **j** in 1 million years, and it *might* be so even if there were no minds then),

or

(8) $Sj_{1my}$ & $\sim(\sim M_{1my} \rightarrow \sim Sj_{1my})$
( = the stars will be in state **j** in 1 million years, and the absence of minds *would not* prevent this).

Perhaps, the objection is, Carnap missed the modal or implicative character of (1) because in 1936 he hadn't developed his intensional semantics, yet; so, he did not meet Lewis' challenge to show that scientific statements like (1) passed the test of liberalized verificationism. In any case, because of the modal or implicative character of (7) and (8), he could not accept them (nor their consequences (2) and (MR)), without committing himself to a metaphysical ontology, either of mind-independent bodies, or necessitarian laws, or physically possible worlds.

But Carnap might reply that if the laws of astronomy (3) and observation (4) *entail* that

(5) the stars will be in state j in 1 million years,

then they teach that this will happen *in any case*, hence, *even if there were no minds* in 1 million years (precisely as the formalization brings out: if we know that (5) $Sj_{1my}$, we also know that (6) $\sim M_{1my} \supset Sj_{1my}$. After all, this is also what Schlick argues in the above quotation. And this entails that

(1) If all minds disappeared from the universe, the stars would still go on in their

courses.

But if so, since (1) entails (2) and (MR), how would Carnap avoid being committed to metaphysical realism? In fact, how can metaphysical claims like (2) and (MR) follow just from observation and scientific laws? As we just asked on behalf of Schlick, does really science tell us that *stars themselves* will be there in 1 million years, or does it simply make predictions about possible observations on sense data?

For a phenomenalist like Schlick, science does not quite show that

(3) $Si_{now} \supset Sj_{1my}$ ( = if the stars are in state **i** now, then they will be in state **j** in 1 million years).

In fact, the only law-like correlations we have actually observed are that perceptions of kind K at a time **t**, were followed by perceptions of kind J at a later time **t'**; moreover, science tells us that there might not be perceptions at all in 1 million years. Therefore, observation and induction cannot really warrant (3), but only

(3') $[\exists P(Si_{now}) \& P_{1my}] \supset \exists P(Sj_{1my})$
( = if there are perceptions of the stars as in state **i** now, and *if* there will be perceptions at all in 1 million years, then there will be perceptions of the stars as in state **j** in 1 million years).

Equally, observation does not quite show that

(4) $Si_{now}$ ( = the stars are in state **i** now),

but only that

(4') $\exists P(Si_{now})$ ( = there are perceptions of the stars as in state **i** now)

But while from (3) and (4) there follows

(5) $Sj_{1my}$ (stars will be in state **j** in 1 million years),

from (3') and (4') there follows only

(5') $\exists P_{1my} \supset \exists P(Sj_{1my})$ ( = if there will be perceptions in 1 million years, then there will be perceptions of the stars as in state **j** in 1 million years),

and obviously from (5') one cannot derive

(6) $\sim M_{1my} \supset Sj_{1my}$ ( = if in 1 million years there will be no minds, the stars will be in state **j**),

nor a fortiori

(1) If all minds disappeared from the universe, the stars would still go on in their courses.

This is why Carnap's (5) and (6) did not seem to render what Lewis meant by (1): because on the one hand Lewis was clearly concerned with the question of the mind-independence of stars, *quite independently of* (5), as brought out by (7) or (8). On the other hand, (5) is the crucial step in Carnap's argument, and (6) is his rendering of (1); but one feels that while science can indeed establish (5) and (6), it does not by itself establish their semantics: it is not its business to decide whether they are to be interpreted physicalistically, so to entail the mind-independence of stars, or phenomenalistically, so to be compatible with antirealism.

But while science cannot decide this question, philosophy can, and we already saw Carnap's argument against phenomenalism. Further arguments are given by Sellars [31], Austin [6] and Quine [27] and [28] sense data as such cannot be perceived, conceptualized or remembered, only physical objects can. So, a phenomenalist language is just impossible, we could never learn it.

Thus, phenomenalism is the mistake which explains Schlick's illusion to be able to accept (1) and (2) while rejecting metaphysical realism. But how about Carnap? In 1936 he was already a physicalist, so how could he avoid metaphysical realism? He felt he could because he regarded the choice of physicalism precisely as a *choice of language* (e.g., [11], pp. 69-70; 78-80), which did not settle the *metaphysical* question of whether there really are physical entities beyond sense data, or not. So, in a sense, he refused to interpret his own language, thus feeling entitled to (a) accept (1) and (2); (b) acknowledge that (1) and (2) are established by science; yet, (c) deny that science could support a metaphysical doctrine and (d) reject any commitment to metaphysical realism.

In this way, however, he missed two important points: (I) Quine's point that quantification involves ontological commitment: by quantifying over mind-independent entities I am claiming that they exist; (II) an epistemic point: if the best way to describe, predict and explain experience is by quantifying over mind-independent entities, that is evidence that they *exist*. Physicalism is not just a vocabulary, it is a theory. Phenomenalism is not only impossible as an interpretition of language, but also arguably wrong as a metaphysical doctrine, because it fails where realism succeeds: in explaining not any particular empirical phenomenon, but (i) the determinacy (ii) the order and (iii) the regularity of phenomena in general [4].

Of course the realist arguments (I) and (II) are not strictly scientific, but philosophical, and this is why scientific evidence supports metaphysical realism, but *in and of itself* it is not enough to establish it. Science *as such* is not concerned with the question of its own semantic and metaphysical interpretation. So, it does speak of physical entities and claim that they exist, but it does not advance the claim that they are metaphysically real. It is a task for philosophy to show that the most correct interpretation of science is the realist one.

*Perhaps* there is also another reason why Carnap could accept (1) and (2) without committing himself to metaphysical realism: perhaps even after giving up phenomenalism and the possibility of verification in a strict sense, he still held a *confirmationist* semantics:[9] i.e., he thought that the content of an assertion is just the experiences which confirm it to some extent (no matter whether these experiences concern sense data or physical bodies); in other words, that meanings are not constituted by truth-conditions, but by confirmation-conditions.

If so, he accepted the astronomical laws $L_1 \ldots L_n$, the *sentence* (3) following from them, plus the observation *sentence* (4), hence their joint consequences (5), (6), (1) and (2). But for him their respective content was just *the set of all actual and possible experiences* (about physical bodies) *which would confirm them*: they meant only that in past conditions $C_1, C_2, \ldots C_m$ we had respectively the confirming experiences $E_1, E_2, \ldots E_m$, and in possible conditions $C_{m+1}, C_{m+2}, \ldots C_n$, we *would* have, respectively, the confirming experiences $E_{m+1}, E_{m+2}, \ldots E_n$.

Obviously, we could not have *any* experiences relative to a condition $C_{\sim M}$ in which there existed no minds (let's call these impossible experiences of a world without minds $E_{\sim M}$). So, the obtaining of the astronomical laws $L_1 \ldots L_n$ could not be observed in condition $C_{\sim M}$; hence, the laws $L_1 \ldots L_n$ could not be (completely) *verified*. However, they can be (partially) *confirmed* (in fact, they are *very well* confirmed): this is why in 1936 Carnap, having substituted verification with confirmation, had no problems in accepting them and their consequences (1), (2), (3), (5), etc. But if he thought that their meaning consisted only of the experiences which could confirm them, for him their content did not include any experiences $E_{\sim M}$ of a world without minds: they said nothing about what would happen in such a world, hence they said nothing on the possibility of the existence of stars and material bodies independently of minds. Therefore he could accept (1), (2), (3), (5), etc., without being committed to metaphysical realism.

But it is hard to tell whether Carnap actually relied on this confirmationist semantics, because it conflicted with his own point that theoretical terms cannot be completely defined in observation terms: for this entails that observation cannot supply the whole meaning of propositions, so their content indeed goes beyond all possible observations: confirmationism can offer a criterion of meaningfulness, but not a whole semantics.[10]

Moreover, even if Carnap actually embraced it, confirmationist semantics could not reconcile his acceptance of (1) and (2) with the rejection of metaphysical realism, for it is wrong: in science and in common discourse the astronomical laws and sentences (1), (2), (3), (4), (5), etc., are actually understood as speaking about *stars*, not *experiences*.

---

[9]Somewhat like the semantics which was later adopted by Dummett [15]; [16], pp. 590-1; [18], ch.14, Section 6; etc.

[10]On the other hand, if Carnap didn't hold a confirmationist semantics, but only (more consistently) a confirmationist *criterion* of meningfulness, he might simply have been self-deceived: by assuming that any scientific result automatically fulfilled his confirmationist criterion of meaningfulness, and that any metaphysical claims failed to do so, he might have accepted (1) and (2), assuming that they did not commit him to (MR). But if so he failed to see that (a) metaphysical claims can be empirically confirmed, and (b) one can avoid the implication from (1) and (2) to (MR) only by a confirmationist *semantics*.

This however is not shown by *science*: contrary to what claimed by Putnam [26] (pp. 105-109) *science* by itself cannot fix the interpretation of its own claims, and Dummett is right that an antirealist interpretation of the whole of science is possible without contradiction [1]. However, there are various *philosophical* moves by which it can be argued, against Dummett, that meanings are truth-conditional rather than confirmation-conditional: (a) showing that Dummett's manifestation and acquisition challenge ([17]; [19], 13) can be met, i.e., that through the compositionality of language we can learn to assign our statements objective (i.e., confirmation-transcendent) truth-conditions ([2], 368-374; [5] 40-41); (b) pointing out that the confirmationist's claim that the meaning of propositions is not what actual speakers understand by them is absurd, because meaning *is* whatever speakers understand; (c) arguing that if the confirmationist theory of meaning were true, it would be inexpressible, hence one cannot really understand what confirmationist semanticists are actually claiming ([2], 377-387).

Thus, the metaphysical dispute between realism and idealism presupposes the more basic decision between realist and confirmationist semantics ([3], 133): if one chooses a realist semantics, the dispute between metaphysical realism and idealism has sense. If one chooses a confirmationist semantics, the metaphysical question cannot even be expressed, as no objective facts can be expressed. What survives, in this case, is only a pale ghost of the original dispute: the question whether it is *assertible* that stars are mind-independent, or that they are mind-dependent, o neither. If one does not take stand on the semantic question, just like Carnap when he considers it as a merely pragmatic choice, one can indeed accept all scientific and common sense "realist" claims, without any philosophical commitment of any kind, more or less like FIne [20] with his "NOA".

Summing up, both Schlick and Carnap accepted the *sentences* (1) and (2), and in a sense also the *sentence*

(MR) the external world is mind-independent,

but Schlick reinterpreted them phenomenistically, so to yield merely propositions on sense data, while Carnap either declined to interpret them, or interpreted them confirmation-conditionally. Thus, they actually rejected them not only in their metaphysical or philosophical sense, but even the sense they have in science or common sense. The "liberalization of empiricism" was probably a smaller change than sometimes is thought: they abandoned strict verificationism for the more liberal confirmationism, and Carnap also abandoned the phenomenalist *language* for a physicalist one. This certainly helped them to offer a better account of scientific methodology; but they remained basically sceptical on the possibility of moving from subjective experience to the knowledge of reality, thus basically keeping their positivist strictures against philosophical doctrines.

Contra Tarozzi's claim (2), the empirical versions of traditional metaphysical doctrines they accepted in 1936 didn't have any philosophical content more than those they had already accepted since the Twenties: they accepted all the sentences of science, just as before; but Schlick still interpreted them phenomenalistically, while Carnap refrained from interpreting them. Science does not fix the semantics of its own sentences, while philosophy tries to do that, through arguments which are specifically philosophical, not scientific. Since Carnap didn't use or consider any of those arguments, on the question of realism he too didn't take any philosophical step beyond his initial positions.[11]

## BIBLIOGRAPHY

[1] Alai, M. (1988). L'argomento della fallacia idealistica nel vecchio e nel nuovo Putnam. In M.L. Dalla Chiara, M.C. Galavotti (eds.), *Temi e prospettive della logica e della filosofia della scienza contemporanee*, Atti del congresso della S.I.L.F.S. 1987, vol. II, Bologna, CLUEB: 93-96.

[2] Alai, M. (1989). *A Criticism of Putnam's Antirealism*, Ann Arbor, U.M.I.

[3] Alai, M. (2013). Ontologia, conoscenza e significato nel realismo scientifico, in M. Bianca, P. Piccari (eds.) *Ontologia, realtà e conoscenza*, Mimesis, Milano.

[4] Alai, M. (2014a). Realismo, idealismo e agnosticismo. Una prospettiva epistemologica, *Hermeneutica* 2014: 109-126.

[5] Alai, M. (2014b). Neopositivism, Realism, and the Status of Philosophy, in V. Fano (ed.) *Gino Tarozzi Philosopher of Physics*, Milano, Angeli, 2014: 33-64.

[6] Austin, J.L. (1962). *Sense and Sensibilia*, Oxford, Clarendon.

[7] Carnap, R. (1928a). *Der Logische Aufbau der Welt*, Berlin-Schlachtensee, Weltkreis-Verlag.

[8] Carnap, R. (1928b). *Scheinprobleme in der Philosophie. Das Fremdpsychische und der Realismusstreit*. Berlin-Schlachtensee, Weltkreis-Verlag.

[9] Carnap, R. (1931). Die physicalische Sprache als Universalsprache der Wissenschaft, *Erkenntnis* II, 5/6: 432-465.

[10] Carnap, R. (1932). Über Protokollsätze, *Erkenntnis* III, 2/3: 215-228.

[11] Carnap, R. (1936). Testability and Meaning, *Philosophy of Science*, 3, (4): 419-471. Repr. in H. Feigl, M. Brotbeck (eds.), *University of Minnesota Readings in the Philosophy of Science*, New York, Appleton-Century-Crofts, Inc.: 47-92.

[12] Carnap, R. (1963a). Intellectual Autobiography. In P.A. Schilpp (ed.) (1963), vol. I.: 3-84.

[13] Carnap, R. (1963b). Replies and Systematic Expositions. In P.A. Schilpp (ed.) (1963), vol. II.: 859-1013.

[14] Carnap, R. (1967). Preface to the second edition. In Id., *The Logical Structure of the World*, La Salle, III., Open Court.

[15] Dummett, M. (1959). Truth, in *Proceedings of the Aristotelian Society* 59 (1): 141-162.

[16] Dummett, M. (1973). *Frege: Philosophy of Language*, London, Duckworth, and Cambridge Mass., Harvard University Press.

---

[11] Carnap's consistency throughout his career in accepting empirical realism and rejecting metaphysical realism is well documented also in [25].

[17] Dummett, M. (1978). *Truth and Other Enigmas*, London, Duckworth, and Cambridge Mass., Harvard University Press.

[18] Dummett, M. (1991). *The Logical Basis of Metaphysics*, Cambridge, Mass., Harvard University Press.

[19] Dummett, M. (1993). *The Seas of Language*, Oxford, Oxford University Press.

[20] Fine, A. (1984). The Natural Ontological Attitude, in J. Leplin (ed.), *Scientific Realism*, Berkeley, University of California 1984: 83-107.

[21] Hempel, C. (1952). *Fundamentals of Concept Formation in Empirical Science*, Chicago, University of Chicago.

[22] Lewis, C.I. (1934). Experience and Meaning, *The Philosophical Review* XLIII: 125-146.

[23] Neurath, O. (1931). Soziologie im Physicalismus, *Erkenntnis*.

[24] Neurath, O. (1932). Protokollsätze, *Erkenntnis* III, 2/3 1932, 204-214.

[25] Parrini, P. (1994). With Carnap, Beyond Carnap: Metaphysics, Science, and the Realism/Instrumentalism Controversy. In W. Salmon and G. Wolters (eds.) *Logic, Language, and the Structure of Scientific Theories*, Pittsburgh and Konstanz: University of Pittsburgh Press and Universitätsverlag Konstanz: 255-277. It. ed. Con Carnap oltre Carnap. Realismo e strumentalismo tra scienza e metafisica, *Rivista di Filosofia* LXXXII, 3 (1991): 339-367.

[26] Putnam, M. (1978). *Meaning and the Moral Sciences*, Oxford, Routledge & Kegan Paul.

[27] Quine, W. V. (1957). The Scope and language of Science, *British Journal for the Philosophy of Science* 8 (29):1-17.

[28] Quine, W. V. (1960). Posits and Reality, in S. Uyeda (ed.) *Basis of the Contemporary Philosophy*, Tokyo, Waseda University, vol. 5.

[29] Schlick, M. (1932). Positivismus und Realismus, *Erkenntnis* III: 1-31 [references to Italian ed., Positivismo e realismo, in M., Schlick *Tra realismo e neo-positivismo*, Bologna, Il Mulino 1974: 77-111].

[30] Schlick, M. (1936). Meaning and Verification, *The Philosophical Review* 45 (4): 339-369 [Italian ed., Significato e verificazione. In A. Bonomi (ed.) *La struttura logica del linguaggio*, Milano, Bompiani, 1973: 71-101].

[31] , Sellars, W. (1956). Empiricism and the Philosophy of Mind, in H. Feigl, M. Scriven (eds.), *Minnesota Studies in the Philosophy of Science, Volume I: The Foundations of Science and the Concepts of Psychology and Psychoanalysis*, Minneapolis, University of Minnesota: 253-329.

[32] Verein Ernst Mach (hrsg.) (1929). *Wissenschaftliche Weltauffassung. Der Wiener Kreis*, Veröffentlichungen des Vereines Ernst Mach, Wien, Artur Wolf Verlag.

# Realism, Empiricism, and Ontological Relativity: A Happy Ménage à Trois?

Karim Bschir

ABSTRACT. In the debate on scientific realism, empiricists often take an anti-realist stance. This need not be. I argue that it is possible to merge an empiricist methodology with a realist perspective on science under the presupposition that one is ready to bite the bullet of ontological relativity. I will show that ontological relativity is not a predicament, neither for empiricism nor for realism. Quite on the contrary, it allows us to bring both together in a consistent manner.

## 1 Introduction

In the tradition of western philosophy empiricism and realism about unobservables stand as opposing views. Ever since the time of Locke and Hume, empiricists have felt a deep discomfort when it comes to the commitment to experience-transcending or abstract entities. And it is only a slight exaggeration to assert that there exists an almost sectarian chasm between strict empiricists on the one hand, who deny the existence of non-observational entities, and realists or rationalists on the other, who are convinced that there must exist something over and above the empirically given.

In the twentieth century, the old dispute between empiricists and rationalists (the "battlefield of endless controversies" as Kant called it) found its continuation in the debate on scientific realism. However, the debate on scientific realism has a sharper focus than the traditional conflict between empiricism and rationalism. The modern debate centers around the question how to justify our commitment to the existence of the numerous unobservable entities that play important explanatory roles in our well-confirmed scientific theories. Taking into account the heritage of empiricism as well as its many virtues, it is by no means difficult to understand why empiricists have often taken the side of the anti-realists in this debate. The reductive empiricism of the Vienna Circle or Bas van Fraassen's constructive empiricism are among the most important varieties of empiricist anti-realisms that we have seen in the twentieth century.

The opposition between realists and empiricists in the debate on scientific realism is, however, rather unfortunate. The reason for this is simple: Both camps make highly sensible claims! Realism holds that, in general, the unobservable entities posited in scientific theories are part of reality, and empiricism boils down to the methodological postulate that all scientific claims about nature must be justified on empirical grounds. Both of these statements appear perfectly reasonable, in particular from the viewpoint of science itself. It may be assumed that most scientists would happily endorse a philosophical framework that allows for both: a) to take a realist stance on theoretical entities and b) to subscribe to empiricism as the preferred methodology for science. Hasok Chang seems to share this assessment when he writes that "it doesn't make much sense that empiricism and realism have been pitted against each other in debates on scientific realism. Typical scientists, as well as most 'normal' people, are both empiricists and realists, and that is not (only) because they are philosophically unsophisticated" ([8], p. 217).[1]

Unfortunately, a reconciliation between realism and empiricism turns out to be a severe philosophical problem. How can we commit to the basic principles of empiricism while at the same time subscribe to the claim that the numerous unobservable entities in our theories are real in the sense that they exist independently of our descriptions of them? This is the question that empiricists find themselves confronted with when they develop realist ambitions; and this will also be the main topic of the following considerations.

I will argue that it is possible to bridge the gap between an empiricist methodology and a realistic stance on science under the presupposition that one is ready to bite the bullet of Quine's doctrine of ontological relativity, which holds that unearthing the ontological commitments of a scientific theory always requires a background framework against which the theory in question is interpreted. As it turns out, ontological relativity compromises neither empiricism nor realism. Quite on the contrary, the upshot of my argument will be that we can be realists and empiricists at the same time if we are ready to accept the fact that science will never lead us to the one and only true ontology and that ontologies are always and necessarily "relative" in the sense which Quine put forward. This amounts to a relaxed view of realism. But empiricism must also relax in order for the *ménage* to be happy one: Empiricists have to give up the claim that strictly empirical criteria alone are sufficient for theory choices.

I will proceed in three steps. First I will explain the reasons for the tension between empiricism and realism (Section 2). I will then turn to Quine's doctrine of ontological relativity. I will show how endorsing ontological relativity can lead the road towards a reconciliation of empiricism and realism (Section 3). I will end by formulating an account which I tag empirical realism (with best regards to Moritz Schlick). The whole analysis will be iced with a short *reductio* argument against scientific anti-realism (Section 4).

## 2 The Tension Between Empiricism and Realism

### 2.1 Empiricism By and Large

Empiricism, taken as a general doctrine in the philosophy of science, can be characterized by the following features:

---

[1] Stathis Psillos calls an empiricism that explicitly denies the reality of theoretical posits a "revisionary stance to science and, besides, not much less metaphysical than scientific realism" ([18], p. 303).

1) The denial of synthetic knowledge *a priori*.
2) The commitment to empirical testability.
3) An instrumentalist stance on scientific theories.
4) A reductionist/nominalist stance on theoretical terms.

The first feature is deeply rooted in the empiricist tradition. It is the idea that all knowledge must be founded in experience. A wholehearted empiricist quite unambiguously claims that the source as well as the justificatory basis for all knowledge must be experience. This dogma has been expressed most explicitly by John Locke himself:

> Whence has it [the mind] all the materials of reason and knowledge? To this I answer, in one word, from EXPERIENCE. In that all our knowledge is founded; and from that it ultimately derives itself. ([16], Book II, Chapter 1).

This foundationalist attitude was also one of the cornerstones of the logical empiricism of the Vienna Circle:

> We have characterized the *scientific worldview* mainly by *two features*: *Frist* it is *empiricist and positivist*: There is only knowledge from experience. This sets the boundaries for legitimate science. ([6], p. 307, my translation).

For the empiricist, experience constitutes the only source of knowledge about the empirical world, even if the sources of mathematical or logical knowledge lie outside experience, i.e. even if we allow for *a priori* knowledge in those realms. Or, to put it in other words: If there is *a priori* knowledge, it cannot be synthetic.

Closely related to the first feature is the commitment to the empirical testability of scientific theories. All claims about nature have to be testable empirically. This also means that all scientific claims are susceptible to revision in light of new empirical evidence and that they are justifiable only up to a certain limit. Scientific theories can never be verified absolutely, however certain or robust they might appear. The best we can strive for are tentative corroborations.

This leads to the third feature of empiricism: Instrumentalism. Because most scientific theories contain statements that go beyond the immediately observable, we find ourselves confronted with the question of whether those statements should be interpreted literally. The empiricist is inclined to answer negatively. Neither need theoretical statements in science to be interpreted literally in the sense that their theoretical terms refer to existing entities, nor do we require theories to be literally true. All we need is empirical adequacy. As long as we are able to deduce empirically testable statements from a theory and as long as the empirical tests of the theory turn out to be successful, we are entitled to accept the theory. Empirical adequacy is sufficient for the acceptance of a scientific theory.[2]

---

[2] A qualification is in order here. There are forms of empiricism that do subscribe to the literal interpretation of theories. Van Fraassen's constructive empiricism is an example. But for the

The forth feature follows from the third. Because theoretical statements do not have to be interpreted literally, empiricists do not have to be committed to the existence of theoretical entities. Theoretical terms are just names, abbreviations or logical constructs that we use in order to facilitate talk about observable phenomena. This is what came to be known, after Quine, as the second dogma of empiricism: "[T]he belief that each meaningful statement is equivalent to some logical construct upon terms which refer to immediate experience" ([15], p. 20); or what Russell called the supreme maxim of scientific philosophising: "Wherever possible, logical constructions are to be substituted for inferred entities." ([24], p. 115).[3]

Now, what exactly is it that makes empiricism an anti-realist position? To be sure, the fourth feature is straightforwardly anti-realist, because it entails an anti-realist attitude towards theoretical entities. There is, however, another anti-realist element in the empiricist package, which is less immediately visible. It is the thesis that scientific theories are usually underdetermined by empirical evidence. So let us take a closer look on the two features that are responsible for why empiricism cannot be realism: Underdetermination and the denial of the existence of theoretical entities.

## 2.2 Underdetermination

Underdetermination arises as a problem within the empiricist program as a result of the normative claim that experience must be the justificatory basis of all scientific knowledge. The problem was first highlighted by Pierre Duhem. In a famous thesis (which came to be known later as the Quine-Duhem-thesis), Duhem asserted that experience is never sufficient to confirm or refute single statements but only theoretical frameworks as wholes. The thesis holds that instead of refuting a single proposition in light of contradictory evidence, it is always possible to adjust an auxiliary assumption related to that proposition such that one can hold on to it despite contradicting evidence. This is the thesis of confirmation holism: Only entire theories can be confirmed or refuted by empirical evidence (See [9] and [15]).

Confirmation holism à la Quine-Duhem leads to underdetermination. Because we can only confirm or refute theories as wholes and because we can always hold on to a theory in the light of refuting evidence by the adjustment of background hypotheses, it is possible (in theory at least) to construct empirically equivalent but logically incompatible alternatives for any given theory. Empirically equivalent theories entail the same observational consequences and are equally well confirmed by the available evidence, but they make incompatible theoretical claims (e.g. with

---

constructive empiricist, "literal interpretation" simply means a correct understanding of what the theory says. It does neither imply the commitment to the existence of theoretical entities nor to the literal truth of the theory: "After deciding that the language of science must be literally understood, we can still say that there is no need to believe good theories to be true, nor to believe *ipso facto* that the entities they postulate are real." ([27], p. 11).

[3]In older empiricisms, the use of theoretical terms was often associated with an economical function. Ernst Mach in his *Mechanics* provides the most articulate example for this line of thought: "It is the object of science to replace, or *save*, experiences, by the reproduction and anticipation of facts in thought. [...] This economical office of science, which fills its whole life, is apparent at first glance; and with its full recognition all mysticism in science disappears." ([17], p. 481).

respect to the entities or properties they posit). Hence the choice between them is underdetermined by empirical data. When we are confronted with two empirically equivalent theories that make contradictory existential claims, we cannot determine which of the two is the true theory based on experience alone. Therefore the underdetermination thesis, if true, forces the empiricist to an anti-realist conclusion: We can never claim for any empirical theory that it is true in a substantial sense. The best we can hope for is empirical adequacy, and we have to accept the fact that sometimes two theories with incompatible theoretical posits can be equally adequate. This is the reason why many believe that empiricism cannot be realism.[4]

It is important to note, however, that underdetermination does not necessarily undermine realism. We can accept underdetermination and still be realists, because it is always possible to allow for more than strictly empirical criteria when confronted with a choice between two empirically equivalent alternatives.[5] However, applying pragmatic or rational criteria, such as coherence, consistency etc., is usually understood as a move away from empiricism. Whether underdetermination forces the empiricist into anti-realism depends on how radical she chooses to be when it comes to her epistemological foundationalism.

## 2.3 Trouble with Theoretical Entities

Let us now take a look at the second anti-realist feature of empiricism: the reductionist stance on theoretical terms. Reductionism was a crucial ingredient of the logical empiricism in the early decades of the twentieth century. Logical empiricism operated on the basis of a verificationist criterion of meaning according to which the meaning of a statement is given by its truth conditions: A statement is meaningful if and only if there are empirically verifiable consequences that determine the conditions under which the statement is true. Verificationism causes a problem with respect to theoretical sentences, because their truth conditions cannot be easily determined. In order to render statements containing theoretical terms meaningful, the logical empiricists assumed that every theoretical term is reducible to observational terms via correspondence rules. This kind of reductionism in combination with the verificationist criterion of meaning constitutes the core of the so called "reductive empiricism" of the Vienna Circle. In the reductionist view, theoretical terms do not refer to unobservable entities, but only to some logical construct that contains only terms which refer to things given in immediate experience. Here is a quote from Carnap, who expresses the empiricist skepticism towards non-observational entities in the following way:

> As far as possible they [empiricists] try to avoid any reference to abstract entities and to restrict themselves to what is sometimes called a nominalistic language, i.e., one not containing such references. However, within certain scientific contexts it seems hardly possible to avoid them. ([5], p. 20).

---
[4]For a more elaborate discussion of the relationship between holism, underdetermination, empiricism and anti-realism see [10].
[5]See for instance Richard Boyd's realist countermove against underdetermination ([2, 3]).

To be sure, the last sentence in this quote is very important because it nicely expresses the trouble that the logical empiricists found themselves in. On the one hand they were fond of the nominalistic consequences of their empiricist methodology. On the other hand they realized two things: First, that the language of science is obviously full of theoretical terms, and second, that it sometimes turns out to be very hard to eliminate them via logical reduction. That is to say that the reductive empiricist program, as it was originally conceived, actually turned out impossible to implement. And it was of course Carnap himself, who in his *Testability and Meaning* from 1936 was among the first who pointed at the problems of reductive empiricism.[6]

So on the one hand, the nominalist stance towards theoretical terms forces the empiricist to back off from a commitment to the reality of theoretical entities. On the other hand, she has to admit that theoretical posits play an important role in scientific explanations and that it seems hard to eliminate them via logical reduction. Hence, the task that the empiricist finds herself confronted with in this situation, can be formulated in the following way: Is there a way of reconciling empiricism with the acceptance of unobservable entities (or properties or structures) without compromising its anti-metaphysical attitude?

Psillos ([18]) has shown that there exists a line within the empiricist tradition that provides a positive answer to the question whether empiricists can "be committed to the reality of explanatory posits without opening the floodgates of metaphysics" ([18], p. 303). Psillos discusses attempts by Schlick, Reichenbach and Feigl, who all provided interesting suggestions for empiricist moves towards realism. Under a certain reading, even Carnap's *Empiricism, Semantics and Ontology* can be seen as an attempt "to develop the rapprochement between empiricism and scientific realism, as this was developed in the Schlick-Reichenbach-Feigl tradition of empirical realism" ([18], p. 313).[7]

Along the lines of Feigl ([11]), Psillos ([18]) develops his own indispensability argument for scientific realism, in which he echoes an important element in all the empiricist moves towards realism: All these contributions have a pragmatic touch. Whether the adoption of the realist framework is legitimate, is relative to the goal of achieving a coherent causal-explanatory view of the world. If this is the aim, then there is no framework that does better than the realist one.

It is not my aim to assess the feasibility of these approaches. What is important for the context of this article is the fact that all these attempts may be taken as an indication that a reunion of realism and empiricism is in principle a desirable aim.

## 3 Ontological Relativity Revisited

Let us take a step back now, and ask again what it would mean for an empiricist to develop realistic leanings. How would an empiricist's take on ontology look like? It

---

[6] See also [12]. To be sure, the original ideas of the logical empiricists were much more nuanced. It is beyond this paper to discuss the subtleties of the original version or the details of the historical development of reductionism. For a much more refined account see for instance [12].

[7] Alspector-Kelly ([1]) also argues that *Empiricism, Semantics and Ontology* can be read as an attempt to free empiricism from the nominalism that it traditionally included.

seems that any empiricist approach to ontology should be compatible with at least two claims:

1) Because empirical science provides the most robust way of acquiring knowledge about reality, the best way of obtaining ontological knowledge about the configuration of reality is through successful and well-confirmed scientific theories. That is to say that there exists no *a priori* ontology.

2) All scientific theories can only be confirmed up to the limit of induction. They always remain subject to revision in the light of new empirical evidence.

Combining 1) and 2) leads to the conclusion that ontological claims are also, like all knowledge claims about the world, susceptible to revision. The belief in the existence of point-like elementary particles, for instance, is vindicated by the fact that theories that postulate point-like elementary particles are successful. But even those theories will eventually be replaced by new theories with different ontological commitments. When a scientific theory gets replaced, the ontological commitments of the theory have to be replaced too. This leads to a further interesting question that the empiricist who engages in ontology has to answer: How do we sort out novel ontological commitments when a theory change happens?

In order to answer this question, it is helpful to consider a historical example. Take for instance the transition from classical mechanics to quantum mechanics. Quantum mechanics contains several theoretical principles that are absent in classical physics. One of these is the superposition principle according to which any linear combination of two well defined states of a quantum system is itself a possible state of the system. In the formalism of quantum mechanics, superpositions are represented as wave functions and the famous Schrödinger equation describes the dynamics of wave functions. The wave function itself has no direct classical correlate, and it is not straightforwardly clear what, if anything, it represents in physical reality. Should we actually decide to ponder on the question what the wave function corresponds to in reality, we could proceed in at least two ways: We could either try to interpret the new theoretical framework against the background of the old picture, in which there are particles with well-defined classical properties moving in space and time. As it turns out, this interpretation leads into severe problems. The interpretation will not work because there is no one to one correspondence between certain elements in the old and the new theory.

We could also go the other way around and try to "invent" an entirely new ontology that fits the new theory. In this case we might come up with an interpretation that refers to "many worlds", "consistent histories, "spontaneous collapses" or what have you. In order to test the plausibility of our preferred interpretation we would then try to reinterpret the old classical picture against the background of the new putative ontology. This could give rise to interesting conceptual problems regarding locality, identity, causality, and we would probably gain important insights into the deficiencies and limitations of our old picture. In any case, accepting the fact that ontologies depend on empirical theories and that they might change as soon as those theories change, leads to the insight that theory changes generate interesting entanglements between the ontological commitments of different theoretical frameworks. Figuring out the ontology of any given theory always seems to require a background against which the interpretation is done. This very idea, the idea that ontologies are always related to theoretical frameworks, and that the interpretation of a theory always requires a background theory, is precisely the verdict of Quine's doctrine of ontological relativity.

Quine developed his doctrine of ontological relativity alongside his views in the philosophy of language concerning issues like "radical translation" and the "inscrutability of reference".[8] The basic claim of ontological relativity is that we always need a background theory (Quine also calls this the "home theory") in order to sort out the ontological commitments of a particular object theory. He says: "The relativistic thesis to which we have come is this, to repeat: it makes no sense to say what the objects of a theory are, beyond saying how to interpret that theory in another" ([21], p. 55).

Ontological relativity entails that ontological stipulations can never be made in an absolute manner because every fixing of a theory's ontological commitments is relative, not only to the theory itself, but also to the background theory used in the fixing. This sounds very much like an awkward infinite regress, for the very choice of a background theory is a relative matter. Quine again: "If questions of reference of the sort we are considering make sense only relative to a further background language, then evidently questions of reference for the background language make sense in turn only relative to a further background language" ([21], p. 49). There is indeed a regress lurking, but Quine compares the kind of relativity that he has in mind to the one that we encounter when we make coordinate transformations in physics. In physics, it makes no sense to speak of absolute position or velocity, because they are always relative to a frame of reference. Likewise, it makes no sense to fix the interpretation of a theory in an absolute way: "What makes sense is to say not what the objects of a theory are, absolutely speaking, but how one theory of objects is interpretable or reinterpretable in another" ([21], p. 50). Ontological relativity boils down to the insight that we cannot uniquely single out the one and only ontology of a theory, i.e. we can never fully eliminate unintended interpretations.

> My answer is simply that we cannot require theories to be fully interpreted, except in a relative sense, if anything is to count as a theory. In specifying a theory we must indeed fully specify, in our own words, what sentences are to comprise the theory, and what things are to be taken as satisfying the predicate letters; insofar we do fully interpret the theory, relative to our own words and relative to our overall home theory which lies behind them. But this fixes the objects of the described theory only relative to those of the home theory; and these can, at will, be questioned in turn. ([21], p. 51).

---

[8]Quine himself was unsure about where to draw the distinction between the inscrutability of reference and ontological relativity. "Kindly readers have sought a technical distinction between my phrases 'inscrutability of reference' and 'ontological relativity' that was never clear in my own mind" ([23], p. 51).

With respect to scientific theories this means that, strictly speaking, we always have to interpret our theories against each other (or against the background of natural language). The crucial point is not which background theory we choose, but the fact that we are always forced to choose a background theory in order to make explicit the ontological commitments of a given object theory.

## 4 Empirical Realism

Many, among them Quine himself, have taken ontological relativity to be a route leading to instrumentalism. There are passages in Quine (especially in his later works), in which he clearly draws anti-realist conclusions based on ontological relativity.[9] Putnam saw "ontological relativity as a refutation of any philosophical position that leads to it" ([19], p. 180), and he believed that it needs to be refuted in order to maintain even a mild version of realism. I argue, however, that ontological relativity does not force us to abandon the basic claims of realism. Quite on the contrary, it even allows us to unify empiricism and realism in a consistent manner.

With respect to the compatibility with empiricism, we have already seen that ontological relativity is well compatible with belief that there is no synthetic *a priori* and accordingly no *a priori* ontology. Ontological relativity also allows for an updating of our belief systems, including their ontological commitments, in the light of new empirical evidence. After all, Quine himself was an empiricist, and he repeatedly claimed that ontology should be seen as a part of empirical science.[10] In fact, the question whether empiricism and ontological relativity are compatible becomes obsolete if we keep in mind that ontological relativity follows as a consequence if one embarks on the project of ontology from an empiricist angle. An empiricist will not only happily accept the fact that all our ontological commitments must come from empirical theories, but also that the interpretation of those theories is a relative matter, i.e. that new theories are usually interpreted against the background of older ones.

The more difficult problem turns out to be the compatibility of ontological relativity with realism. Accepting ontological relativity forces us to admit that ontological commitments can never be fixed once and for all. But does this really prevent us from keeping the belief that scientific theories, each in its own perspective, capture relevant aspects of a mind-independent reality? I do not think so.[11] There is nothing inconsistent in asserting the mind-independence of reality, the ontology of which is revealed by the relative interpretation of scientific theories, while at the same time remaining true to the empiricist claim that our knowledge of this very ontology is never absolutely certain and always subject to revision in the light of

---

[9] See for instance [22], p. 21: "We can repudiate it [our ontology]. We are free to switch, without doing violence to any evidence. If we switch, then this epistemological remark itself undergoes appropriate reinterpretation too; nerve endings and other things give way to appropriate proxies, again without straining any evidence. But it is a confusion to suppose that we can stand aloof and recognize all the alternative ontologies as true in their several ways, all the envisaged worlds as real."

[10] "Ontological questions then end up on par with questions of natural science" ([15], p. 71).

[11] Readers who sense a resonance of the perspectival realism that has been brought forward by Ron Giere ([14]) and more recently also by Paul Teller ([26]) are justified to do so.

new empirical evidence. Ontological relativity allows us to realize that the reference of theoretical terms is inscrutable *in principle*. To be sure, "inscrutable" is an imprecise term. It should not be understood in the sense that the reference of theoretical terms is beyond our grasp. Rather, "inscrutable" in this context means that we cannot fix the reference of theoretical terms absolutely, independently of any background theory.

The notion of realism intended here is admittedly a rather weak one. It contents itself with the claim that reality must be mind-independent. But what exactly does mind-independence mean? It simply means that although there is no one single way of structuring the world ontologically, there exist an objective basis on which the structuring takes place.

Anjan Chakravartty ([7]) has spelled out a similar idea in terms of what he calls "sociability-based pluralism", according to which a) there is more than one structure of mind-independent entities and processes, and b) the mind-independent content of scientific descriptions is identified with properties that are commonly attributed to particulars in those descriptions. While there are many ways in which properties can be grouped together yielding different categories of particulars, the property distributions in space-time exist independently of us. And it appears that properties are not randomly distributed in space-time, but that they are systematically "sociable".[12]

Now what about underdetermination? As we have seen above, underdetermination is often used as an argument for why empiricism cannot be realism. Because every theory choice is in principle underdetermined by empirical data, we are unable to single out the one and only theory that correctly represents the true configuration of reality (at least if we apply strictly empirical criteria). So the argument goes. But bringing ontological relativity into the picture makes it clear that the demand for one and only one true theory containing the one and only true ontology was too exorbitant to begin with. Underdetermination turns out to be a red herring once we accept that ontologies are relative *in principle*. Even the ontological commitments of a purely observational theory cannot be fixed in an absolute way, because we can always reinterpret its terms against a different background and we will find that the ontology changes with every reinterpretation.

This even holds for the ontological commitments of natural languages. To use Quine's often quoted example: "Gavagai" can refer to "rabbit" or "undetached rabbit part" or "temporal stage of a rabbit" or "rest of the universe minus rabbit".[13] So there is no fundamental difference between observational and theoretical terms when it comes to questions of reference.[14] Hence there is no special problem of

---

[12] Roman Frigg also argues that reality need not be uniquely structured to be reality enough: "If a system is to have a structure it has to be made up of individuals and relations. But the physical world does not come sliced up with the pieces having labels on their sleeves saying 'this is an individual' or 'this is a relation' [...] Because different conceptualisations may result in different structures there is no such thing as the one and only structure of a system" ([13], p. 56).

[13] This presupposes that the sentences of natural languages do in fact contain ontological commitments. Quine also suggested a physicalistic interpretation of the indeterminacy of Gavagai utterences in terms holophrasic indeterminacy. Taken as such, "Gavagai" does not refer to anything but some directly observable pattern of sensory stimuli.

[14] Quine draws the same conclusion: "I extend the doctrine to objects generally, for I see all

theoretical terms. The same problem arises with the most trivial terms in our natural language. And who would doubt the reality of rabbits just because the reference of the term turns out to be inscrutable? Certainly no empiricist.

Let us now try to formulate an account, which we may call empirical realism, and which brings together empiricism and realism in a consistent way with the help of ontological relativity.[15] Empirical Realism consists of the following three dimensions:

*1) The methodological dimension: Empiricism*
All scientific knowledge claims must be justified empirically. Empirical testability is a necessary criterion for something to be called "scientific" in the first place. Accordingly, all scientific knowledge claims are susceptible to revision in the light of new empirical evidence.

*2) The semantic dimension: Ontological relativity*
The reference of the terms of any given theory can never be fixed absolutely because they are relative to the background against which the interpretation takes place.

*3) The metaphysical dimension: Realism*
The entities/properties/structures/processes posited in empirically successful scientific theories do represent aspects of a mind-independent reality. Different ontological interpretations can be seen as different perspectives on reality.

The anti-realist might object at this point that this picture is inconsistent because it lacks two important features that are necessary for any philosophical account with a realist leaning. First, the belief that the physical world is structured in a unique way and that there is only one "real" ontology. And second the belief that successful scientific theories are truth bearing descriptions of this uniquely structured reality. The anti-realist might argue that accepting 1) and 2) actually forces us to add *anti-realism* as a third dimension. Precisely because our theories cannot be true in a strict sense and because the ontological commitments of our theories cannot be fixed absolutely, we have to be anti-realist about the entities and structures posited in those theories. A realism without truth and unique structure is not realism enough, the anti-realist will argue. But this claim might backfire at the anti-realist, because accepting 1) and 2) implies that *all* ontological commitments are relative in the way described above, even the ontological commitments of a purely observational theory, or even those of our natural language. So the anti-realist is confronted with a dilemma: Either she becomes anti-realist with respect to observational entities as well, or she accepts the fact that we have no better reason to be anti-realist about theoretical entities than about observable ones. But scientific anti-realists typically

---

objects as theoretical. [...] Even our primordial objects, bodies, are already theoretical [...]" ([22], p. 20).

[15]The name "empirical realism" goes back to Schlick ([25]). Schlick also speaks of "consistent empiricism". Psillos also refers to the realist position developed by Schlick and Feigl as "empirical realism" ([18]).

want to be realists with respect observable things. Taking ontological relativity seriously dissolves the *special* problem of scientific realism, because it confronts us with a more fundamental choice: Either we are anti-realist all the way down and we end up with radical skepticism even in the realm of the observable, or we retain a relaxed realist stance, one that no longer presupposes absolute truth or the belief in a uniquely structured reality. By asserting that realism is compatible with relativity concerning ontological matters, Empirical Realism chooses the second option.

The position that I have just outlined leads to pluralism. But, to be sure, it is not an ontological or metaphysical kind of pluralism. In fact, no conclusion about reality itself follows, except the one that, should the talk of "reality" be sensible in the first place, reality must mean something mind-independent. The position also implies that there exists no single unified scheme that represents the structure of reality exhaustively. The latter claim, however, is an epistemic claim rather than a metaphysical one. Accordingly, the pluralism intended here pertains to our theories and representations of reality rather than to reality itself. Because ontology is a relative matter we must take a permissive stance on theories and we can not accredit an exquisite claim to truth to any single one of them. As has been brilliantly argued by Hasok Chang ([8]), and as I hope to have shown above, this sort of epistemic pluralism does not preclude realism. Certainly the realism that emerges out of this picture is not – as Chang calls it – the "truth-realism" ([8], p. 222) that was the issue in the traditional debate on scientific realism. It is a form of realism that is closer to the practice of real science and that does justice to the history of science. What we end up with is a philosophical account of science which is less revisionary, not only than full-blown empiricist anti-realism, but also than all sorts of overambitious realisms.

## Acknowledgements

I thank the audiences at the SILFS conference 2014 and the philosophy colloquium of the University of Berne in April 2015 for their questions and comments. Claus Beisbart, Matthias Egg, Reto Gubelmann, John Norton and Paul Teller have provided particularly helpful inputs and criticisms. Financial support by Society in Science – The Branco Weiss Fellowship is gratefully acknowledged.

## BIBLIOGRAPHY

[1] Alspector-Kelly, M. (2001). On Quine on Carnap on Ontology, *Philosophical Studies* 102(1), pp. 93–122.
[2] Boyd, R. N. (1973). Realism, Underdetermination, and a Causal Theory of Evidence, *Nous* 7(1), pp. 1–12.
[3] Boyd, R. N. (1983). On the Current Status of the Issue of Scientific Realism, *Erkenntnis* 19(1-3), pp. 45–90.
[4] Carnap, R. (1936). Testability and Meaning, *Philosophy of Science* 3(4), pp. 419–471.
[5] Carnap, R. (1950). Empiricism, Semantics, and Ontology, *Revue Internationale de Philosophie* 4, pp. 20–40.
[6] Carnap, R., H. Hahn, and O. Neurath (1979). Wissenschaftliche Weltauffassung: Der Wiener Kreis, in O. Neurath and R. Hegselmann (eds.), *Wissenschaftliche Weltauffassung, Sozialismus und Logischer Empirismus*, pp. 299–317, Frankfurt: Suhrkamp.
[7] Chakravartty, A. (2011). Scientific Realism and Ontological Relativity, *The Monist* 94, pp. 157–180.

[8] Chang, H. (2012). *Is Water H2O? Evidence, Realism and Pluralism*, Dordrecht: Springer.
[9] Duhem, P. (1914/1991). *The Aim and Structure of Physical Theory*, Princeton: Princeton University Press.
[10] Esfeld, M. (2006). Scientific Realism and the History of Science, in G. Auletta and N. Riva (eds.), *The Controversial Relationships Between Science and Philosophy: A Critical Assessment*, pp. 251–275, Vatican City: Libreria Editrice Vaticana.
[11] Feigl, H. (1950). Existential Hypotheses: Realistic Versus Phenomenalistic Interpretations, *Philosophy of Science* 17(1), pp. 35–62.
[12] Friedman, M. (1987). Carnap's Aufbau Reconsidered, *Nous* 21(4), pp. 521–545.
[13] Frigg, R. (2006). Scientific Representation and the Semantic View of Theories, *Theoria* 55, pp. 49–65.
[14] Giere, R. N. (2006). *Scientific Perspectivism*, Chicago: University of Chicago Press.
[15] Hempel, C. G. (1958). The Theoretician's Dilemma: A Study in the Logic of Theory Construction, in H. Feigl, M. Scriven, and G. Maxwell (eds.), *Minnesota Studies in the Philosophy of Science* Volume II, pp. 37–98. Minneapolis: University of Minnesota Press.
[16] Locke, J. (1690/2004). *An Essay Concerning Human Understanding*, London: Penguin Classics.
[17] Mach, E. (1919). *The Science of Mechanics: A Critical and Historical Account of its Development* (forth ed.), Chicago, London: The Open Court Publishing Company.
[18] Psillos, S. (2011). Choosing the Realist Framework, *Synthese* 180(2), pp. 301–316.
[19] Putnam, H. (1993). Realism Without Absolutes, *International Journal of Philosophical Studies* 1(2), pp. 179–192.
[20] Quine, W. v. O. (1951). Two Dogmas of Empiricism, *Philosophical Review* 60(1), pp. 20–43.
[21] Quine, W. v. O. (1969). *Ontological Relativity, and Other Essays*, New York; London: Columbia University Press.
[22] Quine, W. v. O. (1981). *Theories and Things*, Cambridge; London: Belknap Press.
[23] Quine, W. v. O. (1990). *The Pursuit of Truth*, Cambridge MA.: Harvard University Press.
[24] Russell, B. (1917). The Relation of Sense-Data to Physics, in *Mysticism and Logic*, pp. 108–131. London: Allen and Unwin.
[25] Schlick, M. (1932). Positivismus und Realismus, *Erkenntnis* 3(1), pp. 1–31.
[26] Teller, P. (2015). Pan-Perspectival Realism Explained and Defended, talk given at the EPSA conference 2015 (forthcoming).
[27] Van Fraassen, B. C. (1980). *The Scientific Image*, Oxford: Clarendon Press.

# Content Reduction for Robust Realism

Alberto Cordero

ABSTRACT. Selective realists try to identify truthful parts in successful theories. One recent strategy focuses on compelling derivations of impressive predictions from theory. Its leading trend (variously led by Juha Saatsi, Peter Vickers, and Ioannis Votsis [SVV]) clarifies and refines earlier notions of "theory-part," "success" and "truth-content," but the approach leans worryingly towards a "bare-bones" version of realism that invitites pessimism about the outcome. Section 1 discusses the SVV approach; sections 2 and 3 explore an application that exposes a tension with the augmentative inferential goals of realism. Sections 4 and 5 suggest adjustments that arguably enhance the augmentative prospects while keeping the focus on truth- content; the proposed adjustments enrich the assessment of theory-parts with resources taken from scientific practice. The resulting criterion yields a version of selective realism closer to the selection of theory-parts deemed successful and beyond reasonable doubt by prevailing methodological practices in the natural sciences.

## 1 Background: Bare Bones Realism

To many antirealists the fate of past empirical theories refutes the idea that success betokens truth. Selective realists respond by shifting commitment from whole theories to select theory-parts, seeking to trace empirical success to components with high truth-content. However, which parts are those? Identifying them has proven far from straightforward[1] An intuitive general strategy looks for persistent retention through theory-change, attributing truth-content to parts so retained, estimating that such parts are very likely are either correct or "contain" some abstract version (somehow restricted perhaps) that gets things right. An exemplar case is Fresnel's theory of light, whose assumptions regarding the ether luminiferous are long rejected, yet at intermediate levels the theory contains seemingly reliable content in the form of abstract descriptions rooted in the original proposal. Light is not completely as Fresnel imagined, yet his theory got many things right— e.g. light is made of invisible transversal undulations, and these undulations follow the Fresnel laws of reflection and refraction. Abstracted from reference to the wave substratum, this part of the theory spells out a core (one might call it Fresnel's "broad account") that all subsequent theories of light have retained. As [11] urges, however, retrospective projection of current science both reflects limitations of human imagination as easily as it does truth-content and can be variously misleading and self-serving— and it

---

[1] See [2] and references thereof.

severely weakens realism by giving up the traditional realist goal of identifying the truthful parts of a theory while the theory is still alive. Accordingly, a more developed approach, variously developed by [16], [6], [7], and [8] seeks to trace a theory's empirical success to specific parts responsible for that success. The adequacy of this revamped approach faces serious challenges, however. [5], for example, persuasively argues that features of the caloric theory that were rejected by subsequent physics were central to the success of the theory. An arguably more damaging conceptual charge seemingly applies to Fresnel's theory [2].

In order to convince, these and other critics urge, selective realists need to provide confirmation criteria that specify in advance which parts of a current theory are both empirically successful and likely to survive theory-change. Can one assess the truth of a given partial account while the total theory in which it originates is still in full flight?

In an ongoing theory, at least some parts seem identifiable as very probably truthful by tracking the successful predictions the theory licenses. Recent works in this direction, notably by Juha Saatsi, Peter Vickers, and Ioannis Votsis [SVV] examine particular derivations of impressive predictions, looking for "causally active" posits contained in the steps that lead to those predictions.[2] In explanatory terms, an impressive prediction comes out true because upstream posits invoked in its derivation have truth-content they pass downstream to the prediction. As in the ether case, an upstream proposition might turn out to be not a working posit of the theory, but in that case it will very probably contain a proposition of weaker content that is properly a working posit [9].

The key question is how to tell which of the theory-parts invoked in a derivation count as working posits. To [15], the only parts dependably involved in the logical deduction of impressive predictions are "mathematical parts," chiefly equations and mathematically structured concepts. However, because the latter cannot generate predictions without interpretation, they must be given one, which opens the door to superfluous content that realists must strive to shrink. Interpretation should thus stop at the minimum needed to generate impressive predictions. Accordingly, a prospective realist must (a) drop parts that make no contribution to the empirical success of the theory at hand (e.g. such explanatory accounts of quantum mechanics as Bohmian Theory, many worlds, etc.); and (b) continue reducing content until the process reaches parts that cannot be further trimmed without compromising the theory's power to make impressive predictions. An example of an item surviving such a trimming is the Schrodinger equation, without which whole genres of quantum mechanical phenomena could not be inferred.

Trimming theoretical derivations is problematic, however. The Kinetic Theory of Matter explains and predicts observable properties of bulk matter from the postulated behavior of microscopic molecules. Trimming away the molecular hypothesis in favor of some more parsimonious counterpart severely compromises the kinetic theory's fecundity and fertility (i.e. its power as guide to new results and connections), and so in pragmatic terms also its predictive power, which is why realists generally grant epistemic weight to fertility and fecundity. There are other compli-

---

[2]In particular, [9], [15], and [14]

cations. First, if fecundity and fertility are allowed to provide warrant for the molecular hypothesis, why not also for (in their respective heydays) the posits caloric, phlogiston, Fresnel's ether, and the boundary conditions in Kirchhoff's Theory of Diffraction, along many other posits seriously off the mark? Relaxing minimalism easily lets in seriously wrong posits, it seems. Secondly, a given prediction might, in principle, be derivable from a theory without invoking this or that posit, but to whom the derivation in question will be available depends on the state of background knowledge, which (modulo fallibilism) never rests on ideal epistemic conditions. For example, there are seemingly strong grounds for claiming that 19th century physical theorists proceeded from a metaphysical framework in which the ether was not (and arguably could not be) an optional posit, and so to those theorists all derivations invoking optical waves implicated the ether whether or not they mentioned it[3].

Thirdly, minimalist responses arguably don't work as advertised. [10] critically present Kirchhoff's Theory as an illustration of how bare-bones derivational analysis of impressive predictions can compel realist commitment to "bad" posits. Their analysis is worth a detour.

## 2 Impressive Prediction and Prospective Truth: Kirchhoff's Theory

Kirchhoff's Theory describes how an opening A in an opaque screen disturbs a monochromatic spherical wave of light passing through it. The theory allows calculation of the electromagnetic field U at an arbitrary point P past the opening, the squared module corresponding to the light intensity (Fig. 1). Central to the approach is Green's theorem, used by Kirchhoff to solve Maxwell's homogeneous wave equation at P in terms of conjectured boundary values for the field and its first order derivative at all points on an arbitrary surface that encloses P. The boundary conditions are introduced as assumptions together with some specified approximations. For Kirchhoff's choice of enclosing boundary, considerations from Maxwell's theory indicate that the only non-negligible contributions should come from points within the aperture, resulting in:

$$U(P) = \frac{1}{4\pi} \int\int_A \left\{ U \frac{\partial}{\partial n}\left(\frac{e}{8}\right) - \left(\frac{e}{8}\right)\frac{\partial U}{\partial n} \right\} dA$$

One key assumption—call it "posit H"— hypothesizes that the screen does not perturb waves within the aperture. The resulting predictions are both impressive and come true with astounding accuracy.

But here comes a twist in the story. Recent developments in computer speed and reliable methods of numerical integration now allow for direct calculation of U from Maxwell's equations, removing the need for intermediary conjectures about light's progress through the aperture. The resulting calculations expose a fly in the selectivist ointment: they yield U values over the aperture that are at variance from

---
[3]While today the predicates 'being a wave' and 'requiring a medium' are separable in principle, they were not always so. Waves were thought of as perturbations, requiring that something be perturbed. [2].

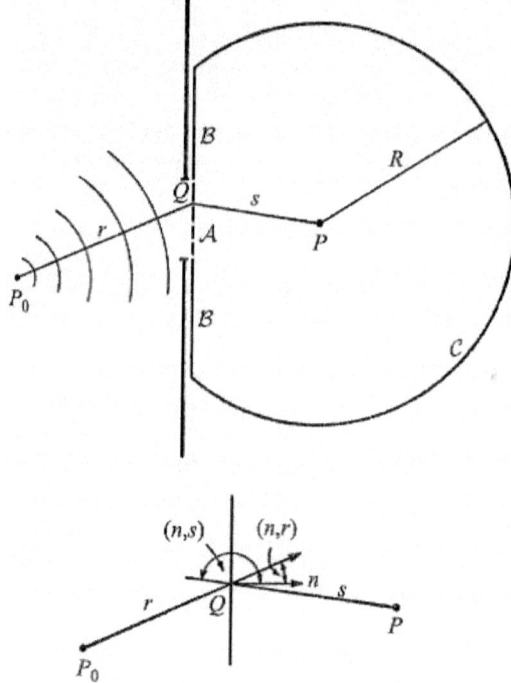

Figure 1. taken from ( [10] p. 35): Kirchhoff's method of determining diffraction at an aperture. P0 is the source of the light, and P is the point beyond the screen at which we want to know the light intensity. Q is a point in the aperture whose contribution we are considering at a given time, r is the distance from P0 to Q, s is the distance from Q to P. An imaginary surface of integration S is comprised of A (the aperture), B (part of the screen), and C (part of a circle of radius R which has P at its center). n is a normal to the aperture, (n, r) is the angle between this normal and the line joining P0 to Q, and (n, s) is the angle between this normal and the line joining Q to P

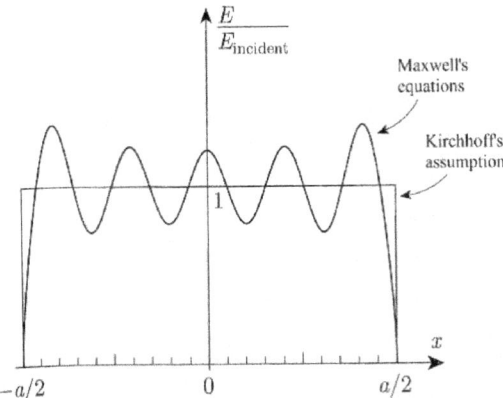

Figure 2. Comparison of Kirchhoff's assumption of a 'flat' amplitude function across an aperture of width a with the amplitude function derived from Maxwell's equations (Adapted by [10] from Figure 3.20(a) in [1], p. 71.

Kirchhoff's conjecture (Fig. 2). To [10] these results show that, although extremely successful, Kirchhoff's theory is not "approximately true." As [14] cautions, that a posit plays an explicit role in a derivation does not make it a 'working posit' to which realists must commit.

Addressing the case of untruthful posits with clear roles in the derivation of impressive predictions, [14] builds on the standard selectivist approach, proposing that untruthful but otherwise successful posits have nonetheless some significant truth to them: logically contained within, they might lodge a more modest, leaner posit with just enough content to make the relevant predictions go through. The realist claim about Kirchhoff's boundary conditions would then be this: taken as a whole the conditions are false, yet the criterion just proposed predicts that a subset of the conditions (or leaner versions of them) will prove both approximately true and sufficient to secure derivations of the theory's impressive predictions. If this could be shown, a serious problem for the realist would turn into a victory for the selectivist strategy. Vickers is clear about the fallibilist background here: "It remains possible that some of what remains, even concerning the boundary conditions, is also idle. Even if the mathematics expressing those conditions must remain, certain features of the interpretation of the relevant equations may be idle." (p. 197) As already noted, one shortcoming of bare bones trimming is a lack of clarity regarding what content is "absolutely necessary." As Vickers realizes, pursuit of interpretive minimalism pushes recipes for theory-part selection into enemy territory—the "absolutely necessary" might seem no more than the step's empirical projection or substructure—the step freed of commitment to any non-empirical content in the original. Vickers pessimistically concludes "the realist is still some distance from prospectively identifying (even some of) the working posits of contemporary science." So, according to Vickers, the realist should commit to just some unspecified parts of what remains once the posits acknowledged as idle are removed.

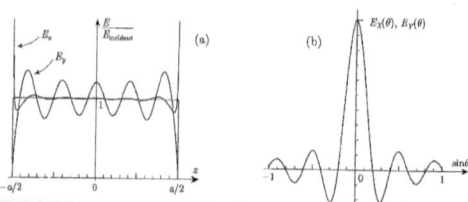

**Brooker:** (a) The electric-field amplitude in the plane of an infinitely conducting slit; the slit length lies in the y-direction and its width in the x-direction. The slit is illuminated normally with light whose E-field lies either along ($E_y$) or across ($E_x$) the length of the slit. For the particular case shown the slit width $a = 5\lambda$. The top-hat curve shows what the field would be according to the usual Kirchhoff assumptions. (b) The amplitude diffracted to the far field, plotted as a function of $\sin\theta$ where $\theta$ is the angle to the 'straight-through' direction. Within the thickness of the printed line, it is the same for both choices of polarisation direction, and also coincides with the Kirchhoff expression $\sin(\frac{1}{2}ka\sin\theta)/(\frac{1}{2}ka\sin\theta)$. All curves (except the top-hat) have been calculated 'properly' from Maxwell's equations.

This stance, he reasons, at least does a good job of restricting the target of realist commitment. Vickers exudes pessimism, however:

> In all this, we find ourselves—even 30+ years after Laudan's confutation of convergent realism—unsure of the extent to which the divide et impera strategy can succeed. Even if the 'working posits' of contemporary science cannot be prospectively identified, it remains possible that we might develop a recipe for identifying certain idle posits. This would be a significant achievement, even if not quite what the realist originally had in mind." [14] p. 209)

This rather gloomy conclusion is at odds with the expectation that realist commitment should encompass much of the array of theory-parts sanctioned as successful and beyond reasonable doubt by the best current confirmational practices in the natural sciences. I will suggest that it is also at odds with Kirchhoff's and other provocative cases Saatsi and Vickers have pointed to. The remaining sections argue that selectivism (the divide et impera strategy) has much better prospects than the above qualms suggest.

## 3   Some Realist Rejoinders

Selectivist, I suggest, can both question the above pessimistic analysis of Kirchhoff's Theory and propose more informative ways of identifying prospective truthful posits. This section considers the first line, leaving the second for the next. [10] concentrate on a case in which the opening is a slit on a conducting material, focusing on values of the electric-field amplitude in the open region of the slit, calculated directly from Maxwell's equations. They point to a particular case of extreme discrepancy whose significance, I will suggest, is debatable in light of the full results obtained by the mentioned calculations. In particular, as Fig. 3 (taken from [1]) illustrates, the direct calculation yields results more nuanced than Saatsi and Vickers suggest.

In general, light reaching the screen comprises waves with electric field components along the slit's length ($E_y$) and also across the slit's length ($E_x$). Saatsi and Vickers consider just Ey. In the particular case Brooker analyzes, the slit's width (parameter $\alpha$) is five times the wavelength, and the discrepancy with Kirchhoff's assumption is maximal for fields oriented along the slit and minimal for fields across the slit. The faulty assumption that plays a role in the derivation of impressive predictions (posit H) matches the Electric field with errors ranging from less than about 2% ($E_x$ in central regions of the slit) to about the 25% error emphasized by Vickers and Saatsi (for $E_y$ at the slit edges). It would thus seem unfair to say that H is drastically off the markin most realizations of posit H. As far as the electric-field amplitude is concerned, H generally provides the kind of approximately correct hypothesis a reliabilist might accept. This is a debatable matter, however. For the sake of argument, then, let us suppose that, in Kirchhoff's Theory, posit H is seriously off the mark. This allowance is independently relevant if (as I agree) at more fundamental levels theories often land in underdetermination and/or high-level conceptual tension. The SVV approach emphasizes truth-content and minimal commitment in reverse derivation of predictions, providing two welcome features: (a) Given a derivation, the relationship between the minimally interpreted parts and the corroborated prediction is entirely an inferential matter, indifferent to historical context. (b) Working parts can be specified while the theory in question is still in full flight. The offer seems promising, but under closer scrutiny issues additional to those highlighted by Saatsi and Vickers become apparent, particularly three. Firstly, the primary goal of selectivism is to identify parts that very likely latch unto what is real and active in the world. If so, the sought identifications seem only modestly helped by searches focused on derivational analyses of just single-case inferential chains: claiming that somewhere in a theory some abstract versions of some part or other gets what is real somewhat right seems below target. What would make a "proper target" here? To non-Kantian realists, what is epistemologically accessible reaches beyond the "phenomenal," into the "noumenal"— as per realism's mind-independence thesis. (Scientific realism is at odds with German-idealist stances). Some realists avoid the term "noumenon," however, arguably to the detriment of their positions.[4]

Secondly, mathematical structures help to make a scientific narrative refutable and warrantable, but so too do other features (e.g. consilience, internal coherence, external support). Vickers's pessimistic passage cited at the end of the previous section is well-taken, but a good deal of the trouble he envisages comes, I suggest, from overlooking resources available to selective realists for identifying more fleshy working posits. These include confirmational relations unavailable at the level of single-case derivations but accessible from integrated, diachronic, records of how some theory-parts make an empirical difference as the theory to which they belong

---

[4]Current scientific realist positions divide up largely into two camps: (a) those who claim that at least some of the (unobservable-by-us) entities posited by empirically successful theories exist mind-independently, and (b) those who, in addition, claim that the entities in question are approximately as the respective theories say. Neither kind is committed to Kantian dogma, and so realists are intellectually free to invoke the much-maligned noumenon, which stands for anything existing without regard to phenomena, e.g. electromagnetic waves not being perceived (or even perceivable) by anyone. Everything that was real before phenomena got added to reality by animals counts as noumenon, as does whatever is real independently of the human mind. Realists are thus free to simply contrast 'noumenal' with 'phenomenal'.

plays the field.

Thirdly, we must not conceptually equate selectivism with high theoretical-level realism. Even if Kirchhoff's posit H was utterly off the mark, the theory would still contribute theoretical (noumenal) knowledge at lower levels. Conjoined with laws and auxiliary assumptions accepted as unproblematic, H yields a theoretical description of field U that agrees with what we now take as a correct description over a large and significant portion of the theory's intended range. This level of achievement, I urge, is what selective realists can and should aim to get. Even if posit H had low truth-content and its "good works" were an accident of underdetermination, the field U and the light waves derived from the theory would still ring true at various significant levels of noumenal description (beyond the reach of unaided perception), even if not the deepest level invoked. In this, posit H would be like the ether of old. Like Fresnel's and Maxwell's ether-based proposals, Kirchhoff's Theory yields a great deal of prospectively truthful content— ranging from descriptions of low theoretical level to levels right below the current boundaries imposed by contingent underdetermination, error, and/or high-level conceptual problematicity. The realist claim to make seems then this: impressive predictive success indicates that the theory at hand contains reliable, prospectively identifiable theory-parts that (a) have high truth-content and (b) are also original to the theory. Whether those theory-parts occur at the theory's foundational level, at high derivational levels, or at intermediate levels is immaterial to realism, so long as they contribute noumenal descriptions (as opposed to descriptions retainable at just very low empirical levels). Can one be more precise about the noumenal parts latched upon? Interpretive minimalism has the advantage of offering an easy criterion for selecting inferential components. But, as said, it leads to much too vague determinations of truth-content and concomitant pessimism. Also, it lets in posits of arguably dubious reliability, from posit H (on one reading of the case, e.g. Vickers') to the ether. More importantly, minimalism shifts attention away from the realist task of identifying theoretical content realists can judiciously commit to— an augmentative rather than minimalist project. If so realists should stick to the strategy of content reduction but without too much emphasis on minimalism. In order to keep unnecessary posits out, help would have to be secured elsewhere, most naturally from effective (but not maximally) purgative confirmational resources steadily used in the sciences. Instances in point include convergence of multiple and varied successful theory applications on specific theory- parts, inferential support from sources initially external to the theory at hand (especially independent theories), and non-self-serving post-mortem analyses of the successes of a discarded theory. These resources, I suggest, can strengthen SVV's accent on implied content by broadening the selectivist focus beyond individual derivations and counterproductive minimalism. In scientific practice, deployment of the noted resources is apparent at several levels [4]: episodes of aggressive probing of a theory's central tenets and auxiliary assumptions, arguments from consilience, internal assessments, explanations of a theory's successes after its demise, and cases of external support.

## 4 Theory-Parts for Realists

The typical SVV derivational analysis runs as follows: pick a representative set of individual impressive predictions associated with a theory T—successful and unsuccessful (if available). For each impressive prediction, (a) look for a compelling derivation from T; (b) then, moving from the prediction statement to the premises, reflect on each derivational step, spelling out its component claims, reducing their content to the absolute minimum needed to advance the derivation towards the prediction; (c) finally, declare a given theory-part "indispensable" only if there are strong grounds for claiming that no conceivable part with lesser content could possibly advance the derivation.

How satisfactory is this strategy? Vickers' words at the end of Section 2 express a level of pessimism not borne out by the findings discussed in the previous section, but there are other issues, as already noted. Realism is an augmentative inferential project, and interpretive minimalism pushes firmly towards antirealism. Also, reasonable expectations of underdetermination and conceptual problematicity heavily tax "indispensability" claims made on behalf of of posits that might otherwise seem otherwise efficacious for moving a derivation towards a given prediction. Furthermore, identifying any content as superfluous is complicated by contextualization to the relevant agents' situation—how they understand the conceptual relations involved, as, for example, with the ether posit [2]. Concerns such as these cast doubt on the inferential strategy under consideration. I wish now to suggest how a version less committed to minimalism can do better. Here is the task at hand: given a theory rich in successful impressive predictions, we seek to identify in it truthful posits reliably. Although realists have limited means for making such identifications, scientific confirmational practice does point to relevant resources often overlooked by commentators. Some long-standing ways of reducing error risk have widespread presence in modern science [4]. One approach concentrates, not on single derivations of impressive prediction, but on rich and varied records of them: for each predicted general phenomenon, a representative set of derivations develops as the theory at hand (T) gains applications. Another complementary approach seeks support from sources external to T, including independent backing that may be available for particular theory-parts invoked in the derivation of impressive predictions from T. Yet another approach benefits from "post-mortem" analyses. Deploying these various complementary ways suggests the following four-front approach:

Front 1: Begin the assessment of a successful theory T by picking compelling derivations of its most impressive predictions. Analyza them one by one. In each case start with the step immediately above the prediction and then move up the inferential ladder, as in the SVV approach. Purge each step of overtly superfluous content, but without embracing interpretive minimalism (a problematic goal, as noted). Concentrate the purge on posits that either (a) seemingly make no accessible empirical difference (e.g. absolute space in Newton's cosmology), or (b) are marred by specific compelling doubts (e.g. arguably posit H had this problem in Kirchhoff's time[5]), or (c) are currently spoiled by effective underdetermination, i.e.

---

[5] In the late 19th century background physics cast serious doubts on Kirchhoff's assumption

theory-parts that, while inferentially contributing to the step below, compete for this role with alternatives for which scientific support seems at least as good (e.g. "deep" explanatory accounts of standard quantum mechanics). Having the derivations thus purged of superfluous content, move to the unfulfilled predictions drawn from T, looking for posits on which inferences leading to such predictions converge. Develop two lists of theory-parts—one list (L+) made of parts implicated in cases of impressive predictive success, and list (L-) made of parts found frequently implicated in unfulfilled predictions. At this point a given part may have a place in both lists.

Front 2: With the two lists in hand, assess the impact of each of the selected parts on T's fruitfulness—by estimating the overall effect (in all fields where T has application) of purging the part under scrutiny while keeping the others in place. Recognize as "very probably crucial" only those parts in L+ whose removal clearly leads to T's stagnation, judging from T's extant track record. Recognize as "suspect" those parts in L- whose removal clearly improves T's predictive power and/or frees T from seemingly intractable conceptual conundrums. Sharpen up L+ and L- accordingly. Give each listed part a weight proportional to the number of different successful and unsuccessful prediction lines in which it appears differentially implicated. In the case of general parts, make the weight (positive or negative) reflect the number of parts that instantiate them in the lists. In Kirchhoff's theory, for example, the yield of impressive predictions (e.g. , the field's values at points significantly beyond the screen compared to the aperture's size) plummets if certain posits are removed (e.g. the field U as structured by Kirchhoff's general Green's equation, or claims like "light comprises a microscopic undulation w(x,t)", and so forth). Here the contention is that, judging by the record of manifest retentions across theory-change, posits selected using front 2 have a much higher than average reliability. On the negative side, this front fails to filter out some off-the-mark posits, conspicuously the ether (at least in the context of 19th century physics) and the faulty part of posit H in Kirchhoff's Theory (although this is debatable, as noted earlier). Reliabilists who demand stronger criteria have an extra resource in Front 3.

Front 3: The focus here is on outside support for items in L+ and L- from independently successful empirical theories. L+ backing occurs when claims assumed in a theory T subsequently gain justification from another, initially unrelated, theory T1. Think, for example, of the numerous aspects of cell biology that have gained justificatory elucidation from molecular biochemistry since the 1950s[6]. In Paul Thagard's version of this strategy [12], [13], the emphasis is on explanation: if a theory not only maximizes explanatory coherence, but also broadens its evidence base over time and its assumptions are elucidated by explanations of why the theory's proposed mechanism works, then—in Thagard's view—we can reasonably conclude

---

concerning field distortions created by sharp slits. Posit H assumes something incompatible with Maxwell's equations, namely that the electric and magnetic fields have discontinuities at the aperture.

[6]E.g. neural mechanisms originally introduced as posits are now explained by noting that neurons consist of proteins and other molecules that organize into functional sub-systems such as the nucleus, mitochondria, axons, dendrites, and synapses.

that the theory is at least approximately true. Elucidation has accompanied much of the advance of modern theoretical science. To the extent that external explanatory elucidation springs from an independently supported theory T1, elucidation raises the credibility of the assumptions and narratives it casts light on—hence its interest to realists. Moreover, the initial remoteness of T1 lowers the likelihood that the two theories share conceptual mortgages, giving elucidation independent purgative power on T. In an elucidation instance, the part that gains explanation is an assumption. On the other hand, the explanation received is of limited emancipatory power in that it does not fully help claims drawn from assumptions shared by both theories. For example, to the extent that, in the 19th century, Lagrangian theory and mechanics shared the traditional metaphysics of waves, Lagragian elucidations could not expose the ether as a "dispensable posit" [2]. Also, elucidation seems neither necessary nor sufficient for realism. Unsavory counterexamples give pause to granting a given part high likelihood on the basis of elucidation alone. Here are two cases [3]. When Kepler looked for theoretical support for his Second Law, he derived it from the Aristotelian laws of motion and some principles of optimal action. Kepler elucidated his law, but by invoking as premises some of the wrongest claims of Aristotelian physics. This type of difficulty can be improved by requiring the elucidating theory to be successful in terms of impressive predictions, but this too fails to filter out some lamentable cases (e.g. in the 1940s and 1950s, Freudians claimed to have grounded in thermodynamics such of their principles as the "death instinct"; they did not convince).

Front 4: Construction of L+ and L- often continues after a theory starts to wane, and even after it dies, adding valuable material to the selectivist realist stance. This variety of retrospective elucidation is not a "self-serving" realist strategy. For one thing, it often unveils causal and/or structural justification for a theory's accomplishments, e.g. in the account wave theorists provided for the success of corpuscularian optics regarding the phenomena of reflection, refraction and polarization. Correspondingly, retrospective analyses frequently add precision to specifications of the parts a past theory got right, as can be seen presently in theoreticians' attempts to show why posit H led to correct predictions (e.g. [1]).

## 5 So, Where is the Beef?

List L+, constituted by parts made salient by the combined application of fronts 1 to 4, provides theoretical claims of high scientific reliability, exemplified by what was termed Fresnel's "broad account" in Section 1. L+ items have a superior record of retention through theory-change, and the collections of cumulative cores in L+ have grown over time, progressing in number and variety from trifling rates of growth in early modern science to increasingly steep rates in recent times. Presently, the theory-parts included, along with bridges built between them, provide a thick and highly textured arrangement of noumenal claims about entities, structures, explanatory and historical narratives invoked by increasingly interconnected scientific theories. Theory-parts that pass the screenings of fronts 1, 2 and 3 generally qualify as items deemed both successful and beyond compelling doubt in current scientific practice. The collection they make is vast, comprising—among much else—rich portions of Newtonian theory, pre-quantum chemistry, basic quantum mechanics, basic geology, Darwin's original theory, and much more, recent additions including detailed claims about the character of microscopic systems (bosons, fermions, nuclei, atoms, molecules and more, particularly regarding their energy states, architecture, and dynamical relations, to mention some of the pluses) and remote systems (e.g. the Big Bang, and the universe shortly after it). One reason why Kirchhoff's Theory matters to the realist project, then, is because posit H (in conjunction with the theory's basic laws and auxiliary assumptions) yields structures that describe the intended field correctly over a significant range. This grounds the realist claim that successful theories "yield" parts and narratives that do reliably latch unto what (by the best current accounts) is real and active in the world—in this case, the field U at spatial points removed from the aperture. Significantly, L+ is largely made of claims below the highest theoretical level, also abstract and more coarse-grained than the theories that lodge them, but still rich in content. Accordingly, in the version of selective realism proposed here, the prospective truth-content holders (the "beef") do not easily occur at the highest theoretical levels. Rather L+ is largely made of claims below that, also abstract and more coarse-grained than the theories that lodge them, but still rich in noumenal content. As argued, L+ provides a profuse display of reliable claims ranging from low empirical levels up to heights right below levels epistemically spoiled by current contingent underdetermination, error, and/or conceptual problematicity.

If the above assessment is correct, then selective realism seems in decent shape as a contemporary project. What needs to be accomplished is not bare-bones realism but rather realism with both mathematical bones and as much "healthy meat" as may be reasonably had. How lean should realist commitment be? In this paper my emphasis has has been on naturalist fallibility abetted by such signs of meliorism as growth of scientifically warranted noumenal content. Other than that, to naturalists, further guidance best comes from the confirmational status of scientific theory-parts, but this remains a contended issue. If the suggestions in this paper are on the right track, then the list of prospectively truthful theory-parts realists can presently commit to is already remarkably rich and varied –and growing. If so, more than thirty years after Laudan' pessimistic reading of the history of science, there is reasonable confidence that the divide et impera strategy of selective realism can succeed.

## BIBLIOGRAPHY

[1] Brooker, Geoffrey (2008). Diffraction at a Single Ideally Conducting Slit. Journal of Modern Optics 55 (3): 423–45.
[2] Cordero, Alberto (2011). Scientific Realism and the Divide et Impera Strategy: The Ether Saga Revisited. Philosophy of Science (Vol. 78, 2011): 1120-1130.
[3] Cordero, Alberto (2013b) Theory-Parts for Realists. In V. Karakostas and D. Dieks (eds.), EPSA11 Perspectives and Foundational Problems in Philosophy of Science: European Philosophy of Science Association. Springer Iternational Publishing, Switzerland (2013): 153-165.
[4] Cordero, Alberto (2013a). Naturalism and Scientific Realism, in Reflections on Naturalism (J.I. Galparsoro and A. Cordero, eds.); Boston: Sense Publishers (2013): 61-84
[5] Chang, Hasok (2003). Preservative Realism and Its Discontents: Revisiting Caloric. Philosophy of Science 70: 902–12.
[6] Kitcher, Philip (1993): The Advancement of Science. Oxford: Oxford University Press.

[7] Leplin, Jarrett (1997): A Novel Defense of Scientific Realism. Oxford University Press.
[8] Psillos, Stathis (1999): Scientific Realism. London: Routledge.
[9] Saatsi, Juha (2005). Reconsidering the Fresnel-Maxwell Case Study. Studies in History and Philosophy of Science 36 (3): 509–38.
[10] Saatsi, Juha and Peter Vickers (2011). Miraculous Success? Inconsistency and Untruth in Kirchhoff's Diffraction Theory. British Journal for the Philosophy of Science 62: 29–46.
[11] Stanford, P. Kyle (2006). Exceeding Our Grasp: Science, History, and the Problem of Unconceived Alternatives. Oxford: Oxford University Press.
[12] Thagard, Paul (2000): Coherence in Thought and Action. Cambridge, MA: MIT Press.
[13] Thagard, Paul (2007). Coherence, Truth, and the Development of Scientific Knowledge, Philosophy of Science (74): 28-47.
[14] Vickers, Peter (2013). A Confrontation of Convergent Realism. Philosophy of Science 80:189-211.
[15] Votsis, Ioannis (2011). Saving the Intuitions: Polylithic Reference. Synthese 180 (2): 121–37.
[16] Worrall, John (1989). Structural Realism: The Best of Both Worlds? Dialectica (43): 99-124.

# A Simple Model of Scientific Progress

Luigi Scorzato

ABSTRACT. One of the main goals of scientific research is to provide a description of the empirical data which is as accurate and comprehensive as possible, while relying on as few and simple assumptions as possible. In this paper, I propose a definition of the notion of *few and simple assumptions* that is not affected by known problems. This leads to the introduction of a simple model of scientific progress which is also discussed. An essential point in this task is the understanding of the role played by measurability in the formulation of a scientific theory.

## 1 Introduction

A characterization of scientific progress has proven extremely elusive, since [11] convincingly showed that the naive idea of progress as an *accumulation of empirically verified statements* is untenable[1].

A weaker, but more plausible, view states that there is scientific progress when new theories are discovered that are better than the available ones [15]. But, what does *better* mean? It is certainly not enough to characterize better theories in terms of empirical adequacy. In fact, if we take seriously the idea that only the agreement with the experiments matters, to evaluate scientific theories, then the *bare collection of all the available experimental reports* should always be the best possible "theory"[2]. However, we certainly do not regard such unstructured collection as a valuable theory, if for no other reason, because it enables no prediction.

This suggests a crucial role of *novel predictions*. The idea is very appealing, and it was at the heart of Lakatos' view of progressive research programmes [13]. There is little doubt that successful predictions are exactly what scientists look for. But, how do we use successful predictions for theory selection? Nobody ever formulated a convincing proposal in this sense. There is at least one very good reason for that: *predictions*, by themselves, are not protected against *brute force attacks*. To illustrate this, consider an already empirically adequate theory (that can always be produced by patching together various models, each with limited applicability, and by adding ad-hoc assumptions to account for any remaining anomalies), and imagine that there will be a new experiment soon, for which that theory makes no prediction. A professor could assign to each one of his many students a different extension of

---

[1] See, however, the debate involving [3], [16], [19].
[2] Experimental reports never contradict each others, as long as they bear different dates or locations.

that theory, by introducing many different ad-hoc assumptions. In this way, these students might even cover the whole range of possible outcomes of the upcoming experiment. At least one of such "theories" will hit the result that will be eventually measured! Is this a prediction?! But, how can we tell the lucky student who got it right that it was only by chance? Similarly, how can we tell the clairvoyant who predicted the earthquake that, considering all the failed predictions by *other* clairvoyants, his success is meaningless? *His own* record of successes is 100%. Also Einstein's General Relativity (GR) made only one *impressive* prediction[3].

Science is not defenseless against those brute force attacks and clairvoyant's claims. There is a difference between Einstein's prediction and the clairvoyant prediction: a *good* theory behind. This judgement clearly goes beyond any old or new empirical evidence. Does it mean that it is a totally subjective judgement? No. But to defend this answer we must identify the *non-empirical cognitive values* that we (implicitly) use when we say that Einstein's theory is *good* and the clairvoyant's is not. Unfortunately, there seems to be no agreement on what should count here. The cognitive values of *explanatory power* or the *simplicity of the assumptions*, their *parsimony, elegance,* etc. are often emphasised, but there is no agreement on what these concepts mean[4]. Quite enigmatically, [12] stated that these values are necessarily imprecise. But what does *imprecise* mean? The word *imprecise* differs from the expression *totally arbitrary* only inasmuch the former necessarily assumes a limited range of possibilities (at least with non negligible probability). If that were the case, we could certainly exploit that limited range to justify many cases of theory selection and define scientific progress! But, unfortunately, nobody ever defined that range. On the contrary, according to well known[10], for any theory $T$, and for a wide class of notions of complexity it is always possible to chose a language in which the complexity of $T$ becomes trivially small. Hence, where Kuhn writes *imprecise*, we are apparently forced to read *totally arbitrary and useless*. Indeed, if we cannot restrict the notions of complexity somehow, the resulting arbitrariness in the cognitive values leads inevitably to almost arbitrary theory selection[5].

To illustrate better this key point, consider the example of the Standard Model of particle physics [6]. The model can be defined in terms of a rather lengthy Lagrangian whose terms must also be defined. The Standard Model represents a spectacular unification of a huge variety of phenomena and it currently agrees, with remarkable precision, with all the experiments.

The problems with the Standard Model are indeed non-empirical. They are

---

[3]The bending of light in 1919. After that, it is easy to build alternative theories that share all the successful predictions of GR, but differ from GR.

[4]See, e.g., [25], where *simplicity* is recognized as the most important constraint, but its characterization is not sufficiently precise to tell why a long collection of experimental reports cannot be regarded as simple as any respectable scientific theory, on suitable metrics.

[5]The status of the value of *explanatory power* is not better, since it also needs some notion of simplicity to be defined. See e.g. the notion of *lovelier* in [14], and the discussion in [22]. See also [7]) and [20].

the lack of an elegant unification with General Relativity[6], the lack of *naturalness*[7], and the presence of about thirty free parameters[8] that must be determined from the experiments. Since none of these is a problem of empirical adequacy, it is essential to understand what are the *non-empirical cognitive values* associated to them. It is difficult to answer this question, because, in principle, we could solve all these problems by rewriting our fundamental laws as $\Xi = 0$, where each fundamental equation of the Standard Model and General Relativity is mapped to some set of digits of the variable $\Xi$. In fact, nothing prevents us to define $\Xi$ as a whole set of complex equations. Superficially, this new formulation is the most elegant and parameter-free formulation we can think of! One could object that $\Xi$ is not directly measurable, and that we can only translate it back to the traditional notation, to assign a value to it. But, what does *directly measurable* mean, exactly? The translation of $\Xi$ is certainly not more complex than building the Large Hadron Collider and interpreting its collision events. Shall we call *directly measurable* only what is seen by a human being without instruments? Isn't the human eye already an instrument (and a very sophisticated one indeed)? We must clarify these issues, if we want to show, convincingly, that the goal of improving the Standard Model is not dictated by subjective taste.

These conclusions are by no means restricted to the Standard Model or to particle physics. In fact, for any empirical scientific discipline, we can always produce an empirically impeccable theory by patching partial models together and resorting to ad-hoc assumptions to save any remaining anomalies. In a suitable $\Xi$-like formulation, that patchwork theory would be both the most accurate and the simplest possible theory[9]. What do we need to improve?

Clearly, we feel that the simplification brought about by $\Xi$ is artificial, and that the idea of simplicity — in spite of its ambiguities — is *not totally arbitrary*. Can we make this feeling precise? What is wrong with $\Xi$? Does it have an *intrinsic* problem that limits its usability or its value? And, if so, which one? Or is it just *our subjective taste* (biased by our cultural legacies) that prevents us to appreciate that odd language? And, if so, why not getting used to $\Xi$? How difficult would it be to build a $\Xi$-meter?

As a matter of fact, interpreting $\Xi$ and building a $\Xi$-meter is not only difficult, but *impossible* for a very fundamental reason [21]: the experimental results cannot always be reported in the form[10]: $\Xi = \Xi_0 \pm \Delta$. This seemingly technical observation has profound consequences for philosophy of science. In simple terms, the idea is the following ([21] reviewed in Sec. 2). The postulates of any empirical scientific theory $T$ must mention at least a set $B$ of properties whose measurements are possible

---

[6]The Standard Model is not necessarily in *contradiction* with (classical) General Relativity: a patchwork theory made of both these theories (combined with any prescription resolving the ambiguities that emerge in contexts that are anyway not yet experimentally accessible) is ugly, cumbersome, but neither contradictory nor in conflict with any experiment.

[7]Naturalness is not precisely defined. Two possibilities are discussed by [2].

[8]Some curiously vanishing; some curiously much much bigger than others.

[9]This argument is sometimes taken as evidence of the need of a *semantic* notion of simplicity, rather than a *syntactic* one. Unfortunately, nobody ever defined *precisely* a semantic notion, able to escape this problem. See also [15] about semantic vs. syntactic views.

[10]I.e., they cannot be written as a central value $\Xi_0$ and a *connected* errorbar of some size $\Delta$.

and can be reported as $b = b_0 \pm \Delta$, for all $b \in B$. Furthermore, the properties $B$ that appear in the postulates must be enough to enable — by employing also the laws of the theory — the operative definition of all the other measurable properties of $T$. We will see that the combination of these two requirements precludes an arbitrarily concise formulation of $T$, which, therefore, cannot be shorter than a minimal length, that is — except for this constraint — *language independent* (see Sec. 2), and, hence, it is a well defined property of $T$ (*conciseness*). By analogy with Kolmogorov complexity, such minimal length is presumably not computable exactly, but can be estimated with finite errorbars. Hence, this notion of conciseness represents a well-defined and non-trivial cognitive value, whose determination is necessarily *imprecise* — as expected — but *not arbitrary*.

Now, having defined a notion of conciseness, can we describe real scientific progress as a Pareto improvement[11] that takes into account only empirical adequacy and conciseness? In this paper I define such simple model of progress and discuss a few points. Because of space limitations a full discussion is impossible here, but it can be found in [22].

## 2 Empirical scientific theories and their reformulations

In order to discuss any cognitive values of scientific theories, we first need to say what we mean by scientific theories in this context. Moreover, we need to establish what may or may not count as a valid reformulation of a scientific theory, since we have seen that $\Xi$-like reformulations undermine any attempt to express precisely any interesting cognitive values and goals of science. Hence, the main goal of this section is to define the set $\mathcal{L}_T$ of theories that are equivalent reformulations of $T$.

As a first step, we need to understand what's wrong with $\Xi$-like reformulations. A key observation is that $\Xi$ is not *directly measurable* even in the weak sense defined by the following[12]:

**Postulate 1.** *(Errorbar-connectedness of direct measurements).* The result of a valid direct measurement of a property $X$ with central value $X_0$ and inverse precision $\Delta$ is always expressed as a <u>connected</u> interval as follows: $X = X_0 \pm \Delta$.

This seems a very weak requirement[13]. Do scientific theories normally have properties for which Postulate 1 does not hold? Yes, they have, one example is $\Xi$. The reason why properties like $\Xi$ do not fulfill Postulate 1 is discussed in great

---

[11]$A$ is a Pareto improvement over $B$, according to a set of qualities, iff $A$ is strictly better than $B$ according to at least one quality, and there is no quality according to which $B$ is better than $A$.

[12]Postulate 1 is not a *definition*: it does not attempt to characterize the intuitive idea of *direct measurements*. Postulate 1 identifies only a minimal requirement, which has the advantage of being clearly verifiable.

[13]It is worthwhile noting that Postulate 1 does not apply only to magnitudes assuming *real* values: $X$ may represent any property that can be associated to a value in the course of an observation. For example, in the context of a botanic theory, a typical observation may involve a decision whether an object is a tree or not. In this case, the property *"being a tree"* assumes values 1 (=true) or 0 (=false). I.e., it can be measured as much as the property *"height of a tree"*. In all cases, the errorbar $\Delta$ remains meaningful and important, because the botanic theory, to which the concept of *tree* belongs, may need to account for the probability of failing to recognize a tree. Hence, the theory must assign the proper meaning to $\Delta$, by associating to it a suitable probability of correct recognition.

detail in [21]. In simple terms, this can be understood because $\Xi$ should encode all the possible empirical consequences of the theory (being $\Xi = 0$ its only law). But any sufficiently complex scientific theory entails many consequence that we cannot measure with any precision. So, for such theories, $\Xi$ cannot be measured with finite errorbars[14].

On the other hand, Postulate 1 is certainly a necessary condition for any property that we might consider directly measurable. In fact, each single direct measurement must at least associate to a quantity $X$ a *central value* and an *error-bar*. Hence, Postulate 1 provides a clear and verifiable recipe to exclude all those formulations of scientific theories that achieve high conciseness at the expenses of any admissible empirical interpretation.

We can now use Postulate 1 to characterize scientific theories to the extent that is needed for our goals. In simple terms, an empirical scientific theory is a mathematical theory, that must also, somehow, make reference to some measurable properties. The precise way in which such reference should be formulated has always been controversial. The minimalist approach adopted here consists in saying that (i) at least some properties of the theory $T$ must be directly measurable, at least in the weak sense of Postulate 1, and (ii) the measurements of any other measurable property of $T$ must be expressible in terms of those that are directly measurable[15]. This idea is made precise by the following Def. 1 and Def. 2:

**Definition 1.** *(Scientific theories).* A scientific theory is a quadruple $T = \{P, R, B, L\}$, where

- $P$ is a set of principles[16],

- $R$ is a set of results deduced from $P$ (according to the logic rules included in $P$),

- $B$ is a set of properties that appear in $P$ and are <u>directly measurable</u> in the sense of Postulate 1 (we call them Basic Measurable Properties, or BMPs, of $T$),

- $L$ is the language in which all the previous elements are formulated[17].

---

[14]For example, in the Standard Model, we can imagine many *thought experiments* in which we can predict the behaviour of a few particles. Most of these thought experiments cannot be realised in practice, because we need ways to produce those particles in the wanted states (e.g., through an accelerator) and detect their later behaviours (e.g. through a detector). Very few phenomena are *both* predictable *and* measurable. See [21] for a very simple — but not too simple — example of a scientific theory that shares this property.

[15]Note that a theory typically contains also *non-measurable* properties, for which we put no constraints here. Their role is important to improve the conciseness of the formulation. See Sec. 3.

[16]The principles contain *all* the assumptions needed to derive the results of the theory, from the logical rules of deduction to the modeling of the experimental devices and of the process of human perception. To be clear, what is sometimes called *background science* is regarded here as part of the theory. Note that also the *domain of applicability* of the theory can and must be defined by specifying suitable restrictions on the principles themselves.

[17]The cognitive values we are interested in might be very sensitive to the choice of the language and therefore, assuming a fixed given language is not an option. One of the goals of Def. 3 is to gather in a single equivalence class all those (infinite) scientific theories that differ by a trivial change of the language.

Note that Postulate 1 and Def. 1 cannot *ensure* that the interpretations of the BMPs are fixed: theories face the tribunal of experience as a whole [18], and the assumptions of sufficient unambiguity of their BMPs are part of the theories. The aim of Postulate 1 is not to fix the interpretation of any theoretical expression, its aim is rather to exclude totally implausible interpretations.

Besides the BMPs, a theory can typically define many other (possibly unlimited) measurable properties. These can be characterized as follows:

**Definition 2.** *(Measurable properties).* The measurable properties (MPs) of a theory $T$ are all those properties that can be determined through observations of the BMPs $B$ of $T$, by employing results $R$ of $T$. Their precision is also determined by $T$.

Hence, the BMPs must be sufficient to enable the measurements of all the MPs that the theory needs to describe. In other words, the BMPs provide — together with the principles to which they belong — the basis[18] on which the whole interpretation of the theory is grounded. Thanks to the identification of the BMPs, the principles truly encode *all the assumptions* of the theory, in a sense that goes beyond the logical structure of the theory. This observation deserves to be emphasized:

**Remark 1.** The principles $P$ of a theory $T$ encode all the information needed to fully characterize $T$, in the following sense: the $P$ are sufficient, in principle, to enable anyone to check whether any given derivation of a result $r \in R$ is actually valid in $T$. Moreover, the principles $P$ are sufficient to enable anyone who can interpret the BMPs $B$ to check the agreement between any result $r \in R$ and any experiment.

We can finally address the question that motivated us at the beginning of this Sec. 2: to what extent can we claim that a theory $T'$ is only a reformulation of another theory $T$? According to Def. 1 any translation of $T$ in a different language counts as a different theory. But we obviously want to identify different formulations, as long as their differences are not essential. This is the case when two theories are equivalent both from the logical and from the empirical point of view, i.e., when all their statements concerning any MPs agree. More precisely:

**Definition 3.** *(Equivalent formulations for $T$).* We say that $T$ and $T'$ are equivalent formulations iff:

(i) there is an isomorphism $\mathcal{I}$ between the logical structures of $T$ and $T'$ (*logical equivalence*);

(ii) and for each MP $c$ of $T$ (resp. $c'$ of $T'$), $\mathcal{I}(c)$ (resp. $\mathcal{I}^{-1}(c')$) is also measurable with the same precision and the same interpretation (*empirical equivalence*).

We denote $\mathcal{L}_T$ the set of all pairs $(L, B)$ of available languages and BMPs in which we can reformulate $T$ and obtain a new theory $T'$ that is equivalent to $T$. In the following, the symbol $T$ refers to a scientific theory up to equivalent formulations,

---

[18] It is not a *universal* basis as in [4]. All MPs (basic or not) are completely theory dependent.

while $T^{(L[,B])}$ refers to its formulation in the language $L$ [and basis $B$][19].

In particular, Def. 3 implies that the $\Xi$ formulation is not equivalent to the Standard Model: the translation that makes them logically equivalent cannot realize also an empirical equivalence, because the $\Xi$ is not an acceptable MP, for the Standard Model.

## 3 A measure of complexity of the assumptions

After having introduced a representation for scientific theories and their admissible reformulations, I turn to the original goal of identifying at least one non-empirical cognitive value that can justify theory selection. In particular I want to define what it means that a theory $T$ has *fewer or simpler assumptions* than another theory $T'$.

The notion of *few, simple assumptions* is closely related to the classic idea of simplicity in the philosophy of science (e.g., [8]; [1]; [24]; [9]). The problem of simplicity is often identified with the fact that simplicity has many different, and even conflicting, meanings. But this point of view implicitly assumes that it is not possible to adopt *that notion of complexity which is optimal (i.e. minimal) for $T$*, when evaluating the complexity of $T$. The reason why this *unambiguous* option is never considered is that — if nothing prevents a $\Xi$-like formulation — then the minimal complexity of any $T$ is always trivial (take the length of its $\Xi$-like formulation). *This* forces us to look for alternative, non-optimal, notions of complexity and *then* it becomes arbitrary to decide which one to adopt. But the previous section rules out precisely the general availability of $\Xi$-like formulations, and it becomes meaningful to define the complexity of $T$ as the *minimum* over the *truly equivalent* formulations:

**Definition 4.** *(Complexity of the assumptions; conciseness).* Let $P^{(L,B)}$ be the principles of $T$, when expressed in language $L$ and with BMPs $B$. Let the *complexity of the assumptions* of $T$ be:

$$\mathcal{C}(T) = \min_{(L,B) \in \mathcal{L}_T} \text{length}[P^{(L,B)}] \qquad (1)$$

where the function length counts the number of characters in the language $L$. Let the *conciseness* of $T$ be the inverse of $\mathcal{C}(T)$.

Although the set of equivalent formulations $\mathcal{L}_T$, defined in Def. 3, is expected to be very large, the minimum of Eq. (1) is not trivial, in general, because Postulate 1 rules out $\Xi$-like formulations and improving the conciseness of a formulation becomes a challenging task. Moreover, being the minimum over all equivalent formulations, Def. 4 effectively assigns to $T$ that notion of complexity in which $T$ fares best, under the constraint of measurability of Postulate 1. *I conjecture that Def. 4 represents well — within the limited precision associated to it — the notion of complexity which is relevant for scientific theory selection.* The rest of this paper is devoted to support this conjecture.

---

[19] Note that we do not require that $B' = \mathcal{I}(B)$: two equivalent theories may choose different BMPs, because what is basic for one theory may not be basic for the other. Only the equivalence of the MPs is required.

## 3.1 Analysis of $\mathcal{C}(T)$

The proper justification of the above conjecture can only come from the comparison with real cases of scientific theory selection[20]. This is done in [22]. Here I only comment on some general aspects of $\mathcal{C}(T)$:

1. The first comment concerns the accessible precision in computing $\mathcal{C}(T)$. Although $\mathcal{C}(T)$ is well defined and non-trivial, it is certainly very hard to compute in practice[21]. Even though we are always only interested in comparing the complexity of two alternative theories, that can be expressed by $\delta\mathcal{C}(T, T') := \mathcal{C}(T) - \mathcal{C}(T')$, also $\delta\mathcal{C}$ is often very hard to compute and can be estimated only approximatively. Often, we are not even able to tell whether $\delta\mathcal{C}$ is positive or negative. In fact, modern scientific theories typically combine many assumptions from very different scientific fields. Even when all the assumptions that distinguish $T$ from $T'$ are clearly identified, finding the formulations that minimize respectively $\mathcal{C}(T)$ and $\mathcal{C}(T')$ may require rewriting a substantial part of the body of science. For this reason, we must often rely on an estimate based on the traditional formulation. Moreover, in some cases, the full list of the assumptions of a theory is not entirely specified. This may happen, for example, when a theory is considered in a preliminary form by its very proponents (a status that may last a long time); but it may also happen when old theories contain implicit assumptions whose necessity was overlooked, until an alternative theory reveals them [22]. All this adds further uncertainty on the estimate of $\delta\mathcal{C}$.

But the limited precision of $\delta\mathcal{C}$ is exactly the feature that we expect from a sensible notion of complexity in science. Because, in scientific practice, we do not rely on complexity to discriminate between theories with a roughly similar amount of assumptions, since we know that some overlooked formulation might easily reverse our assessment. In those cases, we need to suspend the judgment on simplicity (i.e. accept $\delta\mathcal{C} \simeq 0$, within errors) and rather look for potential different predictions.

On the other hand, there are also many important cases where it is totally unambiguous that $T$ is simpler than $T'$. This is especially important when $T'$ achieves good accuracy only because it puts little effort toward any synthesis. This is the case, for example, when $T'$ adds new parameters or ad-hoc assumptions; or when $T'$ is built by patching together different models, each capable of describing only a small subset of the data; or, in the extreme case, when $T'$ is just a collection of experimental reports. In these cases, the scientists often do not even consider $T'$ a *theory*, but this can be justified only because they use — *implicitly* but *essentially* — a notion equivalent to $\delta\mathcal{C}$ to rule out $T'$.

This picture is fully consistent with the intuitive idea that the notion of complexity is ambiguous, but only to some extent, because there are many cases in which there is absolutely no doubt that $T$ is simpler than $T'$, in any conceivable and usable language. This *limited precision* without *arbitrariness* cannot be justified by a generic appeal to different *opinions*. But it can be justified by computational limi-

---

[20] In this sense I see philosophy of science as an empirical science itself, whose goal is understanding the rules behind that historical phenomenon that we call science.

[21] And perhaps even impossible to compute in principle, because of its likely relation with Kolmogorov $K$ [5].

tations (under constraints of measurability) of a well defined notion of complexity.

2. The second comment concerns possible alternatives to Def. 4. In particular, one may argue that the usage of the length[] in Eq. (1) is just one arbitrary choice. However, since the minimum is taken over all possible formulations, I argue that Def. 4 effectively takes into account any *plausible* notion of complexity. For example, instead of the function length[], one might assign more weight to some symbols, or combinations of symbols. But this would be equivalent to a formulation in which those (combinations of) symbols are repeated, and we still use the length[]. Hence, this possibility is already included in Def. 4, but it is not minimal. Alternatively, one might wish to count only some kind of symbols (i.e. give zero weight to others). But if we cannot find a formulation where the neglected symbols can be removed or have negligible contribution, it is hard to claim that they should not count! Of course, in principle one might consider any other function of $P^{(L,B)}$, but, when it is too different from any traditional notion of complexity — that inevitably boils down to count something — it becomes very difficult to justify it as a plausible notion of complexity.

These arguments do not intend to justify Def. 4 *a priori*. Def. 4 can only be justified by showing that it reproduces the preferences that we expect in paradigmatic real cases. To challenge my conjecture, one should find at least one case where our best estimate of $\delta\mathcal{C}(T, T')$ for empirically equivalent $T$ and $T'$ gives an intuitively unacceptable result.

Combining observations 1. and 2. leads to a new, very important, observation: two different definitions of complexity that are consistent within errors entail identical theory selections, and it is immaterial to discuss which one is better. In other words, we may well propose different notions of complexity that do not coincide with Def. 4, but as long as these alternative definitions do not produce estimate that differ from Def. 4 beyond the estimated errorbars, their effect on theory selection (and hence on scientific progress, as discussed in the next section) is exactly the same.

## 4  A model of scientific progress: describing more with less

Having formulated a notion of minimal complexity of the assumptions $\mathcal{C}$, in Def. 4, we can combine it with the notion of empirical adequacy[22] to give a tentative meaning to the notion of *better theories*. This leads to a simple model of *scientific progress*, which is based only on these two cognitive values. For space limitations, in this paper I only state the definitions. The comparison with paradigmatic real cases of scientific progress can be found in [22].

Since the role of empirical adequacy in theory selection is undisputed, comparing real cases of progress to this model is the proper way to test the conjecture that the complexity of the assumptions $\mathcal{C}$ represents well the non-empirical cognitive values that actually matter in science.

**Definition 5.** *(Better theories; state of the art; outdated theories; scientific progress).*

---

[22]Contrary to van Fraassen, empirical adequacy refers here only to what has been actually measured. We also assume that it refers to an unspecified fixed set of MPs.

Let a theory $T$ be *better* than $T'$ if $T$ is more empirically adequate or has lower complexity of the assumptions than $T'$, without being inferior in any of these aspects. If there is a theory $T$ better than $T'$, we say that $T'$ belongs to the *outdated theories*. Otherwise, we say that $T'$ belongs to the *state of the art*. Finally, we say that we have *scientific progress*, when a state of the art theory $T$ becomes outdated[23]. We call this model of scientific progress $SP_0$.

Note that the state of the art may include also theories that are extremely simple but empirically empty, and theories that are extremely lengthy but agree very well with the experiments (e.g. the collection of experimental reports). We have no unambiguous way to rule them out (and probably we should not). Nevertheless, we have achieved the important result of justifying the growth of scientific knowledge [15], in the sense that very popular scientific theories are regularly overthrown and superseded by better ones. *Moving the edge of the state of the art is what constitutes scientific progress*, and this is what valuable science does, quite often. But, it does not achieve it trivially: for example, collecting more and more experimental reports with unsurprising results, does not make any old theory outdated, and it does not produce, by itself, progress.

Note that both the estimate of the empirical adequacy of $T$ and the complexity of $T$ are affected by errors. This makes every statement about empirical adequacy, simplicity, better theories, etc., a provisional one. For example, new experiments, or a better estimate of simplicity, may bring back to the state of the art an already outdated theory. This is always possible, in principle. The errorbars tell us how unlikely we estimate such event to be. Similarly to any scientific concept, also the philosophical concept of *scientific progress* can be *precisely defined* even though the assessment of its occurrence is necessarily *approximate* and *revisable*.

The state of the art represents, as a whole, our current scientific image of the world [23]. The theories that belong to it cannot be assembled in a single, logically coherent, best theory of all. But they represent, altogether, the toolkit from which we can chose the tool(s) that are best suited to the given empirical questions and to the given requirements of accuracy. Some theories based on Newton mechanics still belong to the state of the art for those issues where quantum mechanical effects are undetectable or where the relevant results cannot yet be deduced from a more fundamental set-up. Moreover, when we are overwhelmed by surprising experimental results, in which we cannot find any regular pattern, even the collection of all experimental reports may be the best theory we have.

Although I have stressed the important role of the state of the art, outdated theories are not thrown away, since hardly anything is thrown away in science. They might still contain ideas that will eventually prove fruitful in the future. But we would never use them in any application. Nor can we sell ideas that might be, perhaps, fruitful, as actual accomplishments.

Def. 5 does not allow the comparison of any two theories, even on the same topic: quite often theory $T$ is neither better not worst than theory $T'$. This reflects the fact that in many cases it is impossible and/or unnecessary to declare a winner.

---

[23]Note that this can only happen because either a new theory $T'$ appears, that is better than $T$, or because a new experiment causes an existing theory $T''$ to become better than $T$.

However, it is important that, sometimes, some new theories appear that are better than existing state of the art theories. This does not need to happen very frequently: it simply needs to be realistic, to define properly the goals of science. This claim must be supported with real cases of scientific progress, that are discussed in [22], that also examines many possible challenges to the model of progress proposed here.

## BIBLIOGRAPHY

[1] Baker, A. (2004). "Simplicity", in E. N. Zalta (Ed.) *Stanford Encyclopedia of Philosophy*, Winter 2004 Edition ed. Stanford University.
[2] Bardeen, W. A. (1995). "On Naturalness in the Standard Model", FERMILAB-CONF-95-391-T.
[3] Bird, A. (2007). "What Is Scientific Progress?", in *Noûs*, 41, 64—89.
[4] Carnap, R. (1966). "Der Logische Aufbau der Welt" (3rd ed.). Hamburg, Germany: Felix Meiner.
[5] Chaitin, G. (1969). "On the Length of Programs for Computing Finite Binary Sequences: Statistical Considerations", in *Journal of the ACM*, 16, 145–159.
[6] Cottingham, W. and D. Greenwood (2007). "An Introduction to the Standard Model of Particle Physics", (2nd ed.). Cambridge University Press.
[7] Crupi, V. and K. Tentori (2012). "A Second Look at the Logic of Explanatory Power (with Two Novel Representation Theorems)", in *Philosophy of Science*, 79, 365–85.
[8] Fitzpatrick, S. (2014). "Simplicity in the Philosophy of Science", in *The Internet Encyclopedia of Philosophy*, ISSN-2161-0002, 1.
[9] Goodman, N. (1977). "The Structure of Appearance" (3rd ed.). Dordrecht, Holland: D. Reidel.
[10] Kelly, K. T. (2009). "Ockham's Razor, Truth, and Information". In J. van Behthem and P. Adriaans (Eds.), *Handbook of the Philosophy of Information*. Dordrecht: Elsevier.
[11] Kuhn, T. S. (1962), "Structure of Scientific Revolutions". Chicago: University of Chicago Press.
[12] Kuhn, T. S. (1977). "The Essential Tension", in *Selected Studies in Scientific Tradition and Change*. Chicago: University of Chicago Press.
[13] Lakatos, I. (1970). "Falsification and the Methodology of Scientific Research Programmes", in I. Lakatos and A. Musgrave (Eds.), *Criticism and the Growth of Knowledge*, pp. 91–196. Cambridge University Press.
[14] Lipton, P. (2004). "Inference to the Best Explanation", in *International library of philosophy and scientific method*. Routledge/Taylor and Francis Group.
[15] Lutz, S. (2014). "What is Right with a Syntactic Approach to Theories and Models?", in *Erkenntnis*, 10.1007/s10670-013-9578-5, 1–18.
[16] Niiniluoto, I. (2014). "Scientific Progress as Increasing Verisimilitude", in *Studies in History and Philosophy of Science*, 46, 73–77.
[17] Popper, K. (1963). "Conjectures and Refutations: The Growth of Scientific Knowledge". London: Routledge and Kegan Paul.
[18] Quine, W. v. O. (1950). "Two Dogmas of Empiricism", in *The Philosophical Review*, 60, 20–43.
[19] Rowbottom, D. (2015). "Scientific progress without increasing verisimilitude: In response to Niiniluoto", in *Studies in History and Philosophy of Science Part A*, 5, 100–104.
[20] Schupbach, J. N. and J. Sprenger (2011). "The logic of explanatory power", in *Philosophy of Science*, 78, 105–27.
[21] Scorzato, L. (2013). "On the Role of Simplicity in Science", in *Synthese*, 190, 2867–2895.
[22] Scorzato, L. (2015). "From measurability to a model of scientific progress", http://philsci-archive.pitt.edu/11498.
[23] Sellars, W. S. (1963). "Philosophy and the Scientific Image of Man", in *Science, Perception, and Reality*, pp. 35–78. Humanities Press/Ridgeview.
[24] Sober, E. (2002). "What is the Problem of Simplicity?", in H. Keuzenkamp, M. McAleer, and A. Zellner (Eds.), *Simplicity, Inference, and Econometric Modelling*, pp. 13–32. Cambridge: Cambridge University Press.
[25] Thagard, P. R. (1978). "The Best Explanation: Criteria for Theory Choice", in *Journal of Philosophy*, 75(2), 76–92.

# Total Evidence, Uncertainty and A Priori Beliefs

Benjamin Bewersdorf

ABSTRACT.
Defining the rational belief state of an agent in terms of her initial or a priori belief state as well as her total evidence can help to address a number of important philosophical problems. In this paper, I discuss how this strategy can be applied to cases in which evidence is uncertain. I argue that taking evidence to be uncertain allows us to uniquely determine an agent's subjective a priori belief state from her present beliefs and her total evidence. However, this also undermines a common assumption on the independence evidence.

## 1 Prior and A Priori Based Rules of Rational Belief Change

Theories of rational belief change state rules about how agents should change their beliefs. Most of these rules define the rational posterior belief state of an agent at a time $t_n$ by the agents prior belief state at $t_{n-1}$ and the evidence the agent has received between $t_{n-1}$ and $t_n$. I will call such rules *prior based rules*.

A different type of rules defines the rational posterior belief state of an agent at a time $t_n$ by an initial or a priori belief state of the agent at $t_0$ and the total evidence the agent received up to $t_n$. I will call such rules *a priori based rules*.

A priori based rules date back at least to Carnap and are frequently employed under various names to address a large range of topics.[1] Since a priori based rules presuppose knowledge of the agent's total evidence and her a priori belief state, their applications have stronger prerequisites than prior based rules. In return, a priori based rules allow us to make changes to the agent's a priori belief state while retaining the agent's evidence as well as to remove particular pieces of information from the agent's body of evidence. This can be utilized to address topics such as the problem of old evidence, the relation between credence and chance, scientific revolutions and language change, the doomsday problem, evidence loss due to forgetting or undermining, as well as the dynamics of de se beliefs.[2]

---

[1] My terminology is borrowed from Jeffrey [13]. Carnap [3], Lewis [18] and Hall [9] speak of initial credence functions, Glymour [8] as well as Howson and Urbach [11] of counterfactual degrees of belief, Earman [5], Bartha and Hitchcock [1], Weatherson [22] and Meacham [19] of hypothetical priors and Williamson [25] of a conceptually prior probability distribution.

[2] See for example Glymour [8], pp. 85-93, Howson and Urbach [11], pp. 270-271, and Earman [5], pp. 119-135, on the problem of old evidence, Lewis [18] and Hall [9] on the relation of credence and chance, Earman [5], pp. 195-199, and Wenmackers and Romeijn [24] on scientific revolutions and language change, Bartha and Hitchcock [1] on the doomsday problem, Williamson [25], pp. 218-221, and Titelbaum [20] on evidence loss, and Meacham [19] on the dynamics of de se beliefs.

The a priori based rules used in the literature presuppose that evidence comes in form of certainties. It has been argued that this is rarely, if ever, the case.[3] I will therefore discuss the prospects of applying a priori based rules to cases in which evidence is uncertain. It will turn out that taking evidence to be uncertain has two noteworthy consequences for a priori based accounts. On the one hand, it is possible to uniquely determine an agent's subjective a priori belief state from her present belief state and her total evidence if her evidence is uncertain. This is particularly interesting, since the lack of knowledge about the a priori belief state is a major complaint against a priori based accounts. On the other hand, the applications of a priori based rules rely on an assumption about the independence of evidence and this assumption is less plausible for uncertain evidence. Presupposing evidence to be uncertain thus creates a new challenge for the applicability of a priori based rules.

## 2  Certain Evidence

I will presuppose the Bayesian framework for rational belief change in the following. According to the Bayesian, the belief state of an agent can be represented by a probability distribution on an algebra of propositions. The probability values represent the degree to which the agent believes these propositions to be true. A probability of 1 represents the agent being certain in the truth of a proposition.

Evidence is represented as a change in the degrees of belief in one or more propositions. In the simplest case the evidence of an agent consists in the agent changing her degrees of belief in a proposition $q$ to 1. In this case, I will say that the agent receives certain evidence that $q$.

According to the Bayesian, an agent who receives certain evidence should adjust her beliefs in accordance to the following rule.

### Simple Conditionalization

Let $P_{n-1}$ with $P_{n-1}(q) > 0$ be the prior belief state of an agent at $t_{n-1}$. If the agent receives certain evidence that $q$ between $t_{n-1}$ and $t_n$, her belief state at $t_n$ should be $P_n(\cdot) = P_{n-1}(\cdot \mid q)$.

Instead of defining the agents rational belief state at $t_n$ by her belief state at $t_{n-1}$ and the evidence the agent received between $t_{n-1}$ and $t_n$, it is also possible to define the rational belief state of the agent at $t_n$ by her initial or a priori belief state at $t_0$ and the total evidence the agent has received up to $t_n$.

### A Priori Based Simple Conditionalization

Let $P_0$ with $P_0(q^*) > 0$ be the a priori belief state of an agent. If the agent's total certain evidence at $t_n$ is $q^*$, her belief state at $t_n$ should be $P_n(\cdot) = P_0(\cdot \mid q^*)$.

The a priori based rules discussed in the literature are variants of a priori based simple conditionalization, differing mostly in the interpretations of the evidence and

---

[3]See for example Jeffrey [14].

the a priori belief state.[4] The total certain evidence of an agent at $t_n$ is assumed to be the conjunction of the evidence available to the agent at $t_n$.[5]

### Total Certain Evidence
Let $q_1, q_2, ..., q_n$ be individual pieces of certain evidence, then $q_1 \cap q_2 \cap ... \cap q_n$ is the corresponding total certain evidence.

There is no explicit argument given for this account of total certain evidence, but it can be motivated by the following considerations. Assume an agent receives the individual pieces of certain evidence $q_1, q_2, ..., q_n$ between $t_0$ and $t_n$. As it should not matter whether an agent receives all of her evidence at once or in separate pieces, the total evidence $q^*$ of the agent at $t_n$ should be such that receiving $q^*$ results in the same belief state as receiving the individual pieces of evidence $q_1, q_2, ..., q_n$ one after the other. I will call this the *condition of total certain evidence*.

### Condition of Total Certain Evidence
$q^*$ is the total certain evidence of the individual pieces of certain evidence $q_1, q_2, ..., q_n$ iff a belief change of $P$ by $q^*$ via simple conditionalization results in the same belief state as an iterated belief change of $P$ by $q_1, q_2, ..., q_n$ via simple conditionalization for every $P$.

It can easily be shown that total certain evidence fulfills this condition.

**Theorem 1** *Total certain evidence fulfills the condition of total certain evidence.*

All proofs can be found in the appendix.

## 3 Uncertain Evidence

It has been argued that evidence is rarely if ever certain. Jeffrey [14] proposed to represent evidence more generally by a change in the degrees of belief of the elements of a partition $\{q_1, q_2, ..., q_k\}$ to any probabilistically coherent new probability values. I will call evidence uncertain iff no element of this partition receives probability 1. Jeffrey argues that simple conditionalization can be generalized in the following way.

### Jeffrey Conditionalization
Let $P_{n-1}$ represent the prior belief state of an agent. If the agent receives evidence represented by a change in the degrees of belief of the elements of a partition $\{q_1, q_2, ..., q_k\}$ to $P_n(q_i) \in [0,1]$, with $\sum_{i=1}^{k} P_n(q_i) = 1$, her posterior belief state should be $P_n(\cdot) = \sum_{i=1}^{k} P_{n-1}(\cdot \mid q_i) \times P_n(q_i)$.

Instead of representing evidence by a partition of propositions and new degrees of belief, Field [6] proposed to represent evidence by a partition of propositions and update factors which denote how strongly the probabilities for the elements of the partition change. Field proposed to define these update factor as follows.

### Field Update Factor

---
[4] See footnote 1 and 2.
[5] See for example Williamson [25], p. 220.

If an agent changes her degrees of belief in the elements of a partition $\{q_1, q_2, ..., q_k\}$ from $P_{n-1}(q_i)$ to $P_n(q_i) \in (0,1)$ with $\sum_{i=1}^{k} P_n(q_i) = 1$, then the update factors $\alpha_i$ for each $q_i$ representing this change are $\alpha_i = 1/k \times \log \prod_{j=1}^{k} \frac{P_n(q_i)/P_{n-1}(q_i)}{P_n(q_j)/P_{n-1}(q_j)}$.

Uncertain evidence can then be defined in terms of update factors as follows.

### Uncertain Evidence

A piece of evidence $\xi$ is an ordered pair $\langle \wp, \Im \rangle$, with $\wp$ being a partition $\{q_1, q_2, ..., q_k\}$ and $\Im$ a function $\Im : \wp \mapsto (-\infty, \infty)$ assigning each element of the partition $q_i$ an update factor $\alpha_i$.

Uncertain evidence defined in this way includes certain evidence only as a limiting case if $\alpha_i$ goes to $\infty$ for some i.

Jeffrey conditionalization can be combined with Field's definition of update factors to yield the following version of Jeffrey conditionalization for evidence given in terms of update factors.

### Field Conditionalization

Let $P_{n-1}$ represent the prior belief state of an agent. If the agent receives evidence represented by $\langle \{q_1, q_2, ..., q_k\}, \Im \rangle$ with $\Im(q_i) = \alpha_i$ and $\sum_{i=1}^{k} \alpha_i = 0$, her posterior belief state should be $P_n(\cdot) = \frac{\sum_{i=1}^{k} P_{n-1}(\cdot \cap q_i) \times e^{\alpha_i}}{\sum_{i=1}^{k} P_{n-1}(q_i) \times e^{\alpha_i}}$.

As Jeffrey and Field conditionalization are interdefinable via Field's definition of update factors, they are essentially the same update rule. The difference between Jeffrey's and Field's account is that Jeffrey presupposes evidence to be given in terms of new degrees of belief, while Field presupposes evidence to be given in terms of update factors. This difference will become important below.

It has been argued by Garber [7] that Field conditionalization has the following counterintuitive consequence. Assume that an agent looks at a cup of coffee several times in a row. Let $q$ be the proposition that there is a cup of coffee and assume that the evidence the agent receives by looking at the cup of coffee can be represented by the same evidence $\xi = \langle \{q, \neg q\}, \Im \rangle$ with $\Im(q) > 0$ each time. It can easily be shown that as long as the agents prior degree of belief in $q$ is positive, by repeatedly looking at the cup of coffee the agent will soon be justified to be virtually certain that $q$, according to Field conditionalization. This is the case even if $\Im(q)$ is very low, which means that the evidence for $q$ is very weak.

The assumption that the agent's evidence is $\xi$ each time the agent looks at the cup is essential for Garber's argument. As has been pointed out by Wagner [21], this is the very same assumption that leads to the alleged commutativity problem of Jeffrey conditionalization. Since Jeffrey and Field conditionalization are essentially the same update rule, it is unsurprising that an assumption problematic for one is problematic for the other. The easiest response to Garber's objection is to reject this assumption and this is what I will do for the purpose of this paper.[6] If the

---

[6]This is also Lange's response to the alleged commutativity problems of Jeffrey conditionalization, see Lange [17].

agent learns nothing new by looking at the cup a second time, the evidence she receives should not be represented by $\xi$, but by $\xi' = \langle\{q, \neg q\}, \Im\rangle$ with $\Im(q) = 0$ instead.

## 4 A Priori Based Field Conditionalization and Total Uncertain Evidence

As shown above, simple conditionalization can easily be transformed into an a priori based rule of rational belief change. The same holds true for Field conditionalization.

### A Priori Based Field Conditionalization
Let $P_0$ be the a priori belief state of an agent. If the agent's total uncertain evidence at $t_n$ is represented by $\xi^* = \langle\{q_1, q_2, ..., q_k\}, \Im\rangle$ with $\Im(q_i) = \alpha_i$ and $\sum_{i=1}^{k} \alpha_i = 0$, her belief state at $t_n$ should be $P_n(\cdot) = \frac{\sum_{i=1}^{k} P_0(\cdot \cap q_i) \times e^{\alpha_i}}{\sum_{i=1}^{k} P_0(q_i) \times e^{\alpha_i}}$.

As discussed above, the total certain evidence of an agent can be argued to be the conjunction of the agent's individual pieces of evidence. Since uncertain evidence is not represented by a single proposition, it cannot be aggregated as easily. However, we can employ the same reasoning to determine the right way to aggregate uncertain evidence that we used in the case of certain evidence. As with certain evidence, it should not matter whether an agent receives all of her uncertain evidence at once or in separate pieces. Thus, the following analog to the condition of total certain evidence should hold for an agent who uses Field conditionalization instead of simple conditionalization to update her degrees of beliefs.

### Condition of Total Uncertain Evidence
$\xi^*$ is the total uncertain evidence of the individual pieces of uncertain evidence $\xi_1, \xi_2, ..., \xi_n$ iff a belief change of $P$ by $\xi^*$ via Field conditionalization results in the same belief state as an iterated belief change of $P$ by $\xi_1, \xi_2, ..., \xi_n$ via Field conditionalization for every $P$.

The condition of total uncertain evidence is fulfilled by the following aggregation rule for uncertain evidence.[7]

### Total Uncertain Evidence
Let $\xi_1 = \langle\wp_1, \Im_1\rangle$ and $\xi_2 = \langle\wp_2, \Im_2\rangle$ be individual pieces of uncertain evidence of an agent. Then the total uncertain evidence of this agent is $\xi_{1\oplus 2} = \langle\wp_{1\oplus 2}, \Im_{1\oplus 2}\rangle$, with $\wp_{1\oplus 2} = \{q_i \cap r_j \mid q_i \in \wp_1, r_j \in \wp_2, q_i \cap r_j \neq \emptyset\}$ and $\Im_{1\oplus 2}(q_i \cap r_j) = \Im_1(q_i) + \Im_2(r_j)$ for all $q_i \cap r_j \in \wp_{1\oplus 2}$.

**Theorem 2** *Total uncertain evidence fulfills the condition of total uncertain evidence.*

---

[7] Jeffrey [15] offers a similar account of combining evidence in his discussion of commutativity. His account makes use of a variant of Field conditionalization that defines update factors in a different way. As I will show in the following, Field's original account has the advantages that it guarantees a probabilistically coherent posterior belief state for arbitrary update factors and that update factors can simply be added when evidence is aggregated.

## 5 Normalization

Field conditionalization, as stated above, is defined only for uncertain evidence with update factors that sum to 0. I will call this the *normalization requirement* and will call evidence that fulfills this requirement *normalized evidence*.

### Normalized Evidence

A piece of evidence $\xi = \langle \wp, \Im \rangle$, with $\wp = \{q_1, q_2, ..., q_k\}$ and $\Im(q_i) = \alpha_i$ is normalized iff $\sum_{i=1}^{k} \alpha_i = 0$.

It is easy to see that the total of two pieces of normalized uncertain evidence need not be normalized if total uncertain evidence is defined as above. This means that Field conditionalization is not defined for all cases of total uncertain evidence. Fortunately, it is possible to drop the requirement that the update factors have to sum to 0 in the definition of Field conditionalization or alternatively to normalize every piece of total evidence such that the sum of the update factors of the evidence is 0.

The normalization requirement in Field conditionalization is the equivalent to the requirement of Jeffrey conditionalization that the sum of the new probabilities of the partition of the evidence is 1. For Jeffrey conditionalization this condition is central, since the posterior belief state of the agent will not be probabilistically coherent if this condition is not satisfied. Interestingly, this is not the case for Field conditionalization. If $P_{n-1}$ is probabilistically coherent, updating $P$ with $\xi$ via Field conditionalization results in a probabilistically coherent $P_n$ even if $\xi$ is not normalized. This can be seen by noting that there is an equivalent normalized evidence for every non normalized evidence.

### Normalization of Evidence

Let $\xi = \langle \wp, \Im \rangle$, $\wp = \{q_1, q_2, ..., q_k\}$ and $\sum_{i=1}^{k} \Im(q_i) = x$, then $\xi_N$ is the normalization of $\xi$ iff $\xi_N = \langle \wp, \Im_N \rangle$ with $\Im_N(\cdot) = \Im(\cdot) - x/k$.

### Equivalent Evidence

Two pieces of evidence $\xi_1$ and $\xi_2$ are equivalent iff for every belief state $P_{n-1}$, updating $P_{n-1}$ with $\xi_1$ via Field conditionalization and updating $P_{n-1}$ with $\xi_2$ via Field conditionalization result in the same belief state $P_n$.

**Theorem 3** *For every non-normalized evidence $\xi$ there is an equivalent normalized evidence $\xi_N$.*

Since the normalization requirement is not necessary to guarantee a probabilistically coherent posterior belief state, it can be dropped from the definition of Field conditionalization. Alternatively, a normalization step can be added to the definition of the total evidence.

## 6 Consequences of Uncertain Evidence

Taking evidence to be uncertain has two noteworthy consequences for the applications of a priori based rules. On the one hand, it is possible to uniquely determine the a priori belief state of an agent from the agent's present belief state and the agent's present total evidence if the latter is uncertain and given in terms of update factors. On the other hand, taking evidence to be uncertain undermines the assumption that individual pieces of evidence are independent of each other and the a priori belief state. Since this assumption is crucial for applications of a priori based rules, taking evidence to be uncertain creates a new challenge for a priori based accounts.

To show how the a priori belief state of an agent can be determined, define inverse evidence as follows.

**Inverse Evidence**
Let $\xi = \langle \wp, \Im \rangle$ with $\wp = \{q_1, q_2, ..., q_k\}$ be a piece of evidence, then $\xi_I = \langle \wp, \Im_I \rangle$ is the respective inverse evidence iff $\Im_I(q_i) = -\Im(q_i)$ for all i.

It can be shown that the a priori belief state of an agent is the result of updating her present belief state with the inverse of her present total evidence by Field conditionalization.

**Theorem 4** *Let P be the present belief state and $\xi$ the present total evidence of an agent, then updating P with $\xi_I$ via Field conditionalization results in the a priori belief state of the agent.*

It might come as a surprise that the a priori belief state of an agent can be determined so easily. The crucial assumption for this result is that the agent's evidence is given in terms of update factors. I will call evidence characterized this way *Field evidence* and evidence characterized by new probabilities *Jeffrey evidence*. Both certain evidence and Jeffrey evidence partly determine the new belief state of an agent independently of the agent's old beliefs. If an agent receives certain evidence that q, her new degree of belief in q will be 1, no matter what her old degree of belief in q has been. The same holds for Jeffrey evidence except that the new degree of belief for q can have any value. This implies that we cannot determine the prior belief state of an agent from her present belief state and her certain evidence or Jeffrey evidence. As we have seen, this is different for Field evidence. This might suggest that Field evidence contains more information than Jeffrey evidence, but that is not the case. Field evidence and Jeffrey evidence contain different information: the relation between old and new degrees of belief in the first case and the new degrees of belief in the second. It is an open question whether evidence is best understood in terms of update factors or new degrees of belief.[8] I cannot enter this debate here, but it seems to depend on the underlying conception of evidence. I will say a bit more on this below. For now, I will be content with the conclusion that the a priori belief state of an agent can be determined by her total evidence

---

[8] See for example Hawthorne [10] for a defense of Field evidence and further references on this debate.

and her present belief state if her total evidence is uncertain and given in terms of update factors. This conclusion is of interest since the alleged inaccessibility of the agent's a priori beliefs is a major complaint against a priori based accounts.

In the first section of this paper I mentioned that the applications of a priori based rules involve changes to the agent's a priori belief state or to some of her pieces of evidence. It is commonly assumed without argument that doing so does not affect the agent's remaining pieces of evidence. While this independence assumption might be plausible for certain evidence, it seems less plausible for uncertain evidence. The classic example of uncertain evidence for which the independence assumption is supposed to fail is color perception. Perceptual evidence about the color of an object is argued to depend on the perceiver's background beliefs about the color of the light shining on the object.[9] Apart from these concerns, the independence assumption also causes the problems for Jeffrey and Field conditionalization mentioned in section 3. Thus, taking evidence to be uncertain seems to force us to reject this assumption.

Since the applications of a priori based rules rely on changing the agent's a priori belief state or part of the agent's evidence, rejecting the independence assumption requires us to provide an account on how such changes do affect the remaining evidence of the agent. The prospects of providing such an account depend on the presupposed conception of evidence. Jeffrey [12] takes evidence to be the causal effect of sensory stimulations. According to Jeffrey, evidence can be measured by observing the agent's degrees of belief after such stimulations occurred. This conception of evidence favors evidence understood in terms of new degrees of belief. According to this view, an account of how pieces of evidence depend on each other would be a hard to establish causal law of cognitive psychology. Carnap in contrast treats evidence as the justification provided by our experiences.[10] This conception of evidence favors evidence understood in terms of update factors. According to this view, an account of how pieces of evidence depend on each other would be part of a normative account of rational belief change. I discuss the prospects of providing such an account elsewhere.[11]

# 7 Appendix

**Proof of Theorem 1**

Let $P$ be a belief state and $q$ and $r$ individual pieces of certain evidence. Let $P_{q\cap r}$ be the result of a belief change of $P$ via simple conditionalization with $q \cap r$, $P_q$ the result of a belief change of $P$ via simple conditionalization with $q$ and $P_{q,r}$ the result of a belief change of $P_q$ via simple conditionalization with $r$. To prove theorem 1 it is sufficient to show that $P_{q\cap r} = P_{q,r}$.

$$P_{q,r} = \frac{P_q(\cdot \cap r)}{P_q(r)} = \frac{P(\cdot \cap q \cap r)}{P(q \cap r)} = P_{q\cap r}. \quad \square$$

---
[9] See for example Christensen [4] and Weisberg [23].
[10] See Carnap's letter to Jeffrey in Jeffrey [13].
[11] See Bewersdorf [2].

**Proof of Theorem 2**

Let $P$ be a belief state and $\xi_1$ and $\xi_2$ two pieces of uncertain evidence such that $\xi_1 = \langle \wp_1, \Im_1 \rangle$ with $\wp_1 = \{q_1, q_2, ..., q_k\}$, $\Im_1(q_i) = \alpha_i$ and $\xi_2 = \langle \wp_2, \Im_2 \rangle$, with $\wp_2 = \{r_1, r_2, ..., r_m\}$, $\Im_2(r_j) = \beta_j$. Let $P_{\xi_1 \oplus 2}$ be the result of updating $P$ via Field conditionalization with $\xi_{1 \oplus 2}$, $P_{\xi_1}$ the result of updating $P$ via Field conditionalization with $\xi_1$ and $P_{\xi_1, \xi_2}$ the result of updating $P_{\xi_1}$ via Field conditionalization with $\xi_2$. To prove theorem 2 it is sufficient to show that $P_{\xi_1 \oplus 2} = P_{\xi_1, \xi_2}$.

By the definition of total uncertain evidence and Field conditionalization we get

$$P_{\xi_1 \oplus 2}(\cdot) = \frac{\sum_{i=1}^k \sum_{j=1}^m P(\cdot \cap q_i \cap r_j) \times e^{\alpha_i + \beta_j}}{\sum_{i=1}^k \sum_{j=1}^m P(q_i \cap r_j) \times e^{\alpha_i + \beta_j}}.$$

By applying Field conditionalization twice we get

$$P_{\xi_1, \xi_2} = \frac{\sum_{j=1}^m \frac{\sum_{i=1}^k \frac{P(\cdot \cap q_i \cap r_j) \times e^{\alpha_i}}{\sum_{i=1}^k P(q_i) \times e^{\alpha_i}}}{1} \times e^{\beta_j}}{\sum_{j=1}^m \frac{\sum_{i=1}^k \frac{P(q_i \cap r_j) \times e^{\alpha_i}}{\sum_{i=1}^k P(q_i) \times e^{\alpha_i}}}{1} \times e^{\beta_j}}$$

$$= \frac{\sum_{i=1}^k \sum_{j=1}^m P(\cdot \cap q_i \cap r_j) \times e^{\alpha_i + \beta_j}}{\sum_{i=1}^k \sum_{j=1}^m P(q_i \cap r_j) \times e^{\alpha_i + \beta_j}} = P_{\xi_1 \oplus 2}(\cdot). \quad \square$$

**Proof of Theorem 3**

Let $\xi = \langle \wp, \Im \rangle$, with $\wp = \{q_1, q_2, ..., q_k\}$, $\Im(q_i) = \alpha_i$ and $\sum_{i=1}^k \Im(q_i) = x$. Let $\xi_N = \langle \wp, \Im_N \rangle$ be the normalization of $\xi$. Thus, $\sum_{i=1}^k \Im_N(q_i) = 0$. Let $P_\xi$ be the result updating $P$ with $\xi$ via Field conditionalization and let $P_{\xi_N}$ be the result updating $P$ with $\xi_N$ via Field conditionalization. To prove theorem 3 it is sufficient to show that $P_\xi = P_{\xi_N}$ for all $P$.

By Field conditionalization and normalization of evidence we get

$$P_{\xi_N} = \frac{\sum_{i=1}^k P(\cdot \cap q_i) \times e^{\alpha_i + x/k}}{\sum_{i=1}^k P(q_i) \times e^{\alpha_i + x/k}} = \frac{\sum_{i=1}^k P(\cdot \cap q_i) \times e^{\alpha_i}}{\sum_{i=1}^k P(q_i) \times e^{\alpha_i}} = P_\xi. \quad \square$$

**Proof of Theorem 4**

Let $\xi = \langle \wp, \Im \rangle$, with $\wp = \{q_1, q_2, ..., q_k\}$ be the present total evidence and $P$ be the present belief state of an agent. Let $P_{\xi_I}$ be the result of updating $P$ with $\xi_I$ via Field conditionalization, and let $P_{\xi_I, \xi}$ be the result of updating $P_{\xi_I}$ with $\xi$ via Field conditionalization. Since for all probability distributions $Q$ and $Q'$, $Q_\xi = Q'_\xi$ only if $Q = Q'$, $P_{\xi_I}$ is the unique a priori belief state of the agent iff $P = P_{\xi_I, \xi}$. Thus, to prove theorem 4 it is sufficient to show that $P = P_{\xi_I, \xi}$.

Let $\Im(q_i) = \alpha_i$ and thus by inverse evidence $\Im_I(q_i) = -\alpha_i$. By applying Field conditionalization twice we get

$$P_{\xi_I, \xi}(\cdot) = \frac{\sum_{i=1}^k P(\cdot \cap q_i) \times e^{\alpha_i} \times e^{-\alpha_i}}{\sum_{i=1}^k P(q_i) \times e^{\alpha_i} \times e^{-\alpha_i}} = \frac{\sum_{i=1}^k P(\cdot \cap q_i)}{\sum_{i=1}^k P(q_i)} = P(\cdot). \quad \square$$

## 8 Acknowledgments

My work has been supported by the Netherlands Organisation for Scientific Research under the project number 360-20-282.

## BIBLIOGRAPHY

[1] Bartha, P. and Hitchcock, C. (1999). "No one knows the date or the hour: An unorthodox application of rev. Bayes's theorem", *Philosophy of Science*, 66 (Proceedings), pp. S339-S353.
[2] Bewersdorf, B. (unpublished). "Experience, Belief Change and Confirmational Holism".
[3] Carnap, R. (1971). "Inductive Logic and Rational Decisions", in Carnap, R. and Jeffrey, R. (eds.). *Studies in Inductive Logic and Probability*, Volume 1, pp. 5-31, University of California Press.
[4] Christensen, D. (1992). "Confirmation Holism and Bayesian Epistemology", *Philosophy of Science*, 59, pp. 540-557.
[5] Earman, J. (1992). *Bayes or Bust? A Critical Examination of Bayesian Confirmation Theory*, MIT Press.
[6] Field, H. (1978). "A Note on Jeffrey Conditionalization", *Philosophy of Science*, 45, pp. 361-367.
[7] Garber, D. (1980). "Field and Jeffrey Conditionalization", *Philosophy of Science*, 47, pp. 142-145.
[8] Glymour, C. (1980). *Theory and Evidence*, Princeton University Press.
[9] Hall, N. (1994). "Correcting the Guide to Objective Chance" *Mind*, 103, pp. 505-517.
[10] Hawthorne, J. (2004). "Three Models of Sequential Belief Updating on Uncertain Evidence", *Journal of Philosophical Logic*, 33, pp. 89-123.
[11] Howson, C. and Urbach, P. (1989). *Scientific Reasoning: The Bayesian Approach*, Open Court.
[12] Jeffrey, R. (1968). "Probable Knowledge", *Studies in Logic and the Foundations of Mathematics*, 51, pp. 166-190. Reprinted in Jeffrey 1992.
[13] Jeffrey, R. (1975). "Carnap's Empiricism". *Minnesota Studies in Philosophy of Science*, 6, pp. 37-49
[14] Jeffrey, R. (1983). *The Logic of Decision*, 2 ed., University of Chicago Press.
[15] Jeffrey, R. (1988). "Conditioning, Kinematics, and Exchangeability", in Skyrms, B. and Harper, W. (eds.). *Causation, Chance and Credence*, Volume 1, Kluwer Academic Publishers, pp. 221-255. Reprinted in Jeffrey 1992.
[16] Jeffrey, R. (1992). *Probability and the Art of Judgement*, Cambridge University Press.
[17] Lange, M. (2000). "Is Jeffrey Conditionalization defective by virtue of being Non-Commutative? Remarks on the Sameness of Sensory Experiences", *Synthese*, 123. pp. 393-403.
[18] Lewis, D. (1980). "A Subjectivist's Guide to Objective Chance", in Jeffrey, R. (ed.). *Studies in Inductive Logic and Probability*, Volume 2, University of California Press, pp. 263-293.
[19] Meacham, C. (2008). "Sleeping Beauty and the Dynamics of de se Beliefs", *Philosophical Studies*, 138, pp. 245-269.
[20] Titelbaum, M. (2013). *Quitting Certainties: A Bayesian Framework modeling Degrees of Belief*, Oxford University Press.
[21] Wagner, C. (2002). "Probability Kinematics and Commutativity", *Philosophy of Science*, 69, pp. 266-278.
[22] Weatherson, B. (2007). "The Bayesian and the Dogmatist", *Proceedings of the Aristotelian Society*, 107, pp. 169-185.
[23] Weisberg, J. (2015). "Updating, Undermining, and Independence", *The British Journal for the Philosophy of Science*, 66, pp. 121-159.
[24] Wenmakers, S. and Romeijn, J.-W. (forthcoming). "A New Theory about Old Evidence", *Synthese*.
[25] Williamson, T. (2000). *Knowledge and its Limits*, Oxford University Press.

# The Pragmatics of Self-Deception

Nevia Dolcini

ABSTRACT. The current philosophical debate on self-deception is characterized by a divide between intentionalist and anti-intentionalist views. Despite the differences between the two competing approaches, especially with respect to whether subjects deceive themselves intentionally or unintentionally, I will argue that they tend to converge on, among other aspects, the interpretation of self-deception as a process by which the subjects *fail* to acquire knowledge. The condition for self-deception that the subject's acquired (self-deceptive) belief is a *false* belief gathers wide and transversal agreement from both the sides of the divide. I will provide criticism about the validity of such a condition by showing that it doesn't match our common intuitions; some positive consequences of cutting the false-belief condition out of the set of conditions for self-deception will be explored. Finally, I suggest that self-deceivers manifest a *deviant* doxastic behavior with respect to the wider (doxastic) context, and I introduce a novel condition - the 'Deviation Condition' - which grasps the social and pragmatic dimension of self-deception. The proposed 'tridimensional' account is a unified model applying to both individual and collective self-deception.

## 1 The Imaginary Invalid

Monsieur Argan, the hypochondriac main character of Molière's *The Imaginary Invalid* [17], urges his daughter Angelique to marry the doctor-to-be Thomas Diaforious against her will. Toinette, Argan's maid and Angelique's best confidant, tries to dissuade him by raising doubts about his phantomatic illness.

> Argan: *My reason is, that seeing myself infirm and sick, I wish to have a son-in-law and relatives who are doctors, in order to secure their kind assistance in my illness, to have in my family the fountain-head of those remedies which are necessary to me, and to be within reach of consultations and prescriptions.*
> Toinette: *Very well; at least that is giving a reason, and there is a certain pleasure in answering one another calmly. But now, Sir, on your conscience, do you really and truly believe that you are ill?*
> Argan: *Believe that I am ill, you jade? Believe that I am ill, you impudent hussy?*
> Toinette: *Very well, then, Sir, you are ill; don't let us quarrel about that. Yes, you are very ill, I agree with you upon that point, more ill even than you think.*
> [17]

Toinette's attempt fails miserably, as Argan '*really and truly*' *believes* on his conscience that he is constantly afflicted with a variety of illnesses. According to the story, this is untrue: Argan is the unaware victim of self-deception. What we commonly mean in claiming someone's self-deceptive state is the failure to recognize what is seemingly obvious to others: while everybody around Argan believes him not to be sick, he holds the contrary belief. His belief that he is sick is intense and resistant to the doubt raised by his relatives and friends, who sometimes would invite him to come to his senses. Self-deceptive states might just be temporarily so, and last as long as the self-deceiver fails to 'face the truth'; once the self-deceiver becomes aware of her being victim of self-deception, typically she exits self-deception and adjusts her doxastic attitudes accordingly.

Monsieur Argan is one of the many self-deceptive characters featured in literary works. Depicted as they are in such a richness of details, these characters make ideal study cases for philosophical analysis. Sahdra and Thagard [26], for example, base their analysis of self-deception on the characters from *The Scarlett Letter* by Hawthorne, while Talbott [29] engages in an analysis of La Fontaine's fable *The Fox and the Grapes*. In fact, literary cases might easily be regarded as mirroring real everyday life, where our experience of self-deception is an 'undeniable fact' ([31], 9). Despite a few voices dismissing the very existence of self-deception[1], recent empirical findings suggest that people are often prone to deceive themselves about a variety of subjects, in different contexts, and for different purposes; for example, self-deception seems to occur in denial of physical illness [11], it may account for cases of unrealistic [1], or for positive illusions in self-evaluation tasks [12]. Moreover, self-deception occurs at various degrees in mental conditions classified as 'pathological', such as depression and schizophrenia [18]. As a result, a great variety of phenomena appears to be comprised in the heterogeneous category of self-deception: poor insight, unrealistic optimism, wishful thinking, feigning memories, delusions, *akrasia*, and other irrational beliefs all falling on a broad spectrum ranging from normality to pathology. [2]

Non-fictional and non-clinical people deceive themselves as well. Ordinary cases of self-deception entail the idea that the subject enters self-deception in order to either maintain serenity and psychological stability, or avoid pain. A husband might (falsely) believe that his unfaithful wife is faithful, despite compelling evidence of her betrayal; a mother who has evidence at her disposal that her son has robbed a bank, still believes him innocent. By 'lying to themselves', these subjects manage to sway pain, along with truth. The lying-to-oneself mechanism is intuitively and pre-philosophically regarded as one of the essential traits of self-deception, which shares many traits with cases of other-deception, whereby people deceive the others for their own purposes. Yet, if the subject's gain in interpersonal deception is obvious, the same does not apply to self-deception.

In the philosophical literature, self-deception is regarded as a puzzling mental phenomenon seriously challenging the subject's *doxastic integrity*. Even if there is

---

[1] A few authors claim that self-deception is impossible. In particular, Paluch [19] and Haight [13] are skeptical about self-deception for the reason that self-deceptive states would entail the subject to be in the impossible state of mind of simultaneously having two contradictory beliefs.

[2] For an analysis of the conceptualization of mental insanity as a lack or negation of irrationality, see Bortolotti (2015), pp. 45-80.

little agreement on how exactly to define self-deception, it is common to consider self-deceptive beliefs as acquired and maintained by the subject in face of adverse evidence, under the pressure of motivations, desires, or emotions, which might or might not be hidden to the subject herself. In the attempt of clarifying *how* we acquire self-deceptive beliefs, several accounts have been endorsed, which ultimately loom into two competing approaches: intentionalism *vs.* non-intentionalism[3]. While the non-intentionalist approach appears to currently dominate the scene, there is still little agreement about what exactly the nature and mechanism of self-deception is, and about what cases can be regarded as paradigmatic [30]. Besides, the discourse on self-deception fatally drags in elements, such as belief and evidence - crucial to knowledge analysis - which are *per se* far from being uncontroversial.

In this paper, after a short critical review of the state of art in the philosophical debate, I will highlight the general claims upon which the contrasting voices within the discussion tend to converge. More specifically, I will argue that the various accounts of self-deception, of both intentionalist and non-intentionalist nature, widely agree that the belief acquired by self-deceivers is false. In contrast, I argue that in accounting for self-deception we have good reasons for dropping the condition that the self-deceptive belief is false, and for extending the analysis to the social and pragmatic dimension of the phenomena. I identify such dimension in the doxastic *tension* between the self-deceiver and the 'spectators' to the given self-deceptive occurrence. The proposed account reshapes self-deception as a 'tridimensional' phenomenon, which is dependant on the subject's motivations, the belief *vs.* evidence tension, and the self-deceiver *vs.* spectators (doxastic) tension. By highlighting the social and pragmatic dimension, the account presented here has the advantage of applying not only to individual self-deception, but also to the less explored phenomenon of collective self-deception.

## 2 The philosophical debate: divergences and convergences

The current philosophical discussion is characterized by a variety of proposals of both intentionalist and non-intentionalist kind, all aiming at describing and explaining a phenomenon, which still remains under-defined. Intentionalism[4] about self-deception is the traditional position, quite popular since the 90s, which rests upon what I call the 'analogy' assumption ($A_A$), and on the 'intention' assumption ($A_I$).

($A_A$) Self-deception is the intrapersonal *analog* of interpersonal deception (or 'other-deception'): self-deception and other-deception share the same structure, yet in self-deception the self-deceived subject is simultaneously the deceiver and the deceived. Thus, the following applies to both self- and other-deception: person $X$ deceives person $Y$ (where $Y$ may or may not be the same as $X$) into believing that $p$ only if $X$ knows, or at least believes truly, that non-$p$ (i.e., that $p$ is false) and causes $Y$ to believe that $p$. In the case of self-deception, obviously $X = Y$.

---

[3] A comprehensive and updated survey of the various positions within the current debate on self-deception is provided by [2].
[4] Talbott, Rorty [23] [25], Pears [21], Davidson [8] [9], and Bermudez [4] [5] are among the best representatives of intentionalist views.

($A_I$) The act of deceiving is *intentional*: in self-deception the subject *intends* to deceive herself into believing $p$, while knowing that non-$p$. This claim entails that non-intentional self-deception cannot occur. ($A_I$) is derived from assumption ($A_A$); from the analogy between other-and self-deception, it follows that *intention* (built into the very fabric of other-deception) is a necessary feature in self-deception as well.

Non-intentionalism, especially as represented in the account endorsed by Alfred Mele [16], advances one positive and one negative argument against intentionalism. The negative argument is based on the rejection of both ($A_A$) and ($A_I$), whereas the positive one derives from the assumption about psychological features of motivational biases, which are assigned an essential role in the unfolding of self-deceptive phenomena. Non-intentionalist supporters usually regard intention-based accounts as stemming from an over-interpretation of the phenomenon, which in their view should be rather understood as a *motivationally biased judgment*.

The *pars destruens* of the non-intentionalist project identifies ($A_A$) and ($A_I$) as the sources of fatal paradoxes. Mele [15] suggests that these two assumptions lead to a *static puzzle* and a *dynamic puzzle*, respectively. The static puzzle stems from ($A_A$), since if that assumption holds true, then to deceive oneself into believing that $p$ requires that one knows or believes truly that non-$p$ (i.e., that $p$ is false), and causes oneself to believe that $p$. That is, self-deceivers (simultaneously) believe both $p$ and non-$p$, and therefore find themselves in the problematic situation of *holding contradictory beliefs*. The second assumption ($A_I$) is identified as the source of the dynamic paradox: it does not seem possible for a subject to successfully deceive herself, while doing so *intentionally*. Potential (other-)deceivers would miserably fail their goal if their victims become aware of the deceptive plan: how can $X$, while knowing/truly believing non-$p$, successfully deceive $Y$ into believing that $p$, if $Y$ knows exactly what $X$ is up to? By analogy, how can a subject successfully deceive herself while aware of her own deceptive intention?

Typically, the intentionalist defensive strategy mainly consists in attempts to dissolve the two puzzles by means of various strategies[5] (e.g., temporal partitioning, psychological partinioning, etc.). However, the dynamic puzzle in particular still remains a real challenge to intentionalist views, and its solution seems either to require the adoption of problematic doxastic and mental states (implicit intentions, unaware intentions, aliefs, and the like), or the very letting go of the intention itself; the latter move resulting in the total defeat of intentionalism.

Despite the distance between the two sides of the debate, mainly created by their divergence in regard to ($A_A$) and (AI), intentionalists and their opponents still seem to agree on at least three jointly sufficient conditions for the subject to enter self-deception in acquiring a belief $p$. I call the three conditions, the 'Motivation

---

[5] [4] suggests that no paradox stems from the subject's holding of two contradictory beliefs, since they could be inferentially insulated. Another defensive strategy by [7] aims at supporting the idea that the simultaneous possession of conflicting beliefs is neither impossible nor illogical within current models of human cognition. For coping with the dynamic puzzle, partitioning strategies are sometimes used: [24] regards the 'self' ('persona', or the 'I') as a loose sort of committee, so that deceiver and deceived are but two distinct parts constituting the entire configuration; others make recourse to *mental exotica*, such as "belief without awareness of such belief" [?]fing).

Condition' ($C_M$), the 'Tension Condition' ($C_T$), and the 'False Belief Condition' ($C_{FB}$)[6].

$C_M$ motivational biases (desires and emotions) favored the acquisition of $p$.

$C_T$ the acquired belief that $p$ is in tension with features either internal (subject's doxastic repertoire) or external (evidence) to the process of belief formation.

$C_{FB}$ the belief that $p$ is false.

These three conditions are not only commonly shared within the philosophical discussion, but they are also highly compatible with our pre-philosophical and common sense views of self-deceptive phenomena. While $C_M$ has been especially emphasized by non-intentionalists authors (e.g., Mele's core idea of self-deception as a *motivationally biased judgment*), it is still entailed by intentionalist accounts. The second condition CT has more to do with the process of belief formation, and it is revealing of the *tension* accompanying any self-deceptive occurrence. Finally, $C_{FB}$ is a condition about the output of the self-deceptive process, which ends with the subject holding a false belief. I understand this last condition as providing the natural terrain for the discussion of the epistemological status of self-deceptive beliefs, as well as of the epistemological nature of self-deception as a whole. In fact, the subject's acquisition of the belief that $p$, which satisfies $C_M$ and $C_T$, but does not satisfy $C_{FB}$, would hardly qualify as an instance of self-deception, precisely because of the subject's acquisition of a *true belief*.

This result directly follows from the jointly sufficient conditions for entering self-deception, and it seems to be revealing of how philosophers look at the very nature of the process of self-deceptive belief acquisition. Indeed, $C_{FB}$ is a condition about the truth value of $p$, so that it shifts the observer's attention from a doxastic level to an epistemological level of analysis: entering self-deception looks like a process by which the subject fails to acquire *knowledge*, where knowledge is intended in the traditional sense of (at least) justified *true* belief. In the following, I will show how such result is not desirable, and counterintuitive, too.

## 3 Two scenarios for Argan

The account of self-deception as an instance of failed knowledge - at least, within the framework of the traditional tripartite analysis of knowledge - can be better clarified by looking at the relations between the conditions for self-deception and knowledge conditions, which go as follows: a subject $S$ knows that $p$ iff (i) $p$ is true, (ii) $S$

---

[6] Note that the here proposed conditions differ significantly from the four sufficient conditions for self-deception as formulated by Alfred Mele: "1. The belief that $p$ which $S$ acquires is false; 2. $S$ treats data relevant, or at least seemingly relevant, to the truth-value of $p$ in a motivationally biased way; 3. This biased treatment is a non-deviant cause of $S$'s acquiring the belief that $p$; 4. The body of data possessed by $S$ at the time provides greater warrant for non-$p$, than for $p$." ([15], 95). In particular, Mele's fourth condition is loosened in $C_T$ so as to incorporate both intentionalist and non-intentionalist approaches. The core of the condition which constitutes the point of convergence is exactly the *tension* that is considered to accompany any self-deceptive act.

believes that $p$, and (iii) $S$ is justified in believing that $p$. Each of these conditions echoes or insulates the conditions for self-deception to various degrees: condition (i) opposes $C_{FB}$, condition (ii) is preassumed by each of the three conditions, and (iii) is relevant to $C_T$ as long as evidence and/or the subject's set of beliefs have something to do with her justification for believing $p$. The satisfaction of $C_{FB}$ necessarily entails that the knowledge condition (i) is not met: the subject who enters a state of self-deception does not know that $p$.

Thus, self-deception amounts to *failed knowledge*, and such a view offers indications as to what sort of psychological phenomena might fall within or outside of the category of self-deception. However, the analysis of self-deception through the lenses of the epistemologist might improperly fixate our attention on elements, which are *not*, after all, structural to its nature and mechanism.

Back to the *The Imaginary Invalid* case, let's consider the two following scenarios: in *Scenario 1* we find Monsieur Argan and his adventures as narrated in the original story by [1]ère, whereas *Scenario 2* is an invented continuation of Argan's story.

*Scenario 1*: Monsieur Argan acquires and maintains - 'really and truly believes' - that he is seriously sick in face of compelling evidence to the contrary, and it is not the case that he is sick; in fact, Argan is in perfect health[7].

*Scenario 2*: As in *Scenario 1*, Monsieur Argan acquires and maintains ('really and truly believes') that he is sick in face of compelling evidence to the contrary. However, unbeknownst to him, as well as to his relatives, doctors and friends, Argan is sick with a fatal yet totally symptomless disease.

Argan from *Scenario 1* successfully enters self-deception given that the three conjointly sufficient conditions are met: Argan's acquisition of the belief that he is sick is motivationally favored (possibly, his fear for death favored his acquisition of $p$)[8]; there is tension between his belief that he is sick and the evidence at his disposal; he holds a false belief. Not only he is self-deceived, but (if that is at all relevant in making sense of self-deception) he also has no knowledge that he is sick.

The analysis of *Scenario 2* leads to an entirely (at least temporarily) different conclusion. Both $C_M$ and $C_T$ are satisfied: as in Scenario 1, Argan acquires the belief that he is sick under some motivational bias, and his belief contrasts the evidence at his disposal. However, $C_{FB}$ is not met, since Argan is fatally sick with a disease. As a result, by adopting the three conditions ($C_M$, $C_T$, and $C_{FB}$) Argan's belief that he is sick does not fully qualify as a self-deceptive belief, and we shall conclude that Argan, after all, has not fallen victim of self-deception. We might want to further investigate on whether Argan knows or does not know that he is sick, but this is a controversial matter, and widely depends upon how we understand justification *per se* and its role in knowledge attribution. Within the context of the debate on self-

---

[7]For the sake of clarity, let's assume that here 'being healthy' is not a matter of degrees, but a property which the subject might either possess or not possess.

[8]Monsieur Argan's example is a case of twisted self-deception, in which Argan acquires the unpleasant and undesirable belief that he is sick. At a first glance, cases of twisted self-deception might be challenging, since it is not obvious what would be the motivations of the subject in acquiring an unpleasant belief. However, it has also been suggested that 'fears' and 'anxiety' qualify as motivations [20], [3]: as a general principle a unified account of self-deception, one which can also explain cases of twisted self-deception, is highly desirable.

deception, some authors have discussed justification. Annette Barnes, for example, argues that the motivational biases favoring self-deception automatically rule out any possibility of belief-justification: "As a result of this bias in the self-deceiver's belief-acquisition process, the self-deceiver's belief that $p$ is never justified" ([3], 78)[9].

In sum, the two scenarios only differ in regard to the fact that in the first story Argan holds a false belief, whereas Argan in the second story holds a true belief, all other conditions being the same. The three conditions for self-deception are satisfied in *Scenario 1*, yet stay unfulfilled in *Scenario 2*, since Argan holds a true belief and $C_{FB}$ is not met. Such a result is counterintuitive, as both scenarios appear to revolve around self-deception: why should we conclude that Argan in *Scenario 2* is not deceiving himself, given that Argan's (internal and external) behavior, as well the other features in the story, perfectly matches *Scenario 1*? A hypochondriac subject, who firmly believes that she is seriously ill despite evidence to the contrary, does not cease to be hypochondriacal after her belief becomes accidentally true. Moreover, if we consider the two scenarios from the perspective of the other subjects populating the stories - Argan's relatives and friends - they would still think of Argan as a victim of self-deception in *Scenario 2* as they do in *Scenario 1*; they dispose of the same evidence in both scenarios, therefore there is no reason for assuming that they would believe that Argan is sick in *Scenario 2*, and that Argan is healthy in *Scenario 1*[10].

If the truth and falsity of Argan's acquired belief in *Scenario 1* and *2*, respectively, is not relevant for determining the self-deceptive state, then self-deception itself seems to behave as a phenomenon insensitive to the belief's truth-value. Thus, I suggest to drop $C_{FB}$, and substitute it with a novel condition, the 'Deviation Condition', which better identifies the way in which self-deception attributions are typically made. From this move we get at least three results: first, $C_{FB}$-free accounts of self-deception better match our common intuitions about the phenomenon; second, it rules out the idea that self-deception is a special instance of 'failed knowledge' (this lessens any temptation to treat genuine cases of self-deception as Gettier-like stories); third, it suggests that the understanding of self-deceptive subjects as *epistemic agents lacking (self-) knowledge* is likely inaccurate. Monsieur Argan in *Scenario 2* is deceiving himself in believing that $p$, and $p$ is 'accidentally' true; however, as shown by Argan's two scenarios example, in accounting for self-deception the truth-value of $p$ appears to be fully irrelevant as self-deception occurs independently of the truth value of the self-deceptive belief.

## 4 The Pragmatic Tension of Self-Deception

The proposed $C_{FB}$-free account of self-deception also includes a further condition to enhance the set of conjointly sufficient conditions for entering self-deception:

---

[9]On the negative relation between motivational bias and justification, see also [14].

[10]Note that Argan's friends and relatives potential doubts about Argan's poor health ('what if he is sick?') given the evidence at their disposal would constitute a typical skeptic maneuver, one which we would expect to occur in a philosophical context, but not in everyday ordinary life (see, [22]).

in addition to $C_M$ and $C_T$, I suggest a third condition that I call the 'Deviation Condition' ($C_D$).

1. the subject's acquired belief $p$ is in tension with the spectators's acquired belief non-$p$.

This further condition ideally grasps the *pragmatic* and *social* dimension of self-deception, which remains underdeveloped in the literature on the topic. What the social and pragmatic dimension of self-deception amounts to, is revealed by the common ways in which people attribute self-deceptive beliefs to others, and (sometimes) to themselves. The social and pragmatic character of self-deception is shown in *The Imaginary Invalid* example, where Argan's friends, relatives and doctors would have the function of 'spectators': notwithstanding the shared evidence, Argan and the spectators acquire opposite beliefs. In other words, when the process of belief formation of the spectators and of Argan are taken into account, then Argan's process represents *a deviation from the standard*.

As happens in Argan's case, people deceiving themselves seem to (doxastically) *fail* in ways that the others *succeed*. In other words, in a given circumstance, the self-deceived subject apparently processes information in a significantly different way from how she (or the others) would do if she were in the position of a mere spectator. The sort of tension stemming from the self-deceived subject *vs.* spectators contrast differs significantly from the tension involved in $C_T$, as it is of a social and pragmatic nature, and it is registered over an extended doxastic context, which includes not only the self-deceived subject - as in the case of $C_T$ - but also the spectators. An extended version of $C_D$ can be formulated as follows: let $S_{SD}$ be the self-deceived subject, $S_S$ the spectators, and $E$ the evidence available to any subject within the wider doxastic context, then given $E$, $S_{SD}$ acquires $p$, whereas $S_S$ acquire non-$p$.

Here one essential aspect of self-deception is highlighted: self-deceivers manifest an abnormal doxastic behavior significantly diverting from the behavior of the others. What does such 'abnormality' amount to? I intend 'abnormality' as a deviation from the norm, where the norm is the standard doxastic behavior usually observed in human reasoning. Thus, self-deceptive behaviors can be understood as deviant doxastic behaviors, which should be 'measured' upon the wider doxastic context constituted by the spectators.

What if the spectators incept the self-deceptive belief? If that happens, the tension between the self-deceived and the wider doxastic context would be dispelled: the self-deceptive belief - now shared by both $S_{SD}$ and $S_S$ - would not qualify as a self-deceptive one anymore (within that very same doxastic context). However, the social and pragmatic tension might characterize the relation between one group of subjects sharing the same self-deceptive belief, and a wider community of spectators. That is, the self-deceiver might be either one single individual or one collection of individuals (a group, a committee, a sect, etc.). In this second case, collective self-deception[11], rather then individual self-deception, is at issue.

---

[11]So far, philosophers have been given little attention to collective self-deception. Typically, accounts of self-deception are tailored to individual self-deception alone, and the analysis of collective

## 5 Conclusion

The proposed 'tridimensional' account of self-deception, *via* the elimination of the 'False-Belief Condition', suggests that occurrences of the phenomenon should not be understood as *epistemic mistakes*, or as instances of failed knowledge. Rather, I have argued, self-deception is better analyzed by attending to its social and pragmatic dimension, since an acquired belief is 'self-deceptive' *also* because of its tension with respect to a wider doxastic context. In order to account for its social and pragmatic dimension, in addition to the 'Tension Condition' and 'Motivation Condition', I include a further condition for entering self-deception, the 'Deviation Condition', which is based on the notion of the self-deceiver's 'doxastic deviation' from a wider doxastic context, with this deviation being a by-product of the pragmatic and social dimension of self-deception. Besides vindicating the social dimension of the tension characterizing self-deception, the proposed account offers a unified model, which applies homogeneously to self-deception, twisted self-deception, as well as collective self-deception.

## BIBLIOGRAPHY

[1] Arabsheibani, G., de Meza, D., Maloney, J., and Pearson, B., 2000, 'And a Vision Appeared Unto Them of a Great Profit: Evidence of Self-Deception Among the Self-Employed', *Economics Letters*, 67, pp. 35–41.

[2] Baghramian, M., Nicholson, A., 2013, 'The Puzzle of Self-Deception', *Philosophy Compass*, 8, 11, pp. 1018-1029.

[3] Barnes, A., 1997, *Seeing Through Self-Deception*, Cambridge, Cambridge University Press.

[4] Bermudez, J., 1997, 'Defending Intentionalist Accounts of Self-Deception', *Behavioral and Brain Sciences*, 20, pp. 107-108.

[5] Bermudez, J., 2000, 'Self-Deception, Intentions, and Contradictory Beliefs', *Analysis*, 60 (4), pp. 309-19.

[6] , A., 2015, *Irrationality*, Cambridge, Polity Press.

[7] Brown, S. L., Douglas, T. K., 1997, 'Paradoxical Self-Deception: Maybe not so paradoxical after all', *Behavioral and Brain Sciences*, 20, pp. 109-110

[8] Davidson, D., 1982, 'Paradoxes of Irrationality', in R. Wollheim and J. Hopkins (Eds.), *Philosophical Essays on Freud*, Cambridge, Cambridge University Press, pp. 289-305.

[9] Davidson, D., 1985, 'Deception and Division', in E. Lepore and B. McLaughlin (Eds.), *Actions and Events*, New York, Basil Blackwell, pp. 138-148.

[10] Fingarette, H., 1969, *Self-Deception*, Berkeley, University of California.

[11] Gordbeck, R., 1997 'Denial in Physical Illness', Journal of Phsychosomatic Research, 43, 6, pp.575-593.

---

self-deception looks peripheral to the mainstream intentionalist *vs.* anti-intentionalist debate (see, for example, [28]; [27]). Yet, it would be desirable for any theoretical account of self-deception to also homogeneously apply to cases of collective self-deception.

[12] Gramzow, R.H., Elliot A.J., Asher E., and McGregor H. A., 2003, 'Self-evaluation bias and academic performance: Some ways and some reasons why', *Journal of Research in Personality*, 37, pp. 41–61.

[13] Haight, M. R., 1980, *A Study of Self-Deception*, Sussex, Harvester Press.

[14] Kornblith, H., 1983, 'Justified Belief and Epistemically Responsible Action', *Philosophical Review*, 92, 1, pp. 33-48.

[15] Mele, A., 1997, 'Real Self-Deception', *Behavioral and Brain Sciences*, 20, pp. 91-137.

[16] Mele, A., 2001, *Self-Deception Unmasked*, Princeton, Princeton University Press.

[17] Moliére (Poqueline), *The Imaginary Invalid*. Project Gutemberg. n.p. n.d. Web.

[18] Moore, O., Cassidy, E., Carr, A., and O'Callaghan, E., 1999, 'Unawareness of illness and its relationship with depression and self-deception in schizophrenia', *European Psychiatry*, 14, 5, pp. 264-9.

[19] Paluch, S., 1967, 'Self-deception,' *Inquiry*, 10, pp. 268–278.

[20] Pears, D, 1984, *Motivated Irrationality*, New York, Oxford University Press.

[21] Pears, D., 1991, 'Self-Deceptive Belief Formation', *Synthese*, 89, pp. 393–405.

[22] Piazza, M., Dolcini N., 2015, 'Possibilities Regained: Neo-Lewisian Contextualism and Ordinary Life', *Synthese*, pp. 1-20.

[23] Rorty, A. O., 1972, 'Belief and Self-Deception', *Inquiry*, 15, pp. 387-410.

[24] Rorty, A. O., 1987, 'Self-Deception, Akrasia and Irrationality', in J. Elster (Ed.), *The Multiple Self*, Cambridge, Cambridge University Press, pp. 115-131.

[25] Rorty, A. O., 1994, 'User-Friendly Self-Deception', *Philosophy*, 69, 268, pp. 211–228.

[26] Sahdra, B. and Thagard, P., 2003, 'Self-Deception and Emotional Coherence', *Minds and Machines*, 13, pp. 213–231.

[27] Surbey, Michele, 2004, 'Self-deception: Helping and hindering personal and public decision making', in C. Crawford and C. Salmon (Eds.), *Evolutionary Psychology, Public Policy and Personal Decisions*, [.1cm], Mahwah, NJ, Lawrence Earlbaum Associates, pp. 109-134.

[28] Ruddick, W., 1988, 'Social Self-Deception', in McLaughlin B. P., and A. O. Rorty (Eds.), *Perspectives on Self-Deception*, Berkeley-Los Angeles, University of California Press, pp. 380-389.

[29] Talbott, W., 1995, 'Intentional Self-Deception in a Single Coherent Self', *Philosophy and Phenomenological Research*, 55 (1), pp. 27-74.

[30] Van Leeuwen, N., 2013, 'Self-Deception', in H. LaFollette (Ed.), *The International Encyclopedia of Ethics*, Wiley-Blackwell.

[31] Wood, A., 1988, 'Self-Deception and Bad Faith', in A.O. Rorty, B.P. McLaughlin (Eds.), *Perspectives on self deception*, Berkeley CA, University of California Press, pp. 207-227.

# Succeeding in the False Belief Test: Why Does Experience Matter?

Marco Fenici

ABSTRACT. I challenge the view—commonly shared among developmentalists—that four-year-olds' success in the false belief test mostly depends on the maturation of either computational resource or cognitive processes specific for mental state attribution. In contrast, available evidence suggests that success on the task is importantly shaped through conversation and social interaction. Adult mindreading is not naturally inscribed in our biological endowment, and social experience has a much more important role than what commonly assumed in its development. KEYWORDS: theory of mind; mindreading; social cognition; false belief.

## 1 Introduction

In everyday life, we are apparently very good at attributing mental states to ourselves as well as to others, and to exploit this 'mindreading' capacity—which has been equated to the possession of a 'Theory of Mind'—to predict behaviour (Dennet 1987) [14]. In the last thirty years, while most of philosophers have debated about how to characterise precisely the possession of a ToM (Davies and Stone 1995 [11]), developmental psychologists have instead focused on mapping the emergence of mindreading in infancy and early childhood. To this aim, they have largely employed the experimental paradigm known as the *false belief test* (FBT) (Baron-Cohen 1985 [5], Wimmer and Perner 1983 [64]). Research employing FBT has found that it is not until age four that children consider others' (false) beliefs to make conscious predictions about others' actions (Wellman et al. 2001 [10] and Wellman and Liu 2004 [62]).

Despite recent findings adopting spontaneous-response methodology (see for a review Baillargeon et al. 2010 [4]), the capacity to pass the traditional (elicited-response) FBT denotes a robust empirical result (Wellman et al. 2001 [10]) that resisted many attempts to reduce the difficulty of the task (Wellman et al. 1996 [61], Woolfe et al. 2002 [1]). Thus, it marks an important developmental acquisition in children's understanding of others' minds, which still awaits an explanation.[1] In this article, I aim to improve towards our understanding of this finding by discussing the contribution of experience to it.

---

[1] Accordingly, where the expression is not ambiguous, I will henceforth use 'FBT' to refer only to the traditional elicited-response FBT.

Following Werker (1989) [63], we can distinguish four models by which experience may underpin the development of a psychological competence such as the capacity to pass FBT: (i) *maturation* characterises the unfolding of a psychological ability independently of the exposure to environmental features; (ii) in *facilitation*, experience affects the rate of development of an ability although it does not influence its endpoint; (iii) *attunement* refers to cases in which experience affects the full development of an ability including at least partially determining the final state, while a more basic level of performance develops by mere maturation; (iv) finally, *induction* characterises those cases in which the development of a psychological capacity is entirely structured by the environmental input.

I discuss some accounts relying on the cognitive maturation of ToM capacities in section 2. Section 3 puts together decisive evidence against both maturation and facilitation views. Finally, in section 4, I discuss the difference between how social experience might attune or rather induce ToM abilities, and conclude in favour of the latter.

## 2 Cognitive maturation alone does not promote false belief understanding

It is often assumed that success in FBT indicates the maturation of some cognitive factor so that younger children's difficulty with FBT masks a performance problem (Bloom and German 2000 [7], Fodor 1992 [23]). Nativists about ToM, in particular, claim that the capacity to attribute mental states has been shaped through natural selection because of its survival efficacy (Humphrey 1976 [26]), and is underpinned by dedicated neural processes (Saxe 2004) specific for the social domain (Baillargeon et al. 2010 [4], Baron-Cohen 1995 [5], Leslie 2005 [29]). This ToM module is supposed to develop in early infancy and to underlie 15-month-olds' looking behaviour in spontaneous-response false belief tasks (e.g., Onishi and Baillargeon 2005 [34]).

Because they think that infants already attribute beliefs, ToM nativists contend that younger children's inability to pass FBT attests performance limitations, and that four-year-olds' success in FBT depends on the emergence of additional computational resources overcoming initial processing constraints. The empirical plausibility of ToM nativism then depends on the possibility to clarify what cognitive impairments prevent younger children from manifesting their psychological understanding. I will consider two proposals by Baillargeon (Baillargeon et al. 2010 [4], Scott and Baillargeon 2009 [48]), and Carruthers (Carruthers 2013 [10]).

Baillargeon and colleagues have proposed that the traditional FBT engages at least three distinct cognitive processes: (i) a process to represent the false beliefs of other agents, (ii) a process to access and select one's own representation of another's false belief when being asked the test question, and crucially (iii) a process to inhibit any prepotent tendency from one's own knowledge to answer questions concerning others. According to their view, the maturation of inhibitory capacities after age four motivates children's late success in the traditional FBT (Leslie 2005 [29]), Scott and Baillargeon 2009 [48]). Against this proposal, however, findings on several populations—autistic children (Ozonoff et al. 1991 [36]), children in Asian countries (see for a review Sabbagh et al. 2013 [44]), deaf children (Schick et al. 2007 [9]), and deaf adults (Pyers and Senghas 2009 [39]) — all demonstrate that possessing mature inhibitory capacities is not sufficient to pass FBT.

Carruthers (2013) [10] also endorses a modularist view about ToM abilities, and proposes that the traditional FBT imposes a "triple burden" on the mindreading system because it requires children (i) to generate the prediction of an action by processing the mental states of the target agent, (ii) to figure out the communicative intention underlying the speech of the experimenter, and (iii) to generate a response that conveys the target agent's mental states to the experimenter. The collapse of any of these components under cognitive load, he argues, entails children's failure in the task. In contrast, success in FBT indicates some improvement in the interactions between the basic domain-specific component of the mindreading system and executive, attentional, and planning mechanisms.

Like Baillargeon and colleagues', also Carruthers' analysis is unconvincing, though. Indeed, Carruthers claims that "it is something about language *production* (or the production of communicative actions generally ...) that disrupts successful performance in verbal false-belief tasks" (Carruthers 2013 [10], p. 153). But if this was the case, any task eliminating children's need to communicate an answer to the experimenter should be easier than the traditional FBT—a suggestion refuted by empirical evidence. For instance, de Villiers and de Villiers (2000) [13] told children a false belief story with the help of a series of pictures. At the end of the story, children were requested to select a proper ending by choosing between two different pictures representing the main character's emotion. This modification relieved children from the need to communicate with the experimenter but—against Carruthers' prediction—did not affect their capacity to pass the task. (See also for additional evidence Call and Tomasello 1999 [9], Figueras-Costa and Harris 2001 [22], and Woolfe et al. 2002 [1]).

## 3 Experience does not merely facilitate false belief understanding

The previous section shows that cognitive factors allegedly responsible for children's maturation of false belief understanding—i.e., inhibitory abilities and the capacity to process communicative intentions—do not account for younger children's difficulties with FBT. While the discussion challenges the considered accounts, this section raises more general and striking doubts that the endogenous maturation of some cognitive factors can be alone responsible for, or even facilitate false belief understanding.

As training studies (Rhodes and Wellman 2013 [41]), and scales (Wellman et al. 2006 [60], Wellman and Liu 2004 [62]) attest, children's knowledge of others' minds evolves through distinct conceptual phases. 18-month-olds already understand that people act on the basis of their desires, which can significantly differ from their own (Repacholi and Gopnik 1997 [40]). After age three, children also understand that people's beliefs may differ from their own and nevertheless guide others' actions (as demonstrated by the acquired capacity to pass the *diverse belief task*, DBT, (Wellman and Bartsch 1988 [58]), and that people may lack epistemic access to a situation (as demonstrated in the *knowledge access test*, KAT, (Pratt and Bryant

1990 [38]).

DBT and KAT apparently tap some understanding of mental states different from false beliefs; they seem to pose less difficulties than FBT, are passed before it and related to children's performance in it (Rhodes and Wellman 2013 [41]). However, while children from the United States and Australia pass DBT first and KAT only later, Chinese and Iranian children reliably pass these two tasks in the reverse order (Shahaeian et al. 2011 [51], Wellman et al. 2006 [60]). This significant cross-cultural variation rejects the possibility that the progressive maturation of processing capacities may bring children to pass DBT and KAT, first, and FBT later: if DBT was less demanding than KAT, as attested from Western children's developmental trajectory, why would Chinese and Iranian children find it so difficult and master it only *after* mastering KAT? The reversed problem appears if we consider KAT, which seems cognitively demanding for Western children but much easier for children from Asian countries.

Other studies also provide a final piece of evidence denying that social experience might merely facilitate children's capacity to pass FBT—that is, the idea that social experience may affect, at most, the development of false belief understanding but not its final acquisition. Rhodes and Wellman (Rhodes and Wellman 2013 [41]) trained Western almost-four-year-olds with false belief situations twice a week over one month and half. Crucially, after the training period, only those children who initially passed KAT (and DBT) also improved their success rate in FBT; in contrast, training was not efficacious for children who did not pass KAT. In addition, control children who initially passed KAT but were not later included in the training sessions did not improve their FBT success rate. This shows that children did not progress in their understanding of false beliefs due to the passage of time alone: experience with false belief situation in training session was crucial for this.

This experiment corroborates data from comparative studies showing that subject lacking access to the proper kind of social and conversational experience about others' mental states never develop a proper understanding of others' beliefs as manifested in FBT. Deaf children raised by hearing parents, for instance, are exposed to limited conversational input, and do not develop complex linguistic abilities as well as false belief understanding; in contrast deaf children from non-hearing parents, who are exposed to typical conversational input, present a typical development of mental state understanding (Schick et al. 2007 [9]). Similarly, Nicaraguan non-signer deaf adults, who had limited conversational abilities did not equally manifest good understanding of false beliefs (Pyers and Senghas 2009 [39]). Significantly, both groups recover their initial limitations as soon as they are thought a sign language, thereby they acquire a method to represent and gain information about mental states.

Overall considered, these data exclude that social experience merely facilitates false belief understanding. If that was the case, one would expect that children would end passing FBT anyway—although they should pass it earlier when provided with proper social experience. The cases of deaf children from hearing families and Nicaraguan non-signer deaf adults however show that the capacity to pass FBT remains profoundly impaired when it is not supported by adequate social interaction. Nevertheless, there is neither predetermined period nor critical threshold to start succeeding in FBT.

To sum up: the developmental trajectory of children's understanding of others mental states is open to, and influenced by cultural influences. This excludes cognitive maturation as a correct model for the process of development underlying children's understanding of false beliefs. Neither is facilitation an adequate model: subjects who do not receive adequate social and conversational experience never get to the point to pass FBT. Therefore, false belief understanding is very likely constrained by cognitive maturation, but it is crucially underpinned by social learning.

## 4 Experience does not attune false belief understanding

Having discarded two models of the development of social understanding based on cognitive maturation and the facilitating role of social experience, it remains to be decided whether social experience attunes or rather induces false belief understanding. Scholars embracing cognitive as well as socio-cultural accounts of the development of social cognition firmly opt for the first possibility (see, for instance, for some examples from the two perspectives Carruthers 2013 [10], German and Leslie 2004 [24], San Juan and Astington 2012 [45], Banaji and Gelman 2013 [57]). However, arguing for the attuning role of social experience on children's capacity to pass FBT requires to make two relevant assumptions: (i) that the cognitive processes specific for belief attribution—for instance, those presupposed by ToM nativists and allegedly manifested in spontaneous-response FBT—exist already before the time that children start passing elicited response FBT, and (ii) that it is the refinement of these processes that specifically promotes false belief understanding at age four. There would be otherwise no reason to contend that social experience attunes innate or biological mindreading competences rather than that a novel capacity to attribute mental states is assembled in the course of development following social input.

Empirical data as well as theoretical reflection opposes both assumptions. As to (i), it is undeniable that infants are hardwired to distinguish intentional agents from physical bodies. Nevertheless, that they *selectively respond* to other agents' (false) beliefs in spontaneous-response tasks does not yet demonstrate that they are also *attributing* representational states. Nor it specifies which properties of the agent infants are sensitive to. Their sensitivity to others' beliefs might depend on the capacity to track some simpler properties or features of action that are coextensive in predictive power with the possession of (false) beliefs.

Following these considerations, some have suggested that infants' performance in spontaneous-response tasks actually does not depend on a capacity to attribute beliefs but on (i) more minimal capacities to track others' beliefs by responding to their observable manifestation in overt behaviour (*belief-like* mindreading accounts, (e.g., Apperly and Butterfill 2009 [1], Butterfill and Apperly 2013 [8])), or (ii) some sensitivity to others' goals and perceptual states (*perceptual* mindreading accounts, (e.g., Fenici 2014 [21])), or even (iii) depend on less sophisticated embodied competences that do not have any direct translation in the vocabulary of folk psychology (non-mindreading accounts, (e.g., Heyes 2014 [25]).

More significantly, even if we grant for the sake of the argument that infants' social cognitive abilities identify a minimal capacity to attribute representational states, a full defence of the attuning over the inducing role of social experience for children to pass FBT is also committed—as previously noted—to the additional claim (ii) that passing elicited-response FBTs essentially exploits the same cognitive processes underlying spontaneous-response FBTs. The attuning role of social experience would be indeed rejected if there was only marginal overlapping between the cognitive processes underlying infants' alleged mindreading abilities and those granting the capacity to pass elicited-response FBTs.

Empirical evidence supports the latter hypothesis, though. As Fenici (2013) [20] argued extensively, the development of social cognitive abilities from infancy to early childhood is discontinuous. Summarizing the discussion, at least three distinct sets of considerations support the conclusion. A first line of reasoning is based on the likely existence of a double dissociation between low-level gaze-tracking and processes tapped in spontaneous-response FBTs and high-level belief-tracking processes assessed in the traditional FBT (as suggested by Senju et al. 2009 [50] and 2010 [49]). A second line of argumentation considers evidence showing that performance on spontaneous- and elicited-response FBTs remain separated even in adulthood (as suggested by considering together data from a series of studies (Surtees et al. 2011 [54] and 2012 [53]). Finally, a last piece of evidence supporting the same conclusion descends from transitional studies assessing the development of social cognitive abilities from infancy to early childhood (Thoermer et al. 2012 [15]).

Therefore, available evidence seems to reject the view that common belief attribution capacities at age four actually extend the cognitive abilities underlying infants' sensitivity to false beliefs as manifested in spontaneous-response FBTs. Rather, two distinct sets of capacities appear to be at work in spontaneous-response and elicited response FBTs. This conclusion rejects strict continuity in social cognitive development from infancy to early childhood thereby denies that social experience promotes four-year-olds performance in elicited-response FBTs by merely attuning infants' basic social cognitive capacities.

## 5 Conclusions

According to the received view, our capacity to attribute mental states has been inscribed in our biological endowment by natural selection in the evolution of our species. It follows that experience has a little role to play in the acquisition of mindreading capacities: either it facilitates their acquisition, or it triggers and attunes their development from more basic pre-existing mindreading abilities. Against such a view, available evidence shows that (i) children and even adults never come to understand that people can possess false beliefs when deprived of proper social and linguistic interaction, and that (ii) the development of social cognition undergoes important discontinuities between infancy and early childhood. The first point excludes that experience merely facilitates children's success in FBT; the second suggests—against the attunement hypothesis—that four-year-olds' capacity to pass FBT does not depend on previous basic mindreading abilities.

From these considerations, I conclude that four-year-olds' success in FBT reflects the acquisition of a novel psychological competence, and that life experience in the first years is fundamental to induce such an important change in children's understanding of the social world. The conclusion allows integrating current knowledge about the development of social cognition with both current discussion about the role of language for the acquisition of a ToM (Astington and Baird 2005 [2], Milligan et al. 2007 [33], Siegal and Surian 2011 [52]) and data from a number of studies indicating that success in FBT is affected by social and conversational experience provided by wider familiar environments (Ruffman et al. 1998 [43]) where parents are inclined to elaborate the child's talk (Ensor and Hughes 2008 [17], Ontai and Thompson 2008 [35]), and frequently discuss about mental states (Dunn et al. 1991 [16] and 2005 [7], Meins et al. 2003 [31], Ruffman et al. 2002 [43], Taumoepeau and Ruffman 2006 [55]).

What remains an open question, instead, is what specifically social and linguistic experience provide to children that enables them to pass FBT. On the one hand, one possibility—compatible with the proposals by Apperly and Butterfill (2009) [1], de Villiers (2005) [12], Miller and Marcovitch (2012) [12], and Perner (1991) [37] — is that social and linguistic experience improve domain-general reasoning capacities and allows new representational abilities (Karmiloff-Smith 1992 [28])(Carruthers 2013 [10]). On the other hand, it may also be that social interaction instructs children about the use and function of mental state concepts, and that only this domain-specific knowledge is necessary to pass FBT (Fenici 2011 [18] and 2012 [19], Hutto 2008 [27]). It is up to future research clarifying which of these options is the most likely.

## BIBLIOGRAPHY

[1] Apperly, I. A., & Butterfill, S. A. (2009). "Do humans have two systems to track beliefs and belief-like states?" Psychological Review, 116(4), 953–970.
[2] Astington, J. W., & Baird, J. A. (Eds.). (2005). "Why Language Matters for Theory of Mind". New York: Oxford University Press.
[3] Back, E., & Apperly, I. A. (2010). "Two sources of evidence on the non-automaticity of true and false belief ascription". Cognition, 115(1), 54–70.
[4] Baillargeon, R., Scott, R. M., & He, Z. (2010). "False-belief understanding in infants". Trends in Cognitive Sciences, 14(3), 110–118.
[5] Baron-Cohen, S. (1995). "Mindblindness: An Essay on Autism and Theory of Mind". Cambridge, MA: The MIT Press.
[6] Baron-Cohen, S., Leslie, A. M., & Frith, U. (1985). "Does the autistic child have a "Theory of Mind"?" Cognition, 21(1), 37–46.
[7] Bloom, P., & German, T. P. (2000). "Two reasons to abandon the false belief task as a test of Theory of Mind". Cognition, 77(1), 25–31.
[8] Butterfill, S. A., & Apperly, I. A. (2013). "How to construct a minimal theory of mind". Mind & Language, 28, 606–637.
[9] Call, J., & Tomasello, M. (1999). "A nonverbal false belief task: the performance of children and great apes". Child Development, 70(2), 381–395.
[10] Carruthers, P. (2013). "Mindreading in infancy". Mind & Language, 28(2), 141–172. doi:10.1111/mila.12014
[11] Davies, M., & Stone, T. (1995). "Folk Psychology: The Theory of Mind Debate" (1st ed.). Wiley-Blackwell.
[12] De Villiers, J. G. (2005). "Can language acquisition give children a point of view?", in J. W. Astington & J. A. Baird (Eds.), *Why Language Matters for Theory of Mind*, pp. 186–219. New York: Oxford University Press.

[13] De Villiers, J. G., & de Villiers, P. A. (2000). "Linguistic determinism and the understanding of false beliefs", in P. Mitchell & K. J. Riggs, *Children's Reasoning and the Mind*, pp. 191–228. Hove, UK: Psychology Press.
[14] Dennett, D. C. (1987). "The Intentional Stance". Cambridge, MA: The MIT Press.
[15] Dunn, J., & Brophy, M. (2005). "Communication, relationships, and individual differences in children's understanding of mind". In J. W. Astington & J. A. Baird (Eds.), *Why Language Matters for Theory of Mind*, pp. 50–69. New York: Oxford University Press.
[16] Dunn, J., Brown, J. R., & Beardsall, L. (1991). "Family talk about feeling states and children's later understanding of others' emotions". Developmental Psychology, 27(3), 448–455.
[17] Ensor, R., & Hughes, C. (2008). "Content or connectedness? Mother–child talk and early social understanding". Child Development, 79(1), 201–216.
[18] Fenici, M. (2011). "What does the false belief test test?" Phenomenology and Mind, 1, 197–207.
[19] Fenici, M. (2012). "Embodied social cognition and embedded theory of mind". Biolinguistics, 6(3-4), 276–307.
[20] Fenici, M. (2013). "Social cognitive abilities in infancy: is mindreading the best explanation?" Philosophical Psychology. doi:10.1080/09515089.2013.865096
[21] Fenici, M. (2014). "A simple explanation of apparent early mindreading: infants' sensitivity to goals and gaze direction". Phenomenology and the Cognitive Sciences, 1–19. doi:10.1007/s11097-014-9345-3
[22] Figueras-Costa, B., & Harris, P. L. (2001). "Theory of mind development in deaf children: a nonverbal test of false-belief understanding". Journal of Deaf Studies and Deaf Education, 6(2), 92–102.
[23] Fodor, J. A. (1992). "A theory of the child's Theory of Mind". Cognition, 44(3), 283–296.
[24] German, T. P., & Leslie, A. M. (2004). "No (social) construction without (meta-)representation: modular mechanisms as a basis for the capacity to acquire an understanding of mind". Behavioral and Brain Sciences, 27(1), 106–107.
[25] Heyes, C. M. (2014). "False belief in infancy: a fresh look". Developmental Psychology.
[26] Humphrey, N. K. (1976). "The social function of intellect". In P. P. G. Bateson & J. R. Hinde (Eds.), *Growing Points in Ethology*, pp. 303–317. Cambridge: Cambridge University Press.
[27] Hutto, D. D. (2008). "Folk Psychological Narratives". Cambridge, MA: The MIT Press.
[28] Karmiloff-Smith, A. (1992). "Beyond Modularity: A Developmental Perspective on Cognitive Science". Cambridge, MA: The MIT Press.
[29] Leslie, A. M. (2005). "Developmental parallels in understanding minds and bodies". Trends in Cognitive Sciences, 9(10), 459–462. doi:10.1016/j.tics.2005.08.002
[30] Leslie, A. M., German, T. P., & Polizzi, P. (2005). "Belief-desire reasoning as a process of selection". Cognitive Psychology, 50(1), 45–85.
[31] Meins, E., Fernyhough, C., Wainwright, R., Clark-Carter, D., Gupta, M. D., Fradley, E., & Tuckey, M. (2003). "Pathways to understanding mind: construct validity and predictive validity of maternal mind-mindedness". Child Development, 74(4), 1194–1211.
[32] Miller, S. E., & Marcovitch, S. (2012). "How theory of mind and executive function co-develop". Review of Philosophy and Psychology, 3(4), 597–625. doi:10.1007/s13164-012-0117-0
[33] Milligan, K., Astington, J. W., & Dack, L. A. (2007). "Language and theory of mind: meta-analysis of the relation between language ability and false-belief understanding". Child Development, 78(2), 622–646.
[34] Onishi, K. H., & Baillargeon, R. (2005). "Do 15-month-old infants understand false beliefs?" Science, 308(5719), 255–258.
[35] Ontai, L. L., & Thompson, R. A. (2008). "Attachment, parent–child discourse and theory-of-mind development". Social Development, 17(1), 47–60.
[36] Ozonoff, S., Pennington, B. F., & Rogers, S. J. (1991). "Executive function deficits in high-functioning autistic children: Relationship to theory of mind". Journal of Child Psychology and Psychiatry, 32(7), 1081–1105.
[37] Perner, J. (1991). "Understanding the Representational Mind". Cambridge, MA: The MIT Press.
[38] Pratt, C., & Bryant, P. (1990). "Young children understand that looking leads to knowing (so long as they are looking into a single barrel)". Child Development, 61(4), 973–982. doi:10.1111/j.1467-8624.1990.tb02835.x
[39] Pyers, J. E., & Senghas, A. (2009). "Language promotes false-belief understanding: evidence from learners of a new sign language". Psychological Science, 20(7), 805–812.
[40] Repacholi, B. M., & Gopnik, A. (1997). "Early reasoning about desires: evidence from 14- and 18-month-olds". Developmental Psychology, 33(1), 12–21.

[41] Rhodes, M., & Wellman, H. (2013). "Constructing a new theory from old ideas and new evidence". Cognitive Science, 37(3), 592–604.
[42] Ruffman, T., Perner, J., Naito, M., Parkin, L., & Clements, W. A. (1998). "Older (but not younger) siblings facilitate false belief understanding". Developmental Psychology, 34(1), 161–74.
[43] Ruffman, T., Slade, L., & Crowe, E. (2002). "The relation between children's and mothers' mental state language and theory-of-mind understanding". Child Development, 73(3), 734–751.
[44] Sabbagh, M. A., Benson, J. E., & Kuhlmeier, V. (2013). "False belief understanding in infants and preschoolers". In M. Bornstein & M. Legerstee (Eds.), *The Developing Infant Mind: Integrating Biology and Experience*, pp. 301–323. New York, NY: Guilford Press.
[45] San Juan, V., & Astington, J. W. (2012). "Bridging the gap between implicit and explicit understanding: How language development promotes the processing and representation of false belief". British Journal of Developmental Psychology, 30(1), 105–122. doi:10.1111/j.2044-835X.2011.02051.x
[46] Saxe, R., Carey, S., & Kanwisher, N. (2004). "Understanding other minds: linking developmental psychology and functional neuroimaging". Annual Review of Psychology, 55(1), 87–124. doi:10.1146/annurev.psych.55.090902.142044
[47] Schick, B., de Villiers, P. A., de Villiers, J. G., & Hoffmeister, R. (2007). "Language and theory of mind: a study of deaf children". Child Development, 78(2), 376–396.
[48] Scott, R. M., & Baillargeon, R. (2009). "Which penguin is this? Attributing false beliefs about object identity at 18 Months". Child Development, 80(4), 1172–1196.
[49] Senju, A., Southgate, V., Miura, Y., Matsui, T., Hasegawa, T., Tojo, Y., ... Csibra, G. (2010). "Absence of spontaneous action anticipation by false belief attribution in children with autism spectrum disorder". Development and Psychopathology, 22(02), 353–360. doi:10.1017/S0954579410000106
[50] Senju, A., Southgate, V., White, S., & Frith, U. (2009). "Mindblind eyes: an absence of spontaneous theory of mind in asperger syndrome". Science, 325(5942), 883–885.
[51] Shahaeian, A., Peterson, C. C., Slaughter, V., & Wellman, H. M. (2011). "Culture and the sequence of steps in theory of mind development". Developmental Psychology, 47(5), 1239–1247. doi:10.1037/a0023899
[52] Slaughter, V., & Peterson, C. C. (2011). "How conversational input shapes theory of mind development in infancy and early childhood". In M. Siegal & L. Surian (Eds.), *Access to Language and Cognitive Development*, pp. 4–22. Oxford University Press.
[53] Surtees, A. D. R., & Apperly, I. A. (2012). "Egocentrism and automatic perspective taking in children and adults". Child Development, 83(2), 452–460.
[54] Surtees, A. D. R., Butterfill, S. A., & Apperly, I. A. (2011). "Direct and indirect measures of level 2 perspective taking in children and adults". British Journal of Developmental Psychology, 30, 75–86.
[55] Taumoepeau, M., & Ruffman, T. (2006). "Mother and infant talk about mental states relates to desire language and emotion understanding". Child Development, 77(2), 465–481.
[56] Thoermer, C., Sodian, B., Vuori, M., Perst, H., & Kristen, S. (2012). "Continuity from an implicit to an explicit understanding of false belief from infancy to preschool age". British Journal of Developmental Psychology, 30(1), 172–187. doi:10.1111/j.2044-835X.2011.02067.x
[57] Wellman, H. M. (2013). "Universal social cognition". In M. R. Banaji & S. A. Gelman (Eds.), *Navigating the Social World: What Infants, Children, and Other Species Can Teach Us*, pp. 69–74. Oxford University Press.
[58] Wellman, H. M., & Bartsch, K. (1988). "Young children's reasoning about beliefs. Cognition, 30(3), 239–277.
[59] Wellman, H. M., Cross, D., & Watson, J. (2001). "Meta-analysis of theory-of-mind development: the truth about false belief". Child Development, 72(3), 655–684.
[60] Wellman, H. M., Fang, F., Liu, D., Zhu, L., & Liu, G. (2006). "Scaling of theory-of-mind understandings in Chinese children". Psychological Science, 17(12), 1075–1081. doi:10.1111/j.1467-9280.2006.01830.x
[61] Wellman, H. M., Hollander, M., & Schult, C. A. (1996). "Young children's understanding of thought bubbles and of thoughts". Child Development, 67(3), 768–788.
[62] Wellman, H. M., & Liu, D. (2004). "Scaling of theory-of-mind tasks". Child Development, 75, 523–541.
[63] Werker, J. F. (1989). "Becoming a native listener". American Scientist, 77(1), 54–59.
[64] Wimmer, H., & Perner, J. (1983). "Beliefs about beliefs: representation and constraining function of wrong beliefs in young children's understanding of deception". Cognition, 13(1), 103–128.

[65] Woolfe, T., Want, S. C., & Siegal, M. (2002). "Signposts to development: theory of mind in deaf children". Child Development, 73(3), 768–778.

# How to Bite the Bullet of Quidditism — Why Bird's Argument against Categoricalism in Physics fails

Andreas Bartels

ABSTRACT. Bird's [1] a priori argument against Categoricalism with respect to fundamental physics properties is shown to be ineffective: First, there are categorical characteristics of fundamental properties of physics which are not fixed by the causal roles of these properties, but contribute to the identities of these properties and are thus legitimate candidates for quiddities. The existence of those *substantive* quiddities does not give rise to any in-principle-limitation of our knowledge of properties, but only to familiar sorts of empirical under-determination. Thus, Quidditism with respect to substantive quiddities does not lead to any unacceptable epistemic consequences, and therefore does not compromise Categoricalism. Second, the same sort of under-determination would apply to the dispositional monist's conception of properties. Thus the dispositional monist has to bite the bullet too, if there is any.

## 1 Introduction

[1] has launched an *a priori* argument against Categoricalism with respect to fundamental properties of physics. The argument is, in short, that Categoricalism with respect to fundamental properties of physics entails Quidditism, according to which the identity of a fundamental property is not fixed by its causal roles. But Quidditism leads, as Bird has argued, to an unacceptable epistemic consequence: if Quidditism were true, then we could not know in principle the fundamental properties of nature. Since this in-principle-limitation of knowledge would be forced upon us, not by any known limitation of human knowledge capacities, but by the metaphysical postulate of Categoricalism, which has no independent empirical support, this alleged limitation cannot be accepted. Therefore, Categoricalism cannot be true. Since Categoricalism is the logical negation of the claim of dispositional monism, this in turn entails strong a priori support for the thesis of dispositional monism.

In the following, I will accept the claim that Categoricalism entails Quidditism. Categoricalism is the thesis that fundamental properties have their causal roles (or: their 'powers'), if there are such, not essentially. If Categoricalism is true, then there might exist two *different* properties in the same world (for example, in our world) that have exactly the *same* powers.[1] Thus Categoricalism implies that the identity of fundamental properties is not completely fixed by their powers, which is exactly the claim of Quidditism.

What I will call into question is the claim that Quidditism leads to the consequence that we cannot know in principle the fundamental properties of nature. That would be true only with respect to *primitive* quiddities, which are defined to be just those characteristics the possession of which makes a property to be exactly *this* property. With respect to *this* sort of Quidditism, Lewis has claimed, "Quidditism is to properties as haecceitism is to individuals."[2] Since different properties which are distinct only by their respective primitive quiddities cannot – because of their non-qualitative character – be discerned by any possible empirical consequences, the existence of properties with primitive quiddities would indeed lead to some in-principle empirical under-determination with respect to properties: We may know that *some* property fulfills a certain causal role defined by a theory, but we could in principle never know *which* property it is that actually fulfills this role. This kind of in-principle limitation of property knowledge has been termed *humility*[3] by Lewis.

Now, Bird's argument against Quidditism is not only directed against *primitive* quiddities, but to *all* possible sorts of categorical characteristics of properties which are not causal powers. In the following, I will argue that there are indeed non-primitive categorical characteristics of fundamental properties of physics which are not fixed by the causal roles of these properties, but contribute to the identities of these properties and are thus legitimate candidates for quiddities. Those characteristics are provided by the mathematical representations of properties within their specific theoretical backgrounds. They are the *substantive* quiddities of fundamental physics properties.

Against Bird, I will argue that, in contrast to primitive quiddities, the existence of substantive quiddities does not give rise to any in-principle-limitation of knowledge. Substantive quiddities could rather be involved in phenomena of under-determination of theoretical properties by their causal effects, which is a familiar phenomenon that would not cut any ice concerning the Categoricalism-Dispositionalism-issue. Thus, no inacceptable epistemic limitation concerning the knowledge of properties follows from the existence of those substantive quiddities. This blocks the negative conclusion with respect to Categoricalism. Finally, the point will be further strengthened by the fact that even the dispositional monist has to face the same sort of empirical under-determination of properties which the Categoricalist is confronted with: if properties have their identity by their causal

---

[1] Cf. [1], 71f.

[2] [6] (209); on the other hand, Lewis argues, "haecceitism leads to trouble in a way that quidditism does not" (cf. [6] (210). According to [8], the disanalogy between haecceitism and quidditism originates mainly from the worldboundedness of individuals, in contrast to the repeatability of properties: "Individuals are not repeatable. They are exhausted in one instantiation. That is why it makes sense to treat them as worldbound. But property types are repeatable. And nothing in how they repeat poses a barrier to transworld repetition. That is why it makes no sense to treat them as worldbound" [8] (15).

[3] Cf. [6] 216. Lewis has commented to this sort of limitation of our knowledge of properties in some rather relaxed way: "Who ever promised me that I was capable in principle of knowing everything?" ([6] p. 211).

roles, then exactly those causal roles are under-determined by their manifestations.

## 2   Why Bird's Argument fails

Let us now first consider the dialectics of Bird's argument, i.e. the way in which judgment about conflicting metaphysical claims is taken to be dependent on their respective epistemological consequences. If any metaphysical thesis condemns us to a necessary lack of knowledge of the fundamental properties of the world, this is seen by Bird as a legitimate reason to reject that thesis. Limits of knowledge should be based on facts about the world, either concerning the nature of objects or the nature of human cognitive capacities, which follow from well-confirmed empirical theories. If we dismiss some commonly accepted epistemological assumption – namely that there are no in-principle limits for knowledge about fundamental properties – then this should not happen because of some metaphysical thesis that has yet to be confirmed by empirical theories. In other words: metaphysical theses, if not rooted in well-confirmed empirical theories, should not be taken as a decisive reason to reject commonly accepted epistemology; quite to the contrary, in cases of conflict, it is commonly accepted epistemology that should decide on the validity of metaphysical theses.

Even if we accept this general lesson about the dependence of metaphysics on epistemology, the question remains whether the application of this lesson to the case of Quidditism is legitimate. Is it really true that, in case the identity of properties is not completely fixed by its powers, we will in principle be unable to know those properties? The claim gets some credibility by the assumption that in general the possibility of knowledge of properties is exhausted by the *causal* characteristics of those properties. Now, one could argue, if the identity of a property is *not* completely fixed by causal characteristics, there is something contributing to its identity that has no causal connection to our cognitive apparatus and thus cannot be known by us.

Indeed, properties of the world have to stand in *any* causal connection to our cognitive apparatus, in order to become possible objects of knowledge. On the other hand, it is a well known fact that there are many properties in the world, for instance the spin of electrons, the radiation intensity of extragalactic radio sources or the Quark colors, of which we have knowledge only in some very indirect way. Therefore, it appears to be inadequate to conceive of the 'causal roles' of such properties as something that could be 'directly' observed. The content of our knowledge of those fundamental properties – what they are and under what conditions they will be instantiated – is essentially determined by their mathematical representations within theories. It is not *constituted* by the observable causal effects, to which those properties may contribute.

The condition that theoretical representations must fulfill, in order to be accounted for as representations of real empirical properties, is the condition of *empirical significance*. The condition requires that the instantiation of some property result in *any* observable effect that would not appear if the property were absent. Now, if Quidditism were true in the sense that there are *substantive* quiddities that individuate fundamental properties, the most epistemically troublesome situations that could appear would be like this: There are two competing theories, $T_F$ and $T_G$, where $T_F$ is exactly identical to $T_G$, with the exception that $T_F$ includes property F, whereas in $T_G$, at all places where F occurs in $T_F$, F is replaced by property G. F and G are supposed to be different just by the substantive quidditas Q: F has characteristic Q, whereas G lacks Q. Since Q is, by definition, not a power, there is no causal role that is ascribed to something by the ascription of Q, and therefore, it may appear that both theories are completely alike with respect to all their observable consequences. The properties F and G are different because of their having or not having characteristic Q, without there being any observable effect that would favor the assumption that F is present instead of G or vice versa.

The epistemically troublesome situation, to which Quidditism might give reason, is that of empirical under-determination: No observable fact following from the respective theories provides any empirical evidence in favor of one of these theories. By now, we don't know whether this under-determination is of an in-principle sort, i.e. of a sort that does not leave open any revision in the future, or of a familiar sort, leaving open the possibility of being revised at some later time as a result of theory developments like discovering new connections to other theories or embeddings into richer theories with respect to which $T_F$ and $T_G$ behave differently. The latter case would imply the possibility that new evidence could become available with respect to which both theories could be distinguished.

If Q were a *primitive* quiddity, then the under-determination could, in principle, never be overcome by any further theory development. The reason is that Q, as a primitive quiddity, could in no way couple to qualitative properties represented by other theories (as much as it cannot couple to the other qualitative properties represented by $T_F$). Thus no observable facts could be made available by means of any further theoretical connections. On the other hand, in-principle under-determination based on primitive quiddities would produce only a rather mild sort of limitation for our knowledge of properties. The aspects of reality that would then in principle escape our knowledge would in no way be involved in the qualitative natures of processes in the world.

Esfeld [4] has argued that the epistemic situation following from Quidditism is "in a certain sense [...] a case of under-determination of theoretical entities by observable phenomena"[4]. But that under-determination, according to Esfeld, is not of the familiar sort; rather, because it rests on the necessary non-observability of the categorical characteristics that make up the difference between the under-determined entities, it entails a final verdict about our resources to gain knowledge about the presence of one or the other entity. Again, Esfeld's claim is uncontroversial with respect to categorical characteristics that are primitive quiddities, but it would not be true of substantial quiddities, if there are such. If there were non-primitive categorical characteristics of properties that contribute to their identities which turn out to be *observable* in principle, then Esfeld's claim of the in-principle status of quidditistic under-determination would be undermined.

Thus, the question that we have to tackle now is: Are there really substantive quiddities? Are there characteristics of fundamental physics properties, beneath

---

[4] [4].

primitive natures, that could provide support to the quidditistic thesis that the identity of a fundamental property is not completely exhausted by its causal roles as defined by the best current theory?

The substantive quiddities we look for should be observable characteristics of properties which are categorical in the sense that ascribing them to objects is not ascribing a causal role. Brian Ellis [3] has argued that there are indeed well known examples of substantive observable quiddities, namely paradigmatic *categorical* characteristics like *localization*. In contrast to dispositional characteristics ("whose identities depend on what they dispose their bearer to do"[5]), localization (where in space a property appears) is a characteristics the identity of which depends on what their bearers *are*[6]. The ascription of a place to something does not ascribe to it a causal role. Thus, localization is clearly a categorical characteristic. Whereas localization is not causally active *by definition*, it is, according to Ellis, nevertheless *observable*.

If, for instance, light is reflected by a surface, a certain particle placed on that surface reflects the light, because it has a capacity to do so. The place of the particle does not have any capacity by itself. But it determines, from what direction the reflected light will reach the eye of the observer. The categorical characteristic *localization* thus modifies the causal effects that have been produced by dispositional characteristics in the first place; it is causally effective in an indirect way, in the sense that its causal effectiveness depends on the presence of capacities which are causally efficient in a direct way. Since the localization of a particle with the capacity to reflect light is connected with causal influences that would not appear if the particle had been at another place, localization is clearly observable.[7]

What Ellis supplies us with, is a case of a clearly categorical, observable characteristic. But it is doubtful whether localization is also a quiddity. Quiddities are, by definition, involved in the individuation of properties. But it seems that the capacity of a particle to reflect light can be completely understood without any mentioning of the characteristic of being localized at a certain place, and thus localization does not participate in the individuation of reflectivity. Furthermore, localization has the status of being categorical only within a classical theory of space. Within General Relativity space, or rather spacetime, is represented by a metrical field that is taken to be essentially causally active. Therefore, we have to search out for other candidates for observable quiddities. But, at least two insights of Ellis' considerations can be preserved in this search: First, the insight that categorical characteristics are in principle observable, and second, that their observability comes about in some indirect way, i.e. it is not the result of some intrinsic activity (as it is the case for dispositional characteristics), but rests on their influencing and modifying the way in which intrinsic activities of properties manifest themselves.

Candidates for quiddities that fulfill both of the criteria mentioned above are the *mathematical* properties that characterize fundamental properties in physical theories. One example is the property of "being represented by a scalar (or by

---

[5] [3], 136.
[6] [3], 136.
[7] Cf. [3], 140.

a vector/tensor)". Such mathematical properties are *categorical* because they are not individuated by any causal roles. But they are also in principle *observable*, if only in some *indirect* way. If, for instance, a fundamental physical property is mathematically characterized by a scalar, then certain observable phenomena may be allowed by the respective theory, which would not be allowed if the correct representation of that property were realized by a vector.

Think, for example, of Descartes' scalar theory of momentum that represents momentum by a scalar and requires conservation for total scalar momentum. Descartes' theory allows that a lighter body is reflected by a heavier one, keeping on its scalar momentum during its movement to the opposite direction (whereas the heavier body does not change its state of motion at all). That sort of phenomenon would not be allowed by a theory representing momentum by a vector quantity and requiring a law of conservation for vector momentum that would entail that momentum is conserved *in all possible directions*.

Mathematical properties figuring in the representation of fundamental properties are thus not 'only' mathematical properties. The way in which they contribute to the shape of property representations corresponds to the observable physical behavior of the respective properties. Thus, characteristics like 'being represented by a scalar' have also a *physical* meaning and they are in principle observable in the same way in which localization in Ellis' example is observable. Since these characteristics, despite of their being not definable by causal roles, contribute to the meaning of physics properties, the meaning of physics properties cannot be exhausted by causal roles. Moreover, since those mathematical characteristics are involved in the *individuation* of fundamental physics properties – what a certain physics property is depends critically on those mathematical characteristics – they are legitimate candidates for substantive *quiddities*. If, as a result of ongoing empirical inquiry, observable consequences of those substantive quiddities show up, then this does not in any way diminish their status as quiddities. Even if they *have* empirical consequences, they will for sure not be *definable* by means of these consequences.

In order to avoid misunderstandings, it should be mentioned that the distinction between *quiddities* versus *powers* (causal role-characteristics) does not coincide with the distinction between *simple* versus *structural* properties. Both, quiddities and powers are 'structural' characteristics in the sense that quiddities (like 'being represented by a scalar') just as powers (e.g. the power of gravity to produce gravitational attraction) turn out to be instantiated by realizations of a given characteristic structure. In the case of substantive quiddities, the respective structure determines by which mathematical object the property would be represented, whereas in the case of powers, the structure determines by which sort of connections to other properties it would be actualized.

Fundamental physical properties are individuated by means of their specific substantive quiddities. The powers that can be ascribed to them depend on how the properties are embedded into specific theory nets. Thus, one and the same property may be connected to different powers, depending on how connections to other properties are formed according to the specific theory or theory formulation in which the property occurs.[8] The case of under-determination of properties F and G within their respective theories $T_F$ and $T_G$, mentioned above, would be a case in which different properties (distinct by a substantive quiddity Q) share all their powers. In the following, pairs of properties which are empirically under-determined in that way I will call *Doppelgänger*-properties.

Since the quiddities which we are concerned with now, are not primitive, but substantial observable quiddities, possible cases of empirical under-determination resulting from the existence of Doppelgänger-properties would be of the familiar sort of under-determination: Later theoretical development could provide some extension of the connections to other properties, which in turn could make new evidence available with respect to which one of those Doppelgänger-properties may be favored against the other. There is no reason to suspect any in-principle empirical under-determination following from the possibility of Doppelgänger-properties. Thus, since no principled limitation of knowledge of properties results from such cases, it would be unreasonable to think that, by allowing them, Categoricalism would be compromised.

## 3 Tu quoque: The Under-determination of Causal Roles

The first part of this paper was about how to bite the bullet of Quidditism. We can bear biting it because the possible epistemological consequences connected to it turn out to be of a quite familiar sort. The second part will now show that the dispositional monist will have to bite this bullet too (or, at least a very similar one) – even if empirical under-determination as a result from the possible existence of Doppelgänger-properties were undesirable, dispositional monists would necessarily face exactly the same obstacle. Thus any a priori reason to favor Dispositional Monism against Categoricalism disappears.

For the sake of argument, let us assume that Dispositional Monism is correct, i.e. every fundamental property is completely individuated by the *causal roles* (powers) characterizing it. But what is it that individuates causal roles?

Take, for example, the fundamental property (within Newtonian gravitation theory) of passive gravitational mass $m_p$. One of the causal roles of this property is to produce the force $W = m_p\, g$ (weight), where $g$ is the gravitational acceleration at the place of the body which has $m_p$. The manifestation of this causal role comes about, when the body with weight $W$ is placed on a balance; a pressure will then be exerted upon the surface of the balance, and its pointer may show a certain value of the weight. What individuates the causal role of $m_p$, its weight $W$? It is not its manifestations, but the specific way in which the causal influence of $m_p$ resulting in the manifestation by the balance is exerted, namely the coupling of the passive

---

[8] For example, the metric in General Relativity is determined by the mathematical ('categorical') characteristics of the metric tensor. The connection of the metric with the affine connection, which is provided by the field equations, determines what tidal forces ('powers') can be ascribed to the metric. Non-standard formalisms of General Relativity use other sorts of connections between the metric and the affine structure. According to the Palatini formalism, for instance, the metric and the affine structure are independent structures. Thus, while the tidal forces can be ascribed as 'powers' to the metric according to the standard theory, this cannot be done according to the Palatini formalism (cf.[7], [5]).

gravitational mass to the gravitational field.

The reason that causal roles are not individuated simply by their manifestations is that causal phenomena like the manifestation of a causal role may be caused in quite different ways. The occurrence of a certain pressure upon the surface of the balance may have different sorts of possible causes. As long as we know only the pressure upon the surface of the balance, it remains absolutely under-determined *which* causal role it is that manifests itself by that pressure. The relation between 'theoretical' and 'observational' properties, where the first are under-determined by the occurrences of the latter, re-appears now for the relation between causal roles and their manifestations.

That the under-determination of causal roles by their manifestations is a possibility that appears within real science will be shown now by the example of a theoretical alternative that exists concerning the production of the pressure exerted upon the surface of the balance. In the Newtonian theory of gravitation, the pressure is the result of a coupling between the passive gravitational mass and the gravitational field. Einstein's[9] thought experiment of a box within gravitation-free space shows that this pressure could be produced, within the frame of Newtonian mechanics, by a quite different mechanism. It could be produced in the absence of a gravitational field as the effect of an acceleration field: If the box were be accelerated by some acceleration equal to $g$ in the upward direction, relative to the person in the box, the Newtonian *inertial mass* of the body of the person inside the box would produce exactly the same quantity of pressure upon the balance that, in the first situation, had been produced by the passive gravitational mass (and its coupling to the gravitational field).

As is well known, Einstein took the fact that the two situations are indistinguishable with respect to any empirical effects as indicating that these situations are not distinct in reality, and thus the different theoretical descriptions corresponding to them ("gravitational mass" versus "inertial mass") should be replaced by only *one* applying to both of them. But his famous inference to the principle of equivalence entailing the unification of gravitation and inertia is not in the focus of my interest at this point.

What the example rather shows is that – within one and the same theoretical frame – two different theoretical mechanisms exist that produce indistinguishable observable effects. The causal role which the passive gravitational mass plays – via the mechanism of coupling to the gravitational field – in producing the pressure upon the balance is different from the causal role that is fulfilled by the inertial mass – via its coupling to an acceleration field. Even if, in the case at hand, a unification program concerning these different causal roles has been successful, it cannot be guaranteed by any a priori reason that the duality of empirically indistinguishable causal roles of fundamental physical properties could be overcome by some later unification in general. The dispositionalist might insist that unification is not an accident, but a necessity. But, with respect to our example, this would amount to the claim that Newtonian gravitation theory represents a physically *impossible* world. If a theory of properties would imply such an exaggerated consequence,

---
[9][2], p. 44f.

this would be strong reason for distrust. Einstein's box thought experiment shows that there can be observationally indistinguishable, but different causal roles corresponding to different physical properties which are involved in specific mechanisms as described by a theoretical frame.

The individuation of causal roles, in other words, is not accomplished by observational effects, but by *ways* on which those effects are produced. If causal roles were individuated by their observational effects, then this would amount to a conception of properties as bundles of causal effects. But the dispositional monist, whose perspective we take for granted here for the sake of argument, could not subscribe to such a conception of properties. Rather, properties are genuine *activities* according to the dispositional monist's view. They have to be individuated by types of activities, e.g. by mechanisms or ways of producing observable effects.

Now, mechanisms are themselves *theory-dependent*: The right answer to the question whether some mechanism is the same or rather different from another mechanism depends on whether a theory representing those mechanisms represents them as being the same or as being different. In our example: what causal role has been manifested by exerting a certain pressure upon the balance depends on the theoretical explanation of this causal phenomenon, and thus it depends on the theoretical concepts that are involved in this explanation.

From that it follows that causal roles may be under-determined by empirical evidence in just the same way in which this may be true of the Doppelgänger-properties the possibility of which the categoricalist has to accept. The case of gravitational versus inertial origin of pressure upon a balance exemplifies this claim. Thus, the dispositional monist has to take into account – as a consequence of his conception of properties – just the same sort of empirical under-determination of our knowledge of properties that the categoricalist has to accept with respect to his/her conception of properties.

## 4 Conclusion

Let us summarize: As the first part of the argument shows, the alleged inacceptable epistemic consequence of Quidditism – and thus of Categoricalism from which it follows – turns out to entail nothing more than the possibility of familiar cases of under-determination of properties by empirical evidence: there are possible cases in which we don't know which property is present given our best empirical evidence. Since there is no reason to suspect any in-principle character of that under-determination, those possible cases cannot ground any basic skeptical conclusion with respect to our possible knowledge of properties and thus don't supply strong a priori reason to reject Categoricalism. The second part of the argument shows that Dispositional Monism and Categoricalism are completely on a par concerning the consequences for our knowledge of properties: The dispositional monist faces, with respect to his/her preferred conception of properties as constituted by causal roles, the same sort of empirical under-determination for knowledge of properties which the Categoricalist has to accept with respect to his/her conception of properties. Thus, as far as our possible knowledge of properties is concerned, there is no a priori reason to favor Dispositional Monism over Categoricalism or vice versa.

# BIBLIOGRAPHY

[1] Bird, Alexander (2007): Nature's Metaphysics, Oxford University Press: Oxford

[2] Einstein, Albert (1988): Über die spezielle und die allgemeine Relativitätstheorie, Vieweg: Braunschweig, 23th Edition (1st Edition 1917)

[3] Ellis, Brian (2010): Causal Powers and Categorical Properties, In: Anna Marmodoro (2010) (Ed.): The Metaphysics of Powers, Routledge: London, 133–142

[4] Esfeld, Michael (2009): The modal nature of structures in ontic structural realism, *International Studies in the Philosophy of Science* 23, 179–194

[5] Ferraris, M., M. Francaviglia, C. Reina (1982): Variational Formulation of General Relativity from 1915 to 1925. Palatini's Method Discovered by Einstein in 1925, *General Relativity and Gravitation* 14 (3), 243–254

[6] Lewis, David (2009): Ramseyan Humility, In: D. Braddon-Mitchell and Robert Nola (Eds.): Conceptual Analysis and Philosophical Naturalism, MIT-Press: Cambridge Mass., 203–222

[7] Palatini, Attilio (1919): Deduzione invariantiva delle equazioni gravitazionali dal principio di Hamilton, *Rendiconti del Circolo Matematico di Palermo* 43 (1), 203–212

[8] Schaffer, Jonathan (2005): Quidditistic Knowledge, *Philosophical Studies* 123, 1–32

# A Real World Semantics for Deterministic Dynamical Systems with Finitely Many Components

Marco Giunti

ABSTRACT. This paper shows in detail how it is possible to develop a *real world* semantics for *models* (in contrast with the usual *possible worlds* semantics for *languages* or theories), in the case of a widely used class of scientific models, namely, deterministic dynamical systems with finitely many components.

## 1 Introduction

In general, we take an *empirical theory* to be any theoretical construct, not necessarily of a linguistic type, which is expressly designed to describe or explain real phenomena. The exact nature of the semantic relations that an empirical theory may bear to the real world then depends on how either the theory or the phenomena are further conceived or analyzed.

According to the syntactic view, an empirical theory consists of an axiomatized theory—a purely formal system, together with a set of correspondence rules—an interpretative system (Hempel [11]; [12], sec. 8). The real world reference of the observational terms is supposed to be fixed, but the interpretative system does not typically suffice to set the reference of the theoretical ones. Thus, on this view, only observational sentences turn out to be true or false of the real world. As a consequence, *empirical adequacy*,[1] and not truth, turns out to be the relevant semantic relation between a theory and the world.

According to standard semantics, a theory is true or false in a possible model, which essentially is a set with an appropriate mathematical structure. Therefore, if a theory has to be true of the world in the standard semantic sense, the *world itself* must be a model of the theory (Balzer, Moulines, and Sneed [2], p. 2; Bickle [3], p. 62) and, consequently, it must have a full blown mathematical structure. However, such a strong Platonistic stance may very well seem too high a price to pay.

For van Fraassen ([18], ch. 3), the syntactic view is not adequate even from the empiricist's viewpoint, because its notion of empirical adequacy is utterly flawed. In his view, theories are better conceived semantically, as sets of models, and empirical adequacy is then analyzed as a relation between a model of the theory and the phenomena it describes. In fact, when a theory is empirically adequate, the structures

---

[1]For the syntactic view, a theory is *empirically adequate* just in case all its observational consequences (the so called *empirical content* of the theory) are true.

of the described phenomena are isomorphic to appropriate substructures *empirical substructures* of a model of the theory. Thus, for van Fraassen, phenomena do have a mathematical structure, but not as rich as the structure of the corresponding model. Furthermore, phenomenal structures are purely empirical or observational, as well as the substructures of the model isomorphic to them. In van Fraassen's view, the isomorphism between phenomenal structures (also called *appearances*, [18], p. 45) on the one hand, and substructures of a model of the theory on the other one, is the ultimate and most fundamental semantic relation between an empirical theory and the real world.

Van Fraassen's suggestion, that the crucial semantic relation is an isomorphism between model substructures and phenomena, seems to be on the right track. However, his view leaves at least three important problems unsolved. (a) Are the mathematical structures of the world to which a model is related *exclusively* empirical or observational, as van Fraassen claims? (b) Are such structures *given* independently of the theory, or are they somehow theoretically *constructed*, as Suppe ([16], pp. 132, 144-147, 150) maintains? And, finally, (c) how are we to precisely identify the *empirical substructures* of a model?

In this paper, we are going to delineate an alternative position that does not presuppose any given mathematical structure of the world. This approach elaborates and develops, within a framework of *constructive realism*, the essential aspects of van Fraassen's view with respect to the relation between models and reality, overcoming its difficulties.

More precisely, this position is *realist* in the sense that the representational relation between a model and the world is intended as a relation of *truth* (and not just empirical adequacy), which is based on an identity relation between the mathematical structure of the phenomenon under investigation and an appropriate substructure of the relative model. However, both the phenomenal structure and the model substructure are not independently given, but they are rather *constructed* by means of an appropriate *interpretation* of the model on the phenomenon. This interpretation, which in general is not merely empirical, presupposes, besides the model, also a low level theoretical element—a functional description, which is constitutive of the phenomenon itself.

In the subsequent sections, we are going to show in detail how it is possible to develop a *real world* semantics for *models* (in contrast with the usual *possible worlds* semantics for *languages* or theories), in the case of an important class of models, namely, deterministic dynamical systems with finitely many components.

In the first place, we are going to define an interpretation $I_{DS_L,H}$ of a dynamical system $DS_L$ on a phenomenon $H$. The interpretation $I_{DS_L,H}$ will then allow us to define what it means, for the interpreted dynamical system $(DS_L, I_{DS_L,H})$, to be a *true model of* $H$. In the second place, we will show how such interpretation induces, on the one hand, a mathematical structure on $H$ and, on the other one, a substructure on $DS_L$. Finally, we will prove that $(DS_L, I_{DS_L,H})$ is a true model of $H$ if, and only if, the structure of $H$ induced by $I_{DS_L,H}$ is identical to the substructure of $DS_L$ induced by $I_{DS_L,H}$.

## 2 Deterministic dynamical systems with $n$ components

We are now going to define a real world semantics for a widely used class of models—deterministic dynamical systems whose state space has a finite number $n \in \mathbb{Z}^{>0}$ of components. In general, a deterministic dynamical system can be identified with a pair $DS_L = (M, (g^t)_{t \in T})$, where $M$ is a *state space* and $(g^t)_{t \in T}$ is a family of functions from $M$ to $M$ (called *state transitions*) that satisfy the two conditions $g^0(x) = x$ and $g^{v+t}(x) = g^v(g^t(x))$. The index set $T$ is called the *time set* and each of its elements is to be thought as the *duration* of the corresponding state transition.

Durations can be added and, according to the (decreasing) richness of the algebraic structure imposed to the addition operation $+$, the *time model* $L = (T, +)$ turns out to be a group or a monoid. The set of the corresponding state transitions $\{g^t : t \in T\}$, together with the usual operation of function composition $\circ$, also turns out to be, respectively, a group or a monoid.

Durations are usually taken to be either *continuous* or *discrete* quantities. In the first case, the time set $T$ is identified with either the set of the reals $\mathbb{R}$ or the non-negative reals $\mathbb{R}^{\geq 0}$, and the operation $+$ of addition over durations is the usual addition of two real numbers. In the second case, $T$ is identified with either the set of the integers $\mathbb{Z}$ or the non-negative integers $\mathbb{Z}^{\geq 0}$, and the operation $+$ of addition over durations is the usual addition of two integer numbers.

The usual definition of a dynamical system (Arnold [1]; Szlenk [17]; Giunti [8]; Hirsch, Smale, and Devaney [13]) intends to formally render the intuitive concept of an arbitrary deterministic system, either reversible or irreversible, with continuous or discrete time or state space. However, Giunti and Mazzola [10] have recently noticed that this definition is not completely general, for it does not fix the minimal algebraic structure on the time set $T$ that can still support an adequate notion of a deterministic dynamics on the state space $M$. The two authors have argued that such a minimal structure is a *monoid* and, consequently, that the most general definition of a deterministic dynamical system is the following.

**Definition 1 (Dynamical system).**

$DS_L$ *is a dynamical system* :=

1. (a) $L = (T, +)$;
   (b) $DS_L = (M, (g^t)_{t \in T})$;
2. (a) $+ : T \times T \to T$;
   (b) $\forall t \in T, g^t : M \to M$;
3. (a) $+$ is associative and
   (b) its unity $0 \in T$ exists;
   (c) $\forall x \in M, g^0(x) = x$;
   (d) $\forall v, t \in T, \forall x \in M, g^{v+t}(x) = g^v(g^t(x))$.

The following are all examples of dynamical systems.

**Example 1 (Dynamical systems with discrete or continuous time set or state space).**

1. Discrete time set ($T = \mathbb{Z}^{\geq 0}$) and discrete state space: finite state automata, Turing machines, cellular automata restricted to finite configurations.[2]

2. Discrete time set ($T = \mathbb{Z}^{\geq 0}$) and continuous state space: many systems specified by difference equations, iterated mappings on $\mathbb{R}$, cellular automata not restricted to finite configurations.

3. Continuous time set ($T = \mathbb{R}$) and continuous state space: systems specified by ordinary differential equations, many neural nets.

Definition 1 is a formal rendition of the most general notion of a deterministic dynamical system. However, in this paper, we are going to develop a real world semantics only for those dynamical systems whose state space can be factorized into a finite number $n \in \mathbb{Z}^{>0}$ of components. For any $i$ ($1 \leq i \leq n$), let $X_i$ be a non-empty set. An $n$-component dynamical system is then defined as follows (Giunti [9], sec. 4.1).

**Definition 2 ($n$-component dynamical system).**
$DS_L$ is an $n$-component dynamical system := $DS_L = (M, (g^t)_{t \in T})$ is a dynamical system and $M \subseteq X_1 \times \ldots \times X_n$.
Furthermore, for any $i$, the set $C_i := \{x_i$: for some $n$-tuple $x \in M$, $x_i$ is the $i$-th element of $x\}$ is called *the $i$-th component of $M$*.[3]

**Example 2 (The 4-component dynamical system $DS_P$).**
A typical example of a 4-component dynamical system is the system $DS_P$ (see (2) below), which is individuated by the equation of motion of a projectile:

$$\left( \frac{dx(t)}{dt} = \dot{x}(t), \quad \frac{dy(t)}{dt} = \dot{y}(t), \quad \frac{d\dot{x}(t)}{dt} = 0, \quad \frac{d\dot{y}(t)}{dt} = -\mathbf{g} \right), \tag{1}$$

where $\mathbf{g} \in \mathbb{R}$ is a fixed positive constant. The solutions of this ordinary differential equation univocally determine the 4-component dynamical system:

$$DS_P = (X \times Y \times \dot{X} \times \dot{Y}, (g^t)_{t \in T}), \tag{2}$$

where $P = (\mathbb{R}, +)$ is the additive group of the real numbers, $X = Y = \dot{X} = \dot{Y} = T = \mathbb{R}$ and, for any $t \in T$, for any $(x, y, \dot{x}, \dot{y}) \in X \times Y \times \dot{X} \times \dot{Y}$,

$$g^t(x, y, \dot{x}, \dot{y}) = \left( \dot{x}t + x, \quad -\tfrac{1}{2}\mathbf{g}t^2 + \dot{y}t + y, \quad \dot{x}, \quad -\mathbf{g}t + \dot{y} \right). \tag{3}$$

---
[2] The state space of a cellular automaton is discrete (*i.e.*, finite or countably infinite) if all its states are finite configurations, that is to say, configurations where all but a finite number of cells are non-empty. If this condition is not satisfied, the state space has the power of the continuum.
[3] Let $proj_i$ be the $i$-th projection map. Then, $C_i = proj_i(M)$.

## 3 Deterministic dynamical phenomena

In general, we take a *deterministic dynamical phenomenon* (for brevity, *phenomenon*) to be any manifestation of the real world that an $n$-component dynamical system can represent.

In more detail, any *phenomen* $H$ can be thought as a pair $(F, B_F)$ of two distinct elements, a *theoretical part* $F$ and *a real part* $B_F$ (Giunti [9], sec. 4.1).

The theoretical part $F$ is a *functional description* which provides a sufficiently detailed specification of:

1. the internal constitution and organization, or functioning, of any real system of a certain *type* $AS_F$;

2. a *causal scheme* $CS_F$ of the external interactions of any real system of type $AS_F$ during an arbitrary temporal evolution. In particular, the description of the causal scheme $CS_F$ must include the specification of:

    (a) the *initial conditions* that an arbitrary evolution of any real system of type $AS_F$ must satisfy;

    (b) the *boundary conditions during* the whole subsequent evolution;

    (c) and, possibly, the *final conditions* under which the evolution terminates.

The real part $B_F$ is the set of all real or concrete systems which satisfy the functional description $F$ or, in other words, $B_F$ is the set of all real systems of type $AS_F$ whose temporal evolutions are all constrained by the causal scheme $CS_F$. $B_F$ is called the *realization domain* (or *application domain*) *of* $H$.[4] Any real system $b_F \in B_F$ is called an *F-realizer*.

**Example 3 (The phenomenon of projectile motion).**
We refer to the *phenomenon of projectile motion* by the symbol $H_{p,\phi\theta} = (F_{p,\phi\theta}, B_{F_{p,\phi\theta}})$, where $p$ is an abbreviation for *projectile*, while $\phi$ and $\theta$ are two non-negative real parameters (on which the functional desciption $F_{p,\phi\theta}$ depends), whose meaning is explained below.

*Theoretical part—Functional description* $F_{p,\phi\theta}$

1. Description of any real system of type $AS_{F_{p,\phi\theta}}$: any medium size body in the proximity of the earth.

2. Description of the causal scheme $CS_{F_{p,\phi\theta}}$ of the external interactions of any real system of type $AS_{F_{p,\phi\theta}}$ during an arbitrary temporal evolution;

    (a) initial conditions: the body is released at an arbitrary instant, with an initial velocity and position such that the body hits the earth surface at a later instant, the maximum vertical distance reached by the body with respect to the earth surface is not greater than $\phi$, and the maximum horizontal distance is not greater than $\theta$;

---

[4]Since the functional description $F$ typically contains several idealizations (see Example 3, 1 and 2), no real or concrete system *exactly* satisfies $F$, but it rather fits $F$ up to a *certain degree*. Thus, from a formal point of view, the realization domain $B_F$ of a phenomenon $H = (F, B_F)$ would be more faithfully described as a *fuzzy* set.

(b) boundary conditions: during the whole motion the only force acting on the body is its weight;

(c) final conditions: the motion terminates immediately after the impact of the body with the earth surface.

*Real part—Realization domain* $B_{F_{p,\phi\theta}}$

$B_{F_{p,\phi\theta}}$ = the (fuzzy) set of all medium size bodies in the proximity of the earth whose motions satisfy the causal interaction scheme $CS_{F_{p,\phi\theta}}$. Any body $b_{F_{p,\phi\theta}} \in B_{F_{p,\phi\theta}}$ is called a *projectile*.

## 4 Interpretation of an $n$-component dynamical system on a phenomenon

Let us now see how a dynamical system $DS_L = (M, (g^t)_{t \in T})$ with $n$ components $C_1, ..., C_n$ can be interpreted on a phenomenon $H = (F, B_F)$. The key point of the interpretation consists in establishing a correspondence between the time set $T$ of the dynamical system and the time magnitude of the phenomenon, as well as between each component $C_i$ ($1 \leq i \leq n$) of the state space and a different magnitude of $H$.

In general, we take a *magnitude of a phenomenon* $H$ to be a property $\boldsymbol{M}_j$ of every $F$-realizer $b_F \in B_F$ such that, at different instants, it can assume different values. The set of all possible values of magnitude $\boldsymbol{M}_j$ is indicated by $V(\boldsymbol{M}_j)$.[5]

We further assume that, among the magnitudes of any phenomenon $H$, there always is *its time magnitude*, which we denote by $\boldsymbol{T}$. The set of all possible values (*instants* or *durations*) of the time magnitude of $H$ is indicated by $V(\boldsymbol{T})$.

An interpretation $I_{DS_L,H}$ of $DS_L$ on $H$ consists in stating that (i) each component $C_i$ of the state space $M$ is included in, or is equal to, the set $V(\boldsymbol{M}_i)$ of the possible values of a magnitude $\boldsymbol{M}_i$ of the phenomenon $H$ and (ii) the time set $T$ of $DS_L$ is equal to the set $V(\boldsymbol{T})$ of the possible values of the time magnitude $\boldsymbol{T}$ of the phenomenon $H$. In other words, an interpretation $I_{DS_L,H}$ can always be identified with a particular set of $n+1$ sentences. We thus define:

**Definition 3 (Interpretation of a dynamical system on a phenomenon).** $I_{DS_L,H}$ is an interpretation of $DS_L$ on $H := I_{DS_L,H} = \{C_1 \subseteq V(\boldsymbol{M}_1), ..., C_n \subseteq V(\boldsymbol{M}_n), T = V(\boldsymbol{T})\}$, where $C_i$ is the $i$-th component of the state space of $DS_L$, $\boldsymbol{M}_i$ is a magnitude of $H$, $\boldsymbol{T}$ is the time magnitude of $H$ and, for any $i, j$ ($1 \leq i, j \leq n$), if $i \neq j$, then $\boldsymbol{M}_i \neq \boldsymbol{M}_j$.

Once an interpretation $I_{DS_L,H}$ has been fixed, the dynamical system $DS_L$ provides us with a representation of the temporal evolutions of the real systems (the $F$-realizers) in the realization domain $B_F$ of phenomenon $H$. Hence, the system $DS_L$ together with the interpretation $I_{DS_L,H}$ can be thought as a model of phenomenon $H$. This idea is precisely expressed by the definition below.

---

[5] It should be noticed that this definition does not require that the magnitudes of a phenomenon be observational, or even measurable. Furthermore, the nature of the possible values of a magnitude is not specified as well. This, in particular, means that there may be magnitudes whose possible values are not real numbers.

**Definition 4 (Model of a phenomenon).**
**DS** is a model of $H$ := $H$ is a phenomenon and $\boldsymbol{DS} = (DS_L, I_{DS_L,H})$, where $DS_L$ is an $n$-component dynamical system and $I_{DS_L,H}$ is an interpretation of $DS_L$ on $H$.

Going back to Example 2, we notice that the 4-component dynamical system $DS_P$ is not usually thought as a pure mathematical system. Instead, it is conceived *together with* a largely implicit *intended interpretation*, which makes it a model of the phenomenon $H_{p,\phi\theta}$ of projectile motion (Example 3). This interpretation is made explicit in the following example.

**Example 4 (The intended interpretation of the dynamical system $DS_P$ on the phenomenon $H_{p,\phi\theta}$ of projectile motion).**
We use the symbol $I_{DS_P, H_{p,\phi\theta}}$ to indicate the intended interpretation of the dynamical system $DS_P$ on the phenomenon $H_{p,\phi\theta}$ of projectile motion. Let us also recall that, at the end of Example 3, we stipulated that any real system (projectile) in the realization domain $B_{F_{p,\phi\theta}}$ of the phenomenon $H_{p,\phi\theta}$ be indicated by $b_{p,\phi\theta}$. In order to simplify notation, from now on we are going to refer to an arbitrary projectile just with $b$.

The interpretation $I_{DS_P, H_{p,\phi\theta}}$ can be made explicit as follows. Let $r_b$ the point where the projectile $b$ is initially released. Let us then consider the plane that contains the initial velocity vector of $b$ and the earth center. On this plane, we fix the axes X and Y of a Cartesian coordinate system with origin in the earth center, and whose Y axis passes through $r_b$. We take the positive direction of the Y axis to be the one from the earth center to the point $r_b$. Accordingly, we call the Y axis *vertical* and the X axis *horizontal*.

Let us then consider the following five magnitudes of $H_{p,\phi\theta}$:
**X** = the horizontal component of the position of $b$,
**Y** = the vertical component of the position of $b$,
**Ẋ** = the horizontal component of the velocity of $b$,
**Ẏ** = the vertical component of the velocity of $b$,
**T** = the time magnitude of $H_{p,\phi\theta}$.

We can now let the four components $X, Y, \dot{X}, \dot{Y}$ of the state space of $DS_P$ and its time set $T$ correspond to these five magnitudes of $H_{p,\phi\theta}$. The intended interpretation of the dynamical system $DS_P$ on the phenomenon $H_{p,\phi\theta}$ of projectile motion is thus the following set of five sentences:

$$I_{DS_P, H_{p,\phi\theta}} = \{X = V(\mathbf{X}),\ Y = V(\mathbf{Y}),\ \dot{X} = V(\mathbf{\dot{X}}),\ \dot{Y} = V(\mathbf{\dot{Y}})),\ T = V(\mathbf{T})\}. \quad (4)$$

Let $\boldsymbol{DS}_{p,\phi\theta} = (DS_P, I_{DS_P, H_{p,\phi\theta}})$. By Definition 4, $\boldsymbol{DS}_{p,\phi\theta}$ is thus a model of $H_{p,\phi\theta}$. $\boldsymbol{DS}_{p,\phi\theta}$ is called *the projectile model* (Giunti [9], sec. 4.2.1).

## 5 True models of phenomena

Once an interpretation $I_{DS_L,H} = \{C_1 \subseteq V(\mathbf{M}_1), ..., C_n \subseteq V(\mathbf{M}_n), T = V(\mathbf{T})\}$ is fixed, we can define as follows the possible states and the state space of the phenomenon $H = (F, B_F)$, relative to that interpretation.

**Definition 5 (Possible state of a phenomenon, relative to an interpretation).**
$x$ is a possible state of $H$ relative to $I_{DS_L,H} := x \in V(M_1) \times, ..., \times V(M_n)$.

**Definition 6 (State space of a phenomenon, relative to an interpretation).**

$M := V(M_1) \times, ..., \times V(M_n)$ is called *the state space of $H$ relative to $I_{DS_L,H}$*.

The interpretation $I_{DS_L,H}$ also allows us to define the instantaneous state of any $F$-realizer of the phenomenon $H$. Let $b_F \in B_F$ be an arbitrary $F$-realizer of $H$, and $j \in T$ an arbitrary instant. Then:

**Definition 7 (Instantaneous state of an $F$-realizer, relative to an interpretation).**
$x$ is the state of $b_F$ at instant $j$ relative to $I_{DS_L,H} := x = (x_1, ..., x_n)$, where $x_i$ is the value at instant $j$ of magnitude $M_i$ of $b_F$ (if, at instant $j$, such a value exists).

Obviously, if $x$ is the state of $b_F$ at instant $j$ relative to $I_{DS_L,H}$, then $x \in M$. Note, however, that, depending on the instant $j$, the value of magnitude $M_i$ of $b_F$ may not exist.[6] If this happens, the state of $b_F$ at instant $j$ relative to $I_{DS_L,H}$ is not defined.

Now, relative to the interpretation $I_{DS_L,H}$, we may define the set $C_F$ of all those possible states of $H$ (if any) that *actually* are initial states of $H$.

**Definition 8 (The set $C_F$ of the initial states of a phenomenon, relative to an interpretation).**
$C_F := \{x : \text{for some } b_F \in B_F, \text{ for some temporal evolution } e \text{ of } b_F, \text{ for some } j \in T,$ $j$ is the initial instant of $e$ and $x$ is the state of $b_F$ at $j$ relative to $I_{DS_L,H}\}$. $C_F$ is called *the set of all initial states of $H$, relative to interpretation $I_{DS_L,H}$*.

Intuitively, the set $C_F$ may be thought as the set of all those states in $M$ that are consistent with the initial conditions specified by the causal scheme $CS_F$ and are in fact initial states of some realizer $b_F \in B_F$.

Also note that, depending on the interpretation $I_{DS_L,H}$, $C_F$ may be empty, or $C_F$ may not be a subset of the state space $M$ of $DS_L$.[7] The definition of an admissible interpretation (Definition 12) will exclude these somewhat pathological interpretations.

**Example 5 (The set $C_{F_{p,\phi\theta}}$ of the initial states of the phenomenon $H_{p,\phi\theta}$ of projectile motion, relative to the intended interpretation $I_{DS_P,H_{p,\phi\theta}}$).**
Let $C_{F_{p,\phi\theta}}$ be the set of the initial states of the phenomenon $H_{p,\phi\theta}$ of projectile

---

[6]If, for some reason, $b_F$ no longer exists at instant $j \in T$, then *a fortiori* the value at $j$ of magnitude $M_i$ of $b_F$ does not exist either. Furthermore, it should be noticed that here we are not making any assumption about the continuous existence of the values of a magnitude during any interval of time. Thus, it is always possible that the value of a magnitude $M_i$ of $b_F$ exists at some instant $j$ of $b_F$'s existence, but does not exist at some other instant $k$ of its existence.

[7]In fact, by Definition 7, $C_F$ is empty if, for any $b_F \in B_F$ and any evolution $e$ of $b_F$, some magnitude $M_i$ does not have a value at the initial instant of $e$. Also recall that, according to interpretation $I_{DS_L,H}$ (see Definition 3), each component $C_i$ of the state space $M$ is in general a subset of $V(M_i)$. Thus, if for some $x \in C_F$, its $i$-th component $x_i \in V(M_i)$ is not a member of $C_i$, then $C_F \nsubseteq M$.

motion, relative to the intended interpretation $I_{DS_P,H_{p,\phi\theta}}$ of the dynamical system $DS_P$ on $H_{p,\phi\theta}$. By Definition 8, such a set turns out to be:

1. $C_{F_{p,\phi\theta}} = \{x :$ for some projectile $b \in B_{F_{p,\phi\theta}}$, for some temporal evolution $e$ of $b$, for some $j \in T, j$ is the initial instant of $e$ and $x$ is the state of $b$ at instant $j$ relative to $I_{DS_P,H_{p,\phi\theta}}\}$.

From the equation above, and by recalling how the interpretation $I_{DS_P,H_{p,\phi\theta}}$ is defined (Example 4), we get:

2. $C_{F_{p,\phi\theta}} = \{x :$ for some projectile $b \in B_{F_{p,\phi\theta}}$, for some temporal evolution $e$ of $b$, for some $j \in T, j$ is the initial instant of $e$ and $x = (0, y, \dot{x}, \dot{y})\}$, where $0, y, \dot{x}, \dot{y}$ are the values, at initial instant $j$, of the horizontal position, vertical position, horizontal velocity, and vertical velocity of projectile $b$.

Also note that the three initial values $y, \dot{x}, \dot{y}$ are not completely arbitrary because, according to the causal scheme $CS_{F_{p,\phi\theta}}$ (Example 3, 2), they depend on the two parameters $\phi$ and $\theta$.

Let $C_F \neq \emptyset$. Let us now define, with respect to interpretation $I_{DS_L,H}$, the set all initial instants of the evolutions of a given $F$-realizer $b_F \in B_F$, whose initial state $x \in C_F$ be fixed. We call this set $J_{b_F,x}$.

**Definition 9 (The set $J_{b_F,x}$ of the initial instants of $b_F$ whose initial state is $x$, relative to an interpretation).**
$J_{b_F,x} := \{j_{b_F,x} : j_{b_F,x}$ is the initial instant of some evolution of $b_F$ and $x$ is the state of $b_F$ at $j_{b_F,x}$ relative to interpretation $I_{DS_L,H}\}$. $J_{b_F,x}$ is called *the set of the initial instants of $b_F$ whose initial state is $x$, relative to $I_{DS_L,H}$*.

Note that, for some $b_F \in B_F$ and $x \in C_F$, $J_{b_F,x}$ may be empty.[8] However, by the definition of $C_F$ (Definition 8), for any $x \in C_F$, there is $b_F \in B_F$ such that $J_{b_F,x} \neq \emptyset$.

As we are assuming that the phenomenon $H$ be deterministic, the existence and identity of the instantaneous state, at any fixed stage of an evolution of any realizer $b_F$, is not intended to depend on either the initial instant, or the identity of $b_F$, but only on the initial state. Thus, any admissible interpretation $I_{DS_L,H}$ should at least ensure that the condition below holds.

**Condition D (Determinism).** For any $b_F, d_F \in B_F$, for any $x \in C_F$, for any $j_{b_F,x} \in J_{b_F,x}$, for any $k_{d_F,x} \in J_{d_F,x}$, for any $t \in T$, if $t + j_{b_F,x}$ is an instant of the evolution of $b_F$ that starts at $j_{b_F,x}$ and the state of $b_F$ at instant $t + j_{b_F,x}$ exists, then $t + k_{d_F,x}$ is an instant of the evolution of $d_F$ that starts at $k_{d_F,x}$, the state of $d_F$ at instant $t + k_{d_F,x}$ exists as well, and the state of $b_F$ at instant $t + j_{b_F,x} =$ the state of $d_F$ at instant $t + k_{d_F,x}$.

Let $C_F \neq \emptyset$. For any initial state $x \in C_F$, let us consider the set of all $F$-realizers whose initial state is $x$. This set, denoted by $B_{F_x}$, is in other words the collection of all $F$-realizers $b_F$ whose set $J_{b_F,x}$ is not empty. Note that also this definition, as the previous ones, depends on the interpretation $I_{DS_L,H}$.

---

[8] In fact, $J_{b_F,x}$ is empty if $x$ is not the state of $b_F$ at the initial instant of any of its evolutions.

**Definition 10 (The set $B_{F_x}$ of the F-realizers whose initial state is $x$, relative to an interpretation).**
$B_{F_x} := \{b_F \in B_F : J_{b_F,x} \neq \emptyset\}$. $B_{F_x}$ is called *the set of the F-realizers whose initial state is $x$, relative to interpretation $I_{DS_L,H}$*.

We noticed above that, for any $x \in C_F$, there is $b_F \in B_F$ such that $J_{b_F,x} \neq \emptyset$. Therefore, by Definition 10, for any $x \in C_F$, $B_{F_x} \neq \emptyset$.

Suppose $C_F \neq \emptyset$. Then, for any $x \in C_F$, for any $b_F \in B_{F_x}$, for any $j_{b_F,x} \in J_{b_F,x}$, we define the following set of durations:

**Definition 11 (The set of durations $q_{b_F,j_{b_F,x}}(x)$, relative to an interpretation).**
$q_{b_F,j_{b_F,x}}(x) := \{t : t \in T, t + j_{b_F,x}$ is an instant of the evolution of $b_F$ that starts at $j_{b_F,x}$, and there is $y \in M$ such that $y$ is the state of $b_F$ at $t + j_{b_F,x}$, relative to interpretation $I_{DS_L,H}\}$.

Note that Definition 11, like the previous ones, is relative to the interpretation $I_{DS_L,H}$. Furthermore, $q_{b_F,j_{b_F,x}}(x) \neq \emptyset$, for $0 \in q_{b_F,j_{b_F,x}}(x)$.

Also note that, whenever Condition D above holds, $q_{b_F,j_{b_F,x}}(x)$ depends on $x$, but does not depend on either $b_F$ or $j_{b_F,x}$; therefore, if Condition D holds, we simply write "$q_F(x)$" instead of "$q_{b_F,j_{b_F,x}}(x)$".

By Condition D and Definition 11, for any $x \in C_F$, $q_F(x)$ is the set of all durations $t$ that transform the initial state $x$ of an arbitrary F-realizer $b_F \in B_{F_x}$ into some other state of $b_F$. More briefly, we call $q_F(x)$ *the set of all durations that trasform the initial state $x$ of H into some other state*.

**Example 6 (The set $q_{F_{p,\phi\theta}}(x)$ of all durations that transform the initial state $x$ of the phenomenon $H_{p,\phi\theta}$ of projectile motion into some other state, relative to the intended interpretation $I_{DS_P,H_{p,\phi\theta}}$).**
We recall (Example 5) that $C_{F_{p,\phi\theta}}$ is the set of the initial states of the phenomenon $H_{p,\phi\theta}$ of projectile motion, relative to the intended interpretation $I_{DS_P,H_{p,\phi\theta}}$. We notice that, by 2 of Example 5, $C_{F_{p,\phi\theta}} \neq \emptyset$.

For any projectile $b \in B_{F_{p,\phi\theta}}$ and any initial state $x \in C_{F_{p,\phi\theta}}$, we indicate with $J_{b,x}$ the set of the initial instants of $b$ whose initial state is $x$, relative to $I_{DS_P,H_{p,\phi\theta}}$ (see Definition 9).

We also notice that the intended interpretation $I_{DS_P,H_{p,\phi\theta}}$ does ensure that Condition D holds. Therefore, by Definition 11, for any $x \in C_{F_{p,\phi\theta}}$, we get:

1. $q_{F_{p,\phi\theta}}(x) = \{t : t \in T, t + j_{b,x}$ is an instant of the evolution of $b$ that starts at $j_{b,x}$, and there is $y \in M$ such that $y$ is the state of $b$ at $t + j_{b,x}\}$, where $b$ is an arbitrary projectile member of $B_{F_{p,\phi\theta_x}}$ and $j_{b,x} \in J_{b,x}$ is any initial instant of $b$ whose initial state is $x$.

Let $l(x)$ be the duration of the evolution of $b$ that starts at instant $j_{b,x}$ in state $x$. It is not difficult to show that such a duration does not depend on either the projectile $b$ or the initial instant $j_{b,x}$, but only on the initial state $x$.[9] Therefore, presumably,

---
[9] Let $u(b, j_{b,x})$ the final instant of the evolution of $b$ that starts at $j_{b,x}$ in state $x$, that is to

Equation 1 above reduces to (5) below (Giunti [9], par. 5.3):

$$q_{F_p,\phi\theta}(x) = \{t : t \in T \text{ and } 0 \leq t \leq l(x)\}. \tag{5}$$

As we are not interested in any interpretation $I_{DS_L,H}$ such that (a) $C_F = \emptyset$, or (b) $C_F \not\subseteq M$, or (c) Condition D does not hold,[10] we define:

**Definition 12 (Admissible interpretation).**
$I_{DS_L,H}$ is an admissible interpretation of $DS_L$ on $H$ := (i) $C_F \neq \emptyset$ and (ii) $C_F \subseteq M$ and (iii) Condition D holds.

**Example 7 (The intended interpretation of $DS_P$ on the phenomenon $H_{p,\phi\theta}$ of projectile motion is admissible).**
We notice first that the intended interpretation $I_{DS_P,H_{p,\phi\theta}}$ entails $C_{F_p,\phi\theta} \subseteq M$, for all its component sentences are identities (see Equation (4) above). Second, we have already seen (Example 6) that $C_{F_p,\phi\theta} \neq \emptyset$ and Condition D holds. It thus follows that $I_{DS_P,H_{p,\phi\theta}}$ is an admissible interpretation of $DS_P$ on $H_{p,\phi\theta}$.

We can now precisely state the conditions for an interpretation $I_{DS_L,H}$ to be correct. The intuitive idea is this. We noticed above (sec. 4) that, as soon as an interpretation $I_{DS_L,H}$ is fixed, the dynamical system $DS_L = (M, (g^t)_{t \in T})$ provides us with a representation of the temporal evolutions of the real systems ($F$-realizers) in the realization domain $B_F$ of phenomenon $H$.

In more detail, we should keep in mind that such a representation is provided by the state transition family $(g^t)_{t \in T}$ of dynamical system $DS_L$. The interpretation $I_{DS_L,H}$ will thus turn out to be correct if the representation, provided by $(g^t)_{t \in T}$, of *all* temporal evolutions of *all* $F$-realizers of $H$ is correct. This intuitive idea is formally expressed by the definition below.

**Definition 13 (Correct interpretation).**
$I_{DS_L,H}$ is a correct interpretation of $DS_L$ on $H$ := (i) $I_{DS_L,H}$ is an admissible interpretation of $DS_L$ on $H$ and (ii) for any $x \in C_F$, for any $t \in q_F(x)$, for any $b_F \in B_{F_x}$, for any $j_{b_F,x} \in J_{b_F,x}$, $g^t(x) =$ the state of $b_F$ at instant $t + j_{b_F,x}$ relative to $I_{DS_L,H}$.

The preceding definition finally allows us to define what it means, for an interpreted dynamical system $(DS_L, I_{DS_L,H})$, to be a true model of $H$:

---

say, the instant at which the projectile $b$ hits the earth surface. We can safely assume that the state of $b$ at $u(b, j_{b,x})$ exists, because, by the causal scheme of projectile motion (Example 3, 2c), the motion terminates only after $u(b, j_{b,x})$. Let then $z(u(b, j_{b,x}))$ be this state and let $l(b, j_{b,x}) := u(b, j_{b,x}) - j_{b,x}$. Let us assume for *reductio* that, for some projectile $d$ and initial instant $j_{d,x}$, $l(b, j_{b,x}) \neq l(d, j_{d,x}) := u(d, j_{d,x}) - j_{d,x}$. Assume $l(d, j_{d,x}) < l(b, j_{b,x})$. Since $z(u(b, j_{b,x}))$ is the state of $b$ at $u(b, j_{b,x}) = l(b, j_{b,x}) + j_{b,x}$, and Condition D holds, $z(u(b, j_{b,x}))$ is also the state of $d$ at instant $l(b, j_{b,x}) + j_{d,x}$. It follows that $u(d, j_{d,x})$ is not the final instant of the evolution of $d$ that starts at $j_{d,x}$ in state $x$. Analogously for the case $l(b, j_{b,x}) < l(d, j_{d,x})$.

[10] If either (a), (b), or (c) does not hold, the interpretation $I_{DS_L,H}$ is obviously not corect, because: if (a) holds, no evolution of any $F$-realizer $b_F$ can be represented by means of the state transition family $(g^t)_{t \in T}$ of $DS_L = (M, (g^t)_{t \in T})$; if (b) holds, some evolution of some $F$-realizer $b_F$ cannot be represented by $(g^t)_{t \in T}$; if (c) holds, some evolution of some $F$-realizer $b_F$ cannot be correctly represented by $(g^t)_{t \in T}$.

## Definition 14 (True model of a phenomenon).
**DS** is a true model of $H := \mathbf{DS} = (DS_L, I_{DS_L,H})$ is a model of $H$ and $I_{DS_L,H}$ is a correct interpretation of $DS_L$ on $H$.

We have seen above how, with respect to a given interpretation $I_{DS_L,H}$, it is possible to define the set $C_F$ of the initial states of $H$ and, when Condition D holds, also the set $q_F(x)$ of the durations that transform the initial state of $H$, $x \in C_F$, into another state.

We recall that, if $I_{DS_L,H}$ is an admissible interpretation, Condition D holds and $C_F$, besides being the set of the initial states of $H$, also turns out to be a non-empty subset of the state space $M$ of $DS_L$. In addition, for any $x \in C_F$, any duration $t \in q_F(x)$ is both a duration of $H$ and a duration of the time set $T$ of $DS_L$, for $T = V(\boldsymbol{T})$, by the interpretation $I_{DS_L,H}$.

We are now going to show how this "double nature" of the initial states $x \in C_F$ and the durations $t \in q_F(x)$ allows $I_{DS_L,H}$ to induce, on the one side, a structure on $H$ and, on the other one, a substructure on $DS_L$. Finally, we will prove that $\mathbf{DS} = (DS_L, I_{DS_L,H})$ is a true model of $H$ if, and only if, the structure of $H$ induced by $I_{DS_L,H}$ is identical to the substructure of $DS_L$ induced by $I_{DS_L,H}$.

Let $I_{DS_L,H}$ be an admissible interpretation of a dynamical system $DS_L = (M, (g^t)_t$ on a phenomenon $H = (F, B_F)$.

## Definition 15 (The structure of a phenomenon induced by an admissible interpretation).
*The structure of $H$ induced by* $I_{DS_L,H} := (h_x^t)_{x \in C_F, t \in q_F(x)}$, *where, for any* $x \in C_F$ *and* $t \in q_F(x)$, $h_x^t$ *is the function from* $\{x\} \to \boldsymbol{M}$ *defined by:*

$$h_x^t(x) := \text{the state of } b_F \text{ at instant } t + j_{b_F,x}, \text{ where } b_F \in B_{F_x} \text{ and } j_{b_F,x} \in J_{b_F,x}. \tag{6}$$

Note that (6) is well given, for it does not depend on either $b_F$ or $j_{b_F,x}$, as $I_{DS_L,H}$ is an admissible interpretation, and thus Condition D holds.

Let $I_{DS_L,H}$ be an admissible interpretation of a dynamical system $DS_L = (M, (g^t)_{t\in}$ on a phenomenon $H = (F, B_F)$.

## Definition 16 (The substructure of a dynamical system induced by an admissible interpretation).
*The substructure of $DS_L$ induced by* $I_{DS_L,H} := (g_x^t)_{x \in C_F, t \in q_F(x)}$, *where, for any* $x \in C_F$ *and* $t \in q_F(x)$, $g_x^t$ *is the restriction to* $\{x\}$ *of* $g^t$, *that is to say, the function from* $\{x\} \to \boldsymbol{M}$ *defined by:*

$$g_x^t(x) := g^t(x). \tag{7}$$

Finally, the preceding definitions allow us to prove the following theorem.

THEOREM 1 (Truth as interpretation induced structure identity).
Let $I_{DS_L,H}$ be an admissible interpretation of a dynamical system $DS_L$ on a phenomenon $H$, let $\mathbf{DS} = (DS_L, I_{DS_L,H})$. Then:
**DS** is a true model of $H$ iff the structure of $H$ induced by $I_{DS_L,H}$ is identical to the substructure of $DS_L$ induced by $I_{DS_L,H}$.

**Proof.** The thesis is a straightforward consequence of Definitions 12–16. ∎

## 6  Galilean models—Empirical correctness and truth

By Definition 3 the magnitudes $M_1, ..., M_n, T$ indicated by an interpretation $I_{DS_L,H}$ not necessarily are measurable, or even observational. In the special case when some of them are measurable, we will talk of an *empirical* interpretation. An explicit definition of this concept is the one below.

An interpretation $I_{DS_L,H} = \{C_1 \subseteq V(M_1), ..., C_n \subseteq V(M_n), T = V(T)\}$ is called an *empirical interpretation of $DS_L$ on $H$* if, and only if, the magnitude time $T$ and at least one of the magnitudes $M_1, ..., M_n$ is measurable.

In addition, a pair $DS = (DS_L, I_{DS_L,H})$ is called an *empirical model of $H$* if, and only if, $DS$ is a model of $H$ and the interpretation $I_{DS_L,H}$ is empirical.

Finally, $DS = (DS_L, I_{DS_L,H})$ is called an *empirically correct model of $H$* if, and only if $DS$ is an empirical model of $H$ and all measurements of the measurable magnitudes of $I_{DS_L,H}$ are consistent with $I_{DS_L,H}$'s being a correct interpretation of $DS_L$ on $H$ (Giunti [9], par. 4.1). An empirically correct model of $H$ is also called a *Galilean model of $H$* (Giunti [7]; [8], ch. 3, [9], sec. 4.1).

Note that the three preceding definitions and Definition 14 entail that $DS = (DS_L, I_{DS_L,H})$ is an empirically correct model of $H$ if, and only if, $DS$ is an empirical model of $H$ and all measurements of the measurable magnitudes of $I_{DS_L,H}$ are consistent with $DS$'s being a true model of $H$.

From a strictly formal point of view, however, the three preceding definitions are not completely adequate, for in fact they are based on three intuitive, not analyzed, concepts: (i) measurable magnitudes, (ii) measurements of such magnitudes, (iii) consistence of such measurements with the correctness of an interpretation or, equivalently, with the truth of an empirical model.

Nevertheless, the underlying intuitions seem to be sufficiently clear. Furthermore, as regards the two concepts (i) and (ii), they may very well be analyzed along lines similar to those proposed by Dalla Chiara and Toraldo di Francia ([4]; [5], ch. 2) for, respectively, operationally defined magnitudes and physical situations. Once this analysis is made explicit, also an exact definition of the third concept (*i.e.*, *consistence* between measurements and correctness of an empirical interpretation or, equivalently, between *measurements* and *truth of an empirical model*) may be provided.

The concept of empirical correctness of a model may call to mind van Fraassen's notion of empirical adequacy ([18], ch. 3). Notwithstanding a superficial terminological similarity, the two concepts are in fact very different because, as remarked above, the definition of empirical correctness presupposes the one of truth, while van Fraassen's empirical adequacy is defined independently of the latter.

Empirical correctness of a model is much more similar to Popper's *corroboration*, than to van Fraassen's empirical adequacy. For Popper, in fact, a hypothesis is corroborated if no severe test falsifies it or, that is the same, if all severe tests are consistent with the hypothesis' being true (Popper [14], sec. 5).

However, giving an adequate definition of consistency between severe tests and hypothesis truth is a very serious problem for Popper's falsificationist methodology.

According to the well known Duhem-Quine thesis (Duhem [6]; Quine [15]) the requirement that all severe tests be consistent with the truth of a given hypothesis can always be trivially satisfied, for any putative inconsistency can be ascribed to some other assumption, taken in conjunction with the hypothesis itself. Since these auxiliary assumptions are always present, Popper's concept of corroboration is not well defined, unless one specifies under what conditions such auxiliary hypotheses are to be ignored (or considered).

It should be noticed that this problem of the falsificationist methodology depends on the fact that consistency is intended as a relation between experimental results and truth of a *hypothesis* (*i.e.*, a *sentence*). If, as we maintain, consistency is instead a relation between measurements and truth of a *model*, the problem does not arise, or at least, not in such a way as to block an adequate, formal, development of this methodology.

## BIBLIOGRAPHY

[1] Arnold V. I. (1977). *Ordinary differential equations*, Cambridge MA: The MIT Press.
[2] Balzer W., Moulines C. U., Sneed J. (1987). *An architectonic for science*, Dordrecht: D. Reidel Publishing Company.
[3] Bickle J. (1998). *Psychoneural reduction*, Cambridge, MA: The MIT Press.
[4] Dalla Chiara M. L., Toraldo di Francia G. (1973). "A logical analysis of physical theories", *Rivista del Nuovo Cimento*, 3, 1:1-20.
[5] Dalla Chiara M. L., Toraldo di Francia G. (1981). *Le teorie fisiche*, Torino: Boringhieri.
[6] Duhem P. (1954). *The aim and structure of physical theory*, Princeton, NJ: Princeton University Press, translation of the II French edition (1914). *La Théorie physique: Son objet, sa structure*, Paris: Marcel Rivière & Cie.
[7] Giunti M. (1995). "Dynamical models of cognition", in Port R. F., van Gelder T. (Eds.), *Mind as motion: Explorations in the dynamics of cognition*, pp. 549-571, Cambridge, MA: The MIT Press.
[8] Giunti M. (1997). *Computation, dynamics, and cognition*, New York: Oxford University Press.
[9] Giunti M. (2014). "A representational approach to reduction in dynamical systems", *Erkenntnis*, 79:943-968.
[10] Giunti M., Mazzola C. (2012). "Dynamical systems on monoids: Toward a general theory of deterministic systems and motion", in Minati G., Abram M., Pessa E. (Eds.), *Methods, models, simulations and approaches towards a general theory of change*, pp. 173-185, Singapore: World Scientific.
[11] Hempel C. G. (1952). *Fundamentals of concept formation in empirical science*, Chicago: University of Chicago Press.
[12] Hempel C. G. (1958). "The theoretician's dilemma: a study in the logic of theory construction", in *Minnesota Studies in the Philosophy of Science*, vol. II, pp. 37-98, Minneapolis: University of Minnesota Press.
[13] Hirsch M. W., Smale S., Devaney R. L. (2004). *Differential equations, dynamical systems, and an introduction to chaos*, Amsterdam: Elsevier Academic Press.
[14] Popper K. R. (1963). "Three views concerning human knowledge", in Popper K. R., *Conjectures and refutations*, London: Routledge and Kegan Paul.
[15] Quine W. V. O. (1951). "Two dogmas of empiricism", *The Philosophical Review*, 60:20-43.
[16] Suppe F. (1972). "Theories, their formulations, and the operational imperative", *Synthese*, 25:129-164.
[17] Szlenk W. (1984). *An introduction to the theory of smooth dynamical systems*, New York: Wiley.
[18] van Fraassen B. C. (1980). *The scientific image*, Oxford: Oxford University Press.

# An Embodied-Extended Approach to the Acquisition of Numerical Skills

Simone Pinna

ABSTRACT. An embodied-extended approach to cognition may be useful for the study of some cognitive activities, like arithmetical skills, which are generally thought as involving only internal resources. In this article, after discussing the theoretical assumptions and scopes of such an approach, I will specify its usefulness for an explanation of some specific arithmetic capacities, *i.e.* algorithmic skills. I will, then, formalize a finger counting like procedure for single-digit additions by using a Bidimensional-Turing machine, which is a computational model specifically designed for the formal description of human algorithmic skills. The formal model analysis suggests an hypothesis on numerical facts learning which will be tested through a neural net model.

## Introduction

The development of human arithmetical skills has been associated with some less specific cognitive capacities such as, among others, spatial skills [27] and finger gnosia [22]. It seems that recognition of geometrical shapes on the one side and, on the other, the use of our hands to point objects to be counted or for working memory offloading — *e.g.*, when we use our fingers to keep trace of the numbers to be counted — do positively affect learning, memorization and retrieval of numerical facts.

These results are consistent with an embodied/extended approach to cognition (EEC), where the role of bodily and/or external features — in our case, hands and/or object shapes — is considered as important in the economy of a cognitive skill as the role of internal features — *i.e.* of the central nervous system. EEC is philosophically connected — but not identical, as we will see later — to the Extended-Mind Hypothesis (EMH), which is discussed in the following section.

## 1 Active externalism

In the famous article "The extended mind" (1998), Andy Clark and David Chalmers proposed the so-called "parity principle" as a conceptual tool for identifying genuine cases of extended cognition.

Parity principle:

> If, as we confront some task, a part of the world functions as a process which, *were it done in the head*, we would have no hesitation in recognizing as part of the cognitive process, then that part of the world *is* [...] part of the cognitive process. Cognitive process ain't (all) in the head! [12, p.8]

In such cases we can consider parts of the environment as having causal roles in cognitive *processing*. But then, the authors ask, what about the *mind*? Are there some cases in which we can say external factors to partly constitute *mental states*? The answer, which represents the philosophical core of the Extended Mind Hypothesis (EMH), is that, sometimes, some mental constituents such as *beliefs* may partly consist of environmental features. This concept is explained through an example that would be thenceforth famous. It tells a story about two persons, Inga and Otto. Inga wants to go to an exhibition at the Museum of Modern Art of New York, and knows that the museum is on 53rd Street. We can definitely consider the fact that the MOMA is on that precise address as one of Inga's beliefs, so the cognitive task she carries out consists of retrieving that address from her long-term memory.

> Now consider Otto. Otto suffers from Alzheimer's disease, and like many Alzheimer's patients, he relies on information in the environment to help structure his life. Otto carries a notebook around with him everywhere he goes. [...] For Otto, his notebook plays the role usually played by a biological memory. Today, Otto hears about the exhibition at the Museum of Modern Art, and decides to go to see it. He consults the notebook, which says that the museum is on 53rd Street, so he walks to 53rd Street and goes into the museum. [12, p.12-13]

The experiment is aimed to convince us that we can think at Otto's notebook as it was a container of his *dispositional beliefs*, namely something that reliably replaces his compromised long-term memory.

As long as Otto performs the same kind of process as Inga's by using his notebook, we have to recognize, in accord to the parity principle, that part of Otto's environment, namely his notebook, has a causal role in the cognitive process that makes him able to access one of his beliefs. Also, we can consider Otto's notebook truly as a container of his dispositional beliefs. Hence, insofar as we consider beliefs as constitutive parts of one's mind, we can say that in Otto's case his mind extends beyond his organism in the sense that it includes, at least, his notebook.

*Extended beliefs* such as those contained in Otto's notebook satisfy four conditions, which must be held in all true cases of extended mind [12, p.17]:

**Portability.** When required, the external resource must be readily accessible.

**Constant employment.** The external resource is constantly consulted by its owner.

**Reliability.** The informational content of the resource is accepted and used with no hesitation.

**Past endorsement.** The contents of the resource have been consciously endorsed in the past and, consequently, externalized.

Clark and Chalmers call the mind's view implied both in EEC and in EMH cases as *active externalism*. The term active is meant to distinguish this variety of externalism from the standard Putnam-Burge style externalism [7, 25], which is typically based on *supervenience thought experiments* [18] as in "Twin Earth" cases.

In these cases, the relevant external features are passive. Because of their distal nature, they play no role in driving the cognitive process in the here-and-now. This is reflected by the fact that the actions performed by me and my twin are physically indistinguishable, despite our external differences. [12, p.9]

On the contrary, in active externalism cases

> [...] the human organism is linked with an external entity in a two-way interaction, creating a coupled system that can be seen as a cognitive system in its own right. All the components in the system play an active causal role, and they jointly govern behavior in the same sort of way that cognition usually does. If we remove the external component the system's behavioural competence will drop, just as it would if we removed part of its brain. Our thesis is that this sort of coupled process counts equally well as a cognitive process, whether or not it is wholly in the head. [12, p.8-9]

## 2 Ontological issues

Critics of the EMH focused their attention mainly on what I call the ontological question of EMH, namely, the following: Can we say that, sometimes, some "chunks of the world" may be regarded as ontologically constitutive of the mind, or should we just consider them as mere auxiliary instruments, given that the true cognitive processes always take place inside the organism?

In a series of target articles [5, 1, 2, 3, 4], Fred Adams and Ken Aizawa argue that the EMH falls into a "coupling-constitution fallacy", where causal relations between mental activity and external objects or processes (in Otto's case, for instance, the association between searching for an information and checking into his notebook) are confused with constitutive relations (e.g, in the same case, the tenet that the notebook is part of Otto's mind). In general, the argument goes, the fact that an object or process X is coupled with an object or process Y does not imply that X is a constitutive part of Y. Thus, the fact that Otto performs a cognitive process that is coupled with the use of his notebook is not sufficient to consider the notebook as a proper part of Otto's mind.

> Question: Why did the pencil think that $2 + 2 = 4$ ? Clark's answer: Because it was coupled to the mathematician. That about sums up what is wrong with Clark's extended mind hypothesis. [3, p.67]

Adams and Aizawa argue that the four conditions given by Clark and Chalmers are not sufficient to decide whether some external resource is a proper part of a cognitive process. In addition to these conditions, we also need some other reliable "mark of the cognitive".

Clark replies to this objection by showing that the introduction of such a criterion is in turn a source of unsolvable problems. For example, in [11] Clark argues that, as it makes no sense to ask whether a pencil or a notebook is or is not a cognitive object, the same can be said of any putative part of a cognitive system (such as a neuron or a set of neurons).

> Consider the following exchange, loosely modeled on Adams and Aizawa's opening "reductio":
> 
> Question: Why did the V4 neuron think that there was a spiral pattern in the stimulus? Answer: Because it was coupled to the monkey.
> 
> Now clearly, there is something wrong here. But the absurdity lies not in the appeal to coupling but in the idea that a V4 neuron (or even a group of V4 neurons, or even a whole

parietal lobe) might itself be some kind of self-contained locus of thinking. It is crazy to think that a V4 neuron thinks, and it is (just as Adams and Aizawa imply) crazy to think that a pencil might think. [11, p.81]

In Clark's view, therefore, the problem of determining the boundaries of cognition arises again and again, no matter where we decide to set those boundaries.

Besides the aforesaid *part destruens* of their position, Adams and Aizawa propose also a *part costruens*, namely a strategy to recognize what may be regarded as a proper part of a cognitive system. To this purpose, they suggest the "Intrinsic Content Condition", according to which properly cognitive states must involve intrinsic, non-derived content, *i.e.* non-representational contents.

This proposal have provoked a complex debate, whose reconstruction is beyond the scopes of this work.[1] However, Clark's ultimate position on this issue is that Adams and Aizawa's objection to EMH draws our attention to a probably useless question, for it is, at least at present, very difficult to be empirically tackled.

> Since what is at issue is [...] whether the notebook might now be part of the local supervenience base for some of Otto's dispositional beliefs [...] the status of the notebook itself, as "cognitive" or "noncognitive", is (to whatever extent that idea is even intelligible) simply irrelevant. By contrast, the precise nature of the coupling between the notebook and the rest of the Otto system seems absolutely crucial to how one then conceives of the overall situation. [11, p.90]

Clark suggests that a more fruitful question on which we should focus concerns the explanation of the cognitive mechanism that links Otto's behaviour to its notebook, and not the ontological status of the single components of the (extended) system comprising Otto and his notebook. In other terms, this question should be: what is the *function* played by the notebook in this kind of extended cognitive system?

## 3  Explanation of functional roles

In "Supersizing the Mind" (2008) Clark tries to shift the focus to what I call the functional-explanatory question:

- What is the role of external (with respect to the organism) instruments and objects for the explanation of the development and performance of specific kinds of cognitive processes?

Clark discusses how this problem is approached in various studies within current cognitive science-related fields, such as robotics, dynamical approach, cognitive psychology, connectionism etc. As a consequence of this "field-test" of the theory, he extends, reviews and re-writes the set of propositions which represents the philosophical core of EMH. Among those propositions, there is one that clearly shows Clark's idea that ontological issues about the mind must be subordinated, at least temporarily, to functional and explanatory questions.

> Hypothesis of Organism-Centered Cognition (HOC): Human cognitive processing (sometimes) literally extends into the environment surrounding the organism. But the organism (and within the organism, the brain/CNS) remains the core and currently the most active element. Cognition is organism centered even when it is not organism bound. [10, p.39]

---

[1] In [21] Menary interestingly reviews this debate from an externalist stance.

This hypothesis may be seen as an attempt to prevent EMH from ontological objections as those seen in the discussion about the alleged couple-constitution error. Indeed, the message contained in HOC is exactly that in the field of philosophy of mind and cognitive science we need, at least temporarily, to abandon ontological issues. Rather, our attention should be focused on the explanation of the role of any cognitively relevant resource, be it internal or external to the organism.

This view, while reducing EMH ontological scope, is fully coherent with an Embodied-Extended approach to Cognition (EEC), according to which the explanation of some cognitive phenomena has to seriously consider the role of bodily features and external objects. In the following section, I will show the usefulness of this approach for the explanation of early arithmetic skills development.

## 4   Use of fingers in early arithmetic and EEC

The use of fingers for counting plays an acknowledged role in the development of early arithmetic skills [8, 9, 15, 19]. A recent research [26] brings evidences that finger gnosia, *i.e.* the correct representation of fingers, is associated on the one side to a greater probability of finger-use in computation and, on the other side, to better arithmetical performance in 5 to 7 years old children. Given that relations between poor finger gnosia and poor arithmetical skills have also been found [22], it should be interesting to inspect the cognitive mechanism on which this phenomenon is grounded. Indeed, we have here a cognitive phenomenon that seems well suited to be investigated through the lenses of EEC.

Nevertheless, some claim that this emphasis on EEC, in the case of numerical skills acquisition, is somehow exaggerated. In [13] Crollen *et al.* report many studies which suggest that, although the use of fingers in computation may help children's development of arithmetical abilities, it is not a necessary stage for that development. So we face here what seems to be a tough problem for EEC. According to this approach, indeed, the relevant bodily and external features for the performance of a given cognitive task are as fundamental for the cognitive system which performs that task as the internal ones. However, in this specific case the weight to be given to the external and internal features is not at all the same: we can totally bypass the external part of the cognitive system and obtain the same result!

An argument like this can really threaten an EEC approach to the explanation of early counting skills[2] if the discussion is focused on what I called the ontological question of EMH. In this case, the problem would be the cognitive status of finger use for counting, *i.e.* whether this use is constitutive for early arithmetic development or has a mere causal role in some subjects

However, as we have seen before, it is doubtful that this is a meaningful question, for this kind of ontological approach takes for granted, at least, the existence of a shared general definition of a cognitive system. But how this definition should be formulated is far from clear, whether or not we confine a cognitive system to its internal features [10].

---

[2]In the relevant cases for *active externalism*, indeed, "if we remove the external component the system's behavioural competence will drop, just as it would if we removed part of its brain." [12, 8-9]

On the other hand, the cited argument against EEC does not seem to threaten another line of investigation, namely that focused on what I called the functional-explanatory question of EMH, which represents the real scientific challenge for the extended-embodied approach to cognition.

I propose a way to answer this last question in the specific case of the acquisition of numerical skills. We can thus reformulate the functional-explanatory question: how does finger-use in counting routines affects learning of arithmetical skills?

The hypothesis I will explore is that the use of fingers for counting provides effective and reliable strategies in order to obtain correct results. Arithmetical results acquired this way have a high probability to be stored in long term memory and, hence, to cause the development of a (correct) set of basic arithmetical facts. In this hypothesis, the crucial role of finger use for counting is identified with its algorithmic meaning, i.e, with the possibility given by the use of fingers to carry out simple and effective computations. But how may we verify this hypothesis?

I propose a twofold method to face this question. First, I will formally describe a finger using counting strategy, in order to isolate its relevant algorithmic features by using a Turing machine-inspired computational model, namely a Bidimensional Turing Machine. Through this analysis it could be possible to inspect the relevant operations at work when performing a finger-using counting routine. The model should also give some indications about the implicit knowledge necessary to perform a given counting routine.

Second, the information obtained by the analysis of the BTM model will be useful for the simulation of learning and retrieval of a set of basic number facts – namely, the results of single-digit additions (SDA) n + m – through a neural network model. The network is given different training-sets in order to simulate different ways of acquisition of SDA results, one of which is modeled on the finger-counting routine.

## 5 Algorithm formalization

The finger-counting strategy I will focus on will be formalized by using a Bidimensional Turing Machine (BTM), which is a computational model expressly designed in order to describe human strategies for computation, where the external memory is able to reliably represent a sheet of squared paper [16]. I cannot give here a full theoretical and formal account of this computational model. Nevertheless, a brief description of its main features is necessary for understanding the following algorithm formalization.

It is supposed that the readers are familiar with the functioning of ordinary Turing machines (TM). The most visible difference between a BTM and a TM is that the the latter's unidimensional tape is replaced in the former with a two-dimensional grid, as in Two-dimensional Turing machines [14]. Other important features are sketched above:

1. Similar to an ordinary TM, each single instruction of a BTM is a sequence of 5 objects (quintuple) of the following types: (i) internal state, (ii) symbol, (iii) symbol, (iv) movement, (v) internal state. However, the machine table (set of instructions) of a BTM is more sophisticated than that of a TM. In

fact, in the general case, each instruction is not just a single quintuple, but a quintuple schema that typically represents a possibly infinite set of quintuples of a specified form.

2. Both internal states and movements of a BTM are more complex than those of a TM, for

   a) a generic internal state of a BTM is not a simple unstructured state $q_i$, but it is an $n+1$-tuple $(q_i, x_1, ..., x_n)$, where $q_i$ is an element of a finite set $Q$ of internal states, and any $x_j$ is a generic object member of a given data-type $X_j$; the special symbol $b$ (blank) is a member of each data-type. Any position of the $n+1$-tuple is called a register. Given a generic internal state of this kind, any $x_j$ may be either full or empty. If each register of a given generic internal state is empty (namely, consists of the symbol $b$), then the internal state reduces to its component $q_i$; all possible movements from the present head location are legal; any legal movement is thus represented by a pair of integer numbers, i.e. the coordinates of the head destination relative to its present location.

   b) In ordinary TMs, the only way to refer to a simple internal state $q_i$, to a symbol $s_j$, or to a movement $R$ (right), $L$ (left), or $H$ (halt) is by means of the corresponding constants "$q_i$", "$s_j$", "$R$" "$L$", or "$H$". In BTMs, by contrast, (complex) internal states, symbols and movements can be referred to also by means of complex terms, which are formed by repeated applications of function terms to symbol constants or variables.

3. The admissible domain of each variable is a specified set of either symbols or nonnegative integers, while each function term stands for a specified function, which must be computable in the intuitive sense.

4. The syntax of quintuple schemas is so designed that each such function works as an auxiliary operation, which is called as needed during computation and executed in just one step.

5. Finally, quintuple schemas may also contain relational terms (n-place predicates) which can be used to express additional conditions. Such conditions may only refer to the presently read symbol or to the internal state terms.

Now I will describe a BTM ($BTM$count-on) that reflects an algorithm for single-digit addition that is normally carried out with the use of fingers. $BTM$count-on performs the sum of two natural numbers $x_1$ and $x_2$ with $0 \leq x_i \leq 9$ by using a counting-on strategy, i.e, by starting from the value of the first addend and counting out the value of the second.

An informal description of the procedure formalized by this machine is the following:

   i. in the initial state, two addends with a value between 0 and 9 are held in internal memory;

| Input | | Condition | Output | | |
|---|---|---|---|---|---|
| internal state | read | | write | move | internal state |
| $q_1, [r_1], [r_2]$ | $s$ | $[s \neq r_2]$ | $\sigma(s)$ | $(0,0)$ | $q_1, [\sigma(r_1)], [r_2]$ |
| $q_1, [r_1], [r_2]$ | $s$ | $[s = r_2]$ | $s$ | $(0,0)$ | $q_1, [r_1]$ |
| $q_1, [r_1]$ | $s$ | $[s = r_2]$ | $s$ | $(0,0)$ | $q_1, [r_1]$ |

Table 1. Machine table $BTM$count-on

ii. a cell of the grid contains a counter with initial value 0;

iii. at each step, both the value of the first addend and of the counter is incremented by 1, until the value of the counter is equal to that of the second addend;

iv. finally, the machine stops, holding the result in its internal memory. If the value of the second addend is 0, the machine shifts directly to a stop instruction.

To construct the $BTM$count-on we need to define:

a) a set of constants $A = \{$"0", "1", "2", "3", "4", "5", "6", "7", "8", "9"$\}$ which constitutes the vocabulary of the grid;

b) two registers $R_1$ and $R_2$, corresponding to two internal states variables $r_1$ and $r_2$. The data-type correspondent to each register includes all non negative numbers and the special symbol $b$ (blank), which stands for an empty position.

c) a variable $s$ for the grid. The range of $s$ is the set $A$;

d) a simple internal state $q_1$;

e) an auxiliary function $\sigma : 0, ..., 17 \to Z^+$ s. t. $\sigma(x) := x + 1$.

f) two auxiliary conditions $x = y$ and $x \neq y$, which are the standard relations of identity and diversity on natural numbers.

The machine table of $BTM$count-on is given in table 1.

This BTM starts with 2 numbers $0 \leq n \leq 9$ held in its internal variables $r_1$ and $r_2$. The head is positioned on an arbitrary cell of the grid, which is used by the machine as a counter with initial value 0, and remains on the same cell during the entire procedure. Then, at each step of the computation, the value of $r_1$ and that of $s$ is increased by 1 through the function $\sigma$, until $s$ is equal to $r_2$. At this point, the machine stops, holding the result of the sum in its internal variable $r_1$.

## 6  Algorithm analysis and net simulation

Assuming that the algorithm carried out by $BTM$count-on reflects correctly the correspondent counting procedure, we can make interesting considerations by paying attention to some features of the formal model. For example, although this machine performs a very simple procedure, it needs to use at least one auxiliary function, namely $\sigma$, which is a successor function defined on a finite domain ($\{0, ..., 17\}$). Thus, we could conjecture that a subject which is able to use a finger-counting procedure for simple additions needs at least to possess this basic knowledge.

Moreover, if a subject is able to use this procedure to perform simple additions, he would apply it to SDAs presented randomly and not according to a given order (*e.g.*, first all the 1-digit sums $1 + n$, then $2 + n$, and so on).

On the basis of these considerations, Pinna and Fumera [24] formulated an hypothesis on the link between use of finger-counting procedures and SDA learning. According to their hypothesis, SDA results retrieval skills are affected by the order of memorization of SDAs results, and this order is influenced by the counting procedure used.

To test this hypothesis, they used a feed-forward backpropagation neural network, designed in order to learn SDAs.[3] The order of the training-set (TR) examples (consisting of all 100 SDAs) were manipulated as to simulate two alternative training conditions.

In the first condition (A), which reflected a "rote learning", the net was trained, at each epoch, on all the TR, from smaller to larger sums $(0+1, 0+2, ..., 0+9, 1+1, ..., 1+9, ..., 9+1, ..., 9+9)$, as if it was following the order of an "addition table".

The second condition (B) was modeled on the counting-on strategy. To simulate the knowledge of the successor function the net was, first, trained on the 20 sums of the form x+1 and 1+x until no error is made on them. Second, all the other TR examples were given, at each epoch, randomly. This second phase was intended to reflect that a subject using a counting-on procedure may apply this strategy on each SDA, independently from the value of its addends.

The results showed an advantage of training condition B in terms of learning efficiency (see fig.1). However, this fact alone was insufficient to exclude trivial explanations of this apparent convenience, **e.g.**, if it is only a consequence of the first training phase of condition B, which gives an initial advantage with respect to the net trained in condition A, or the effect of the number encoding scheme adopted. It was then necessary to assess the influence of learning strategies on the net by checking if and in what conditions it is able to reflect some cognitive phenomenon related to the simulated arithmetic skill, independently of net properties.

To do this, the net has been tested in order to see if and in what conditions it is able to simulate a very robust phenomenon in mathematical cognition, namely the problem-size effect [17, 6, 20, 28, 23]. This effect consists of an increase in reaction time and error rate in response to arithmetic problems with larger solutions (for instance, solving $7 + 8$ takes longer and is more error-prone than solving $4 + 3$). The net used in this simulation was tested only on one side of the problem-size

---
[3] Full technical details of this neural-net model are given in [24].

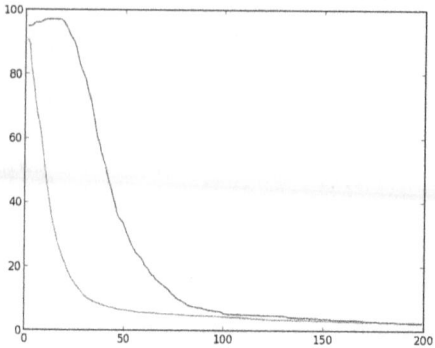

**Figure 1.** Number of errors per epoch. The blue line refers to experiment A, the green line to experiment B. Each curve is the average of 100 curves obtained in the following way: in experiment B the learning procedure is repeated 10 times with 10 different orders of presentation of the training set, randomly chosen, where each time the order of presentation is changed connection weights are also randomly changed; in experiment A the training set presentation order remains always the same. Every 10 repetitions, connection weights are randomly changed, and this procedure is then repeated for 10 times.

effect, namely if, during the training phase, it is more error prone on training set cases where the solution is larger, for temporal features could not be simulated.

The method used for the verification of the problem-size effect on the net is the following:

a) The training set has been divided in two subsets:
Small-size problems: the 49 one-digit sums x + y with x; y less than or equal to 6.
Large-size problems: all the 51 remaining one-digit sums.

b) The percentage of net errors per epoch, with regard of both subsets, has been verified in training conditions A and B.

Figure 2 shows the percentage of errors committed by the net in each subset of the training set. Results are quite interesting. The net shows a clear effect of problem-size in training condition B, while in condition A the effect is not verified. In this case trivial explanations in terms of net properties may be excluded, for the net responds very differently according to the learning strategy used. If, for example, the verification of the problem-size effect in the net were due to the way it encodes numbers, the training condition would have no impact on this effect.

## 7 Conclusion

The link between finger use in early arithmetic procedures and development of advanced arithmetic skills may be investigated through the lenses of an embodied-extended approach to cognition. In this paper, I tried to justify the utility of this

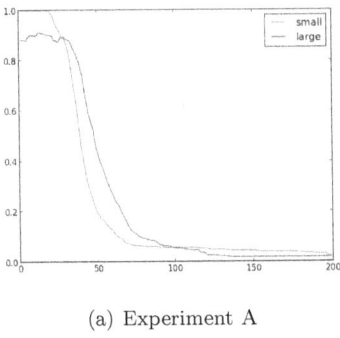
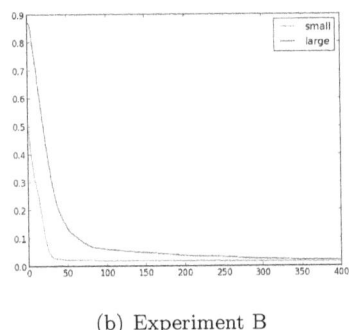

(a) Experiment A    (b) Experiment B

Figure 2. Problem-size effect tested on the same net in different training condition. Curves represent the percentage of errors (y-axis) per epoch (x-axis) in two subsets of the training set (blue: large-size problems; red: small-size problems). a) The net is trained as in experiment A. In the first 200 epochs, no clear effect of problem size is visible. b) The net is trained as in experiment B. The problem-size effect is evident until the 400th epoch, than the two curves overlap.

approach for the explanation of arithmetic skills development. First, I discussed the extended mind hypothesis (EMH), with particular regard to the debate on the "coupling-constitution fallacy". Then, I proposed that the main troubles of EMH may be avoided by distinguishing between the ontological question and the functional-explanatory question about EMH. The arguments presented here were aimed to show that, while the former is (at this moment, at least) a scientifically unfruitful question, the latter is perfectly fitted to be a source of novel and promisingly more thorough cognitive explanations. An embodied-extended approach to cognition, where the role of bodily and/or external features is considered as much importance in the economy of a cognitive skill as the role of internal features, should be focused on functional-explanatory issues and avoid, for the moment, any kind of ontological problems.

On this theoretical basis is grounded the algorithmic approach to the analysis of arithmetic skills, developed in the second part of the paper. I introduced a formal model for the analysis of arithmetic procedures and formalized a specific counting procedure, which is normally used in finger-counting strategies. Then, I proposed a possible answer to a specific instance of the functional-explanatory question with regard to the use of fingers in early arithmetic skills development, i.e, that the memorization of single-digit addition (SDAs) results is affected by the order in which they are processed, and this order is influenced by the counting procedure used.

Lastly, I showed the results of a neural net simulation designed to test the effect of learning strategies on net performances. The training condition suggested by the analysis of the BTM-model is effectively more convenient for a faster reduction of net's errors during the training phase. Also, results on the problem-size effect showed that the training condition modeled on the finger-counting strategy leads not only to a faster reduction of errors during the training phase, but also provides to the net the ability to reproduce a well-known cognitive effect.

Obviously, it would be difficult to clearly state the verisimilitude of the proposed hypothesis on the link between counting procedures and general development of arithmetical knowledge on the basis of the net simulation results. However, these results are encouraging towards future employment of an algorithmic approach to cognitive arithmetic, which seems well fitted to give novel explanations on issues regarding, *e.g.*, the cognitive role of algorithmic schemes (like those generally used in pen and paper procedures) or the relation between space and number representations.

## BIBLIOGRAPHY

[1] Adams F., Aizawa K. (2001). "The bounds of cognition", *Philosophical Psychology*,14:43-64.
[2] Adams F., Aizawa K. (2009). "Why the mind is still in the head", in Robbins P., Aydede M. (Eds.), *The Cambridge handbook of situated cognition*, pp. 78-95, Cambridge: Cambridge University Press.
[3] Adams F., Aizawa K. (2010a). "Defending the bounds of cognition", in Menary R. (Ed.), *The extended mind*, pp. 67-80, London: Bradford Books/MIT Press.
[4] Adams F., Aizawa K. (2010b). "The value of cognitivism in thinking about extended cognition", *Phenomenology and the Cognitive Sciences*, 9:570– 603.
[5] Aizawa K. (2010). "The coupling-constitution fallacy revisited", *Cognitive Systems Research*, 11:332-342.
[6] Ashcraft M. (1992). "Cognitive arithmetic: A review of data and theory", *Cognition*, 44:75-106.
[7] Burge T. (1979), "Individualism and the mental", *Midwest Studies in Philosophy*, 5:73-122.
[8] Butterworth B. (1999). *The mathematical brain*, Macmillan: London, UK.
[9] Butterworth B. (2010). "Foundational numerical capacities and the origins of dyscalculia", *Trends in Cognitive Sciences*, 14:534-541.
[10] Clark, A. (2008). *Supersizing the mind*, New York: Oxford University Press.
[11] Clark A. (2010). "Coupling, constitution, and the cognitive kind: A reply to Adams and Aizawa", in R. Menary (Ed.), *The Extended Mind*, London: Bradford Books/MIT Press.
[12] Clark A., Chalmers D. (1998), "The Extended Mind", *Analysis*, 58: 10–23.
[13] Crollen V., Seron X., Noël M.P. (2011). "Is finger-counting necessary for the development of arithmetic abilities?", *Frontiers of Psychology*, 2, article 242.
[14] Dewdney A. K. (1989). "Two-dimensional Turing machines and Tur-mites", *Scientific American*, 261: 180-183.
[15] Fuson K. (1988). *Children's counting and concepts of number*, Springer-Verlag.
[16] Giunti M. (2009). "Bidimensional Turing machines as Galilean models of human computation", In Minati G., Abram M, Pessa E. (Eds.), *Processes of emergence of systems and systemic properties*, pp. 383-423, Cambridge, MA: World Scientic.
[17] Groen G., Parkman J. (1972). "A chronometric analysis of simple addition", *Psychological Review*, 79: 329–343.
[18] Hurley S. (2010). "The varieties of externalism", in Menary R. (Ed.), *The Extended Mind*, pp. 101-153, London: Bradford Books/MIT Press.
[19] Lakoff G., Núñez R. (2000). *Where mathematics comes from: How the embodied mind brings mathematics into being*, Basic Books.
[20] LeFevre J., Sadesky G.S., Bisanz J. (1996). "Selection of procedures in mental addition: Reassessing the problem size effect in adults", *Journal of Experimental Psychology: Learning, Memory, and Cognition*, 22: 216–230.
[21] Menary R. (2010). "Introduction: The Extended Mind in Focus",in Menary R. (Ed.), *The Extended Mind*, pp. 101-153, London: Bradford Books/MIT Press.
[22] Noël M. (2005). "Finger gnosia: a predictor of numerical abilities in children?", *Child Neuropsychology*, 11: 413–430.
[23] Núñez-Peña M. (2008). Effects of training on the arithmetic problem-size effect: an event-related potential study. Experimental Brain Research, 190: 105-10.
[24] Pinna S., Fumera G. (2016). "Testing Different Learning Strategies on a Simple Connectionist Model of Numerical Fact Retrieval", in Minati G., Abram M., Pessa E. (Eds.), *Towards a Post-Bertalanffy Systemics*, pp.31-40, Springer International Publishing.

[25] Putnam H. (1975). "The meaning of meaning", In Gunderson K. (Ed.), *Language, mind, and knowledge*, pp. 3-52, Minneapolis: University of Minnesota Press.
[26] Reeve R., Humberstone J. (2011). "Five-to-7-year-olds' finger gnosia and calculation abilities", *Frontiers in Psychology*, 2, article 359.
[27] Szucs D., Devine A., Soltesz F., Nobes A., Gabriel F. (2013). "Developmental dyscalculia is related to visuo-spatial memory and inhibition impairment", *Cortex*, 49, 10:2674–2688.
[28] Zbrodov N., Logan, G. (2005). "What everyone finds: The problem-size effect", In Campbell J. (Ed.), *Handbook of mathematical cognition*, pp. 331–346, New York: Psychology Press.

# PART II

# LOGIC AND PHILOSOPHY OF LOGIC

# On an Account of Logicality

Massimiliano Carrara and Ciro De Florio

ABSTRACT. In [13], Linnebo argues that Boolos' interpretation of second-order monadic logic (MSOL) is not logic. His argument starts by proposing some conditions for *logicality* (*ontological innocence*, *cognitive primacy*, and *universal applicability*) and goes on arguing that Boolos' proposal does not satisfy them. The problem is that, in our view, Linnebo's conditions for logicality meet some general difficulties on the very same notion. His formulation cannot be considered a good *test* for demarcating what is logic at all. Take first order logic (FOL): either it does not satisfy the conditions proposed or it satisfies them trivially. A positive result of our analysis is that an account of *logicality* based on *independency* is better evaluable in a *relational* setting.

## 1 Introduction

Is there any way to characterize logicality? Well, there are some well-known accounts of it. A first, standard one, is by *permutation invariance*: logical notions are not altered by arbitrary permutations of the domain of discourse.[1] The idea is to demarcate what is logic by isolating specific logical notions or constants *via* the above criterion. Moreover, such criterion is often presented as the standard charachterization of logicality, even if, as Catarina Dutilh Novaes has recently observed, "It is now widely acknowledged that, in its straigtforward formulation, the criterion is not satisfactory, and many of the analyses propose modifications that allegedly amend its shortcomings"" [6, 82].

A second, perhaps less standard, account ties logicality to certain proof-theoretic properties, such as *proof-theoretic harmony*.[2] Consider introduction and elimination rules for & in FOL:

$$\&\text{-I} \frac{\alpha \quad \beta}{\alpha \,\&\, \beta} \qquad \&\text{-E} \frac{\alpha \,\&\, \beta}{\alpha} \quad \frac{\alpha \,\&\, \beta}{\beta}.$$

They are intuitively sound. Moreover, they are perfectly balanced in the sense that what is required to introduce statements of the form $\alpha \,\&\, \beta$, viz. $\alpha$ and $\beta$, perfectly matches what we may infer from such statements. Following Michael Dummett's terminolgy, the introduction and elimination rules for & in FOL are in *harmony*. Intuitively, a pair of introduction and elimination rules is harmonious if the elimination rules are neither too strong (they don't prove too much), nor

---
[1] For a discussion on this account, see e.g. [21], [20], [12], [2], [6]. For a general introduction to the problem of logical constants, see [10].
[2] On this account see, e.g., [9], [17], [11], [7], [22].

too weak (they don't prove too little). This intuitive idea can be spelled out in a number of ways. Dummett in [7] (p. 250). defines harmony as the possibility of eliminating *maximum formulae* or *local peaks*, i.e. formulae that occur both as the conclusion of an introduction rule and as the major premise of the corresponding elimination rule.

Again, as it is well-known, the account of logicality in terms of *proof-theoretic harmony* meets some difficulties.

There is, in general, a certain skeptcism about the possibility of finding a satisfactory account of the logical/non-logical divide. This skeptcism is motivated by Etchemendy's reasons for thinking that no satisfactory account of the logical/non-logical divide can be forthcoming [8].

For Etchemendy any account of the logical/non-logical divide is, if true, necessarily true. And yet, there are counterfactual situations in which any such account would get things wrong. Etchemendy concludes that any account of the logical/non-logical can at best accidentally get things right: it cannot in general guarantee extensional correctness.

In this paper we consider the problem from a different, restricted, point of view. The aim of this paper is to analyse Linnebo's conception of *logicality*, proposed in [13]. In that paper Linnebo argues that Boolos' reinterpretation of second-order monadic logic (MSOL) is not logic because it does not satisfy some *standards* to be a logic, i.e. MSOL does not satisfy *ontological innocence*, *cognitive primacy*, and *universal applicability*.

In the paper we argue that Linnebo's conditions for logicality meet some general difficulties and cannot be considered a good *test* for demarcating what is logic at all. Following also Etchemendy's criticism we will show that the proposed conditions are so narrow that one can argue that even FOL (first-order logic) does not satisfy them.

Is there a positive result of our analysis? Yes, there is. The moral of our analysis, briefly sketched in the concluding section of the paper, is that accounts of logicality based on *independency* – the *invariance criterion* – are better evaluable in a relational setting. In the paper we briefly outline a way to expand this idea.

We proceed as follows. In section 2 we briefly resume the principal elements of Boolos' reinterpretation of second-order monadic logic in terms of *plural quantification*. Section 3 is devoted to Linnebo's criterion of logicality. Section 4 is on Linnebo's criticisms to Boolos' proposal. In section 5, 6 and 7 we analyse the three general requirements Linnebo proposes for *logicality* (*ontological innocence*, *cognitive primacy*, and *universal applicability*); we argue that they meet some general difficulties and cannot be considered a good *test* for demarcating what is logic at all.

## 2 Boloos on second order monadic logic

Boolos in ([3], [4]) proposed a reinterpretation of MSOL in terms of *plural quantification*.[3] He argued that such interpretation shows – against Quine's criticim (for example in [18]) – that MSOL is a *genuine logic*.

---

[3] For an introduction to the topic see [14].

In a nutshell, his proposal consists in considering second-order variables as ranging not over sets of individuals but over individuals *plurally*. So, no second order entity is involved in MSOL. And second order monadic logic – so interpreted – is logic.

Boolos' basic idea consists of interpreting the atomic formulas of the form:

$Xy$,

as

$y$ is one of the $X$s,

and the existential formulas having the form

$\exists X \ldots$

as

There are some individuals $X$s such that....

Boolos gives no explanation of how to refer to an arbitrary plurality of individuals. He simply treats directly *plural existential quantification* taking as primitive the locution:

There are some objects such that...

used in natural language for referring to an arbitrary plurality.

Observe, *passim*, that this passage is problematic. Indeed the meaning of this locution is somewhat ambiguous, strictly depending on the context of discourse. In some contexts, it has the same meaning of the first-order expression

There is at least an object such that....

But, when it is not reducible to a first-order quantification, as in the famous Geach-Kaplan's proposition

Some critics admire only one another,

it may seem to be just a sloppy way of referring to some *class* of individuals.[4]

Boolos provides also a formal semantics for his language in *Nominalistic Platonism* [4]. It is done restating the Tarskian truth definition by modifying the notion of *assignment*. Given a domain $D$ of individuals, he defines as an assignment any binary relation $R$ between variables and individuals that correlates a unique individual with every first order variable, while it is subject to no constraint for second order variables. So $R$ may correlate a second order variable with no, one or (possibly infinitely) many individuals. The *satisfiability* relation is inductively defined as usual, with the following clauses for atomic formulas and second order existential quantification:

---

[4]For a detailed criticism on this passage see [19]. For a survey of the proposal see [5].

(i) *R* satisfies the atomic formula *Fx iff* the correlate of $x$ is one of the correlates of $F$;

(ii) *R* satisfies $\exists FA$ *iff* there is a relation $R'$, differing from $R$ at most for the correlates of $F$, such that $R'$ satisfies $A$.

Truth is then defined, again as usual, in terms of *satisfaction*. So the *set* of the correlates of $F$ is not involved in the definition of truth. This makes the notion of plural quantification precise and shows how it yields an alternative semantics for second order logic. This semantics turns out to be equivalent to the usual one, according to which the values of second order variables are all sets of individuals. And since the notion of *value of a variable* can be made precise only by the definition of *assignment*, in Boolos' perspective the proposed reformulation shows that, using Quine's slogan that "to be is to be the value of a variable", there is no commitment in second order logic to any entities but individuals.

## 3 Linnebo on Logicality of Plural Quantification

For Linnebo, Boolos' theory of plural quantification could be qualified as a *logic*, if (at least) the following conditions are satisfied:

**Ontological Innocence** The basic axioms are not ontologically committed to any entities beyond those already accepted in the ordinary first-order domain;

**Universal Applicability** The theory of plural quantification can be applied to any realm of discourse, no matter what objects this discourse is concerned with;

**Cognitive Primacy** The theory of plural quantification presupposes no extra-logical ideas in order to be understood, but can be understood directly. Our understanding of it does not consist, even in part, in an understanding of extra-logical ideas, such as ideas from set theory or from other branches of mathematics.[5]

Linnebo, cautiously, avoids to consider the three conditions previously cited as the *last* word on *logicality*. However, for their very nature, these should hold in principle for every system of logic and not only for second order plural logic. Thus, we think that it is not wrong to think that these requirements of logicality are not just *locally valid*. Indirect supports to this thesis are given by Dulith Novaes in [6]. On *permutation invariance* she writes:

> The main philosophical appeal of permutation invariance as a criterion for logicality seems to be the *generality* afforded by it. Another characteristic traditionally attributed to logic, namely its topic-neutrality, is (*prima facie*) also captured by the criterion. indeed, if logic is not concerned with these peculiarities insofar as they are related to different domains of investigation [6, 85].

---

[5][13], 77.

Following the above quotation, it is not difficult to notice that two Linnebo's constraints – *ontological innocence* and *universal applicability* – are in line with the traditionally accepted conditions for logicality.

Now, given the above three conditions can Boloos' proposal of plural quantification considered as a logic? Linnebo's main thesis is that Boolos' interpretation of plural quantification fails to make the *impredicative comprehension principle*:

$$(\text{CP}) \quad \exists X \forall x (Xx \leftrightarrow \phi)$$

(where '$\phi$' is a formula in the language of MSOL that contains '$x$' and possibly other free variables but no occurence of '$X$'. If $\phi$ contains no bound second-order variables, the corresponding comprehension principle is said to be *predicative*; otherwise it is *impredicative*) as a genuine logical principle. Linnebo observes that:

> Adding the theory of plural quantification to an interpreted first-order theory involves adopting the plural comprehension axioms, applied to the domain of his theory. What justifies us adopting these axioms? Because we want the impredictaive plural comphrensions axioms as well as the predicative ones, it is not enough to be justified in taking there to be pluralities corresponding to all *predicative* substitution instances for the substitution instances for the plural variables; that is, in taking there to be pluralities corresponding to the form
> 
> $a_1$ and ... and $a_m$ and the $\phi$s,
> 
> where $m$ is a natural number, the $a_i$'s are singular terms, and $\phi$ contanins no bound plural variable. Rather what we need to justify is that there are pluralities corresponding to *all* expressions of the form 'the $\phi$s', even where $\phi$ contains bound plural variables. But in order to do this, we must understand what these bound plural variables range over. This means that we must understand the notion of a *determinate range of arbitrary sub-pluralities* of the original domain.[6]

According to Linnebo, once such a notion of a *determinate range of arbitrary sub-pluralities of the original domain* is adopted we are commit to allow for collecting together in turn the pluralities so construed, so forming higher-order pluralities. If so, we would be led by the plural interpretation of second-order logic to larger and larger extensions of the domain of individuals. So, if second-order logic were pure logic in virtue of plural quantification, such would be also higher- and higher-order logics. Linnebo concludes that the plural interpretation fails to make the impredicative comprehension principle (CP) a genuine logical principle. Allowing for the iteration of the operation of sub-plurality leads Boolos' plural logic to the inexhaustibility of the layers of higher- and higher-order pluralities. If it is so plural logic would not make us able to talk about all the pluralities there are, violating the condition of *universal applicability*.

---

[6][13], 85.

Moreover, consider the appeal to combinatorics and set theory. It shows, according to Linnebo, that Boolos' *plural logic is not even cognitively prior to mathematics*. Indeed, if we want to understand the notion of arbitrary sub-plurality requires that we need to understand some extra-logical, set-theorethical, ideas. So, Boolos proposal does not satisfy condition of *cognitive primacy* either.

Finally, consider the combinatorial feature of the notion of sub-plurality. It is not problematic *per se*. But it is problematic if we think that *pluralities* are entities of some sort. Linnebo does not take a final stand as for the *ontological status* of pluralities. But, one can easily observe that his argument against the logicality of second-order logic as it is interpreted by Boolos follows only if pluralities are taken to be entities of some kind (*ontological innocence*). Consider the following quotation:

> [A defense of Boolos' position] is based on the idea that only *things* can be collected together. If this idea is right, and if Boolos avoids reifying pluralities, then there will simply be no *things* available to be collected together to form higher pluralities. But this defense too is unconvincing. There is no obstacle to iterating the combinatorial considerations that give content to our talk about arbitrary sub-pluralities; in particular, combinatorics has no ontological qualms about collecting together first-order pluralities so as to form higher pluralities. For instance, from the point of view of combinatorics, it is no more problematic to arrange individual Cheerios in the following way: 00 00 00 than it is to arrange them as: 000000, although the former arrangement is most informatively described as three pairs of Cheerios – which is a higher order plurality – whereas the latter arrangement is a mere first-order plurality based on the same six Cheerios. To whatever extent the more complex arrangement involves additional ontological commitments, these commitments pose no problem to combinatorics.[7]

Here, Linnebo seems to implicitly distinguish between *things*, like for instance concrete objects, and more general *entities*, and consider the non-reification of pluralities as the claim that they are not things. In fact, that combinatorics has no qualms about ontology comes down to the fact that combinatorics is completely indifferent to the nature of the entities it combines: it may combine *things*, like Cheerios in the bowl, and abstract *entities*, like sets, in a very large variety of arrangements. So, regardless of the kind of entities pluralities are, they are capable of being combined. It is clear, therefore, that Linnebo shares with other Boolos' crititcs [8] the view that pluralities are entities of some sort. But, then, plural quantification is not *ontologically neutral*.

Is Linnebo criterion of logicality applicable to FOL too?

---

[7][13, pp. 87–88].

[8]See, for example, Parsons in [15] and [16]. Parsons acknowledges that Boolos proposal has the merit to give a clarification to the notion of *manifold*. However, he argues, this interpretation is not ontologically noncommittal. Parsons in [16] criticizes Boolos' semantics by holding that the appropriate reading of the locution 'There are some $X$s' is 'There is a plurality $X$', which unveils the hidden commitment of plural quantification to pluralities.

## 4 Universal Applicability

According to Linnebo, $MSOL$ – in Boolos' interpretation – has to be *universally applicable*. But, what does it mean 'universally applicable'?

An intuitive, minimal requirement for *universal applicability* seems to be the following:

> A logic is *universally applicable iff* its logical truths do not depend on some extra-logical facts which, inasmuch extra-logical, vary over domains.

If this is what Linnebo thinks about Boolos' interpretation of MSOL then it is a criterion hardly satisfied also by FOL.

Etchemendy in [8] has shown how certain ontological questions about cardinality of the domain are counterfactually relevant for the applicability of FOL.

Let us consider the following sentence:

$$(Fin)\ (\forall x \forall y \forall z\, (Rxy \wedge Ryz \rightarrow Rxz) \wedge \forall x \neg Rxx) \rightarrow \exists x \forall y \neg Rxy$$

$(Fin)$ – in which just $R$ is a non-logical predicate – is true only in models whose domain has a finite number of elements. Now, $(Fin)$ is not logically true from an intuitive point of view and, again, it is not true in the model-thoeretic Tarskian semantics. It is straightforward to find a model where $(Fin)$ is false: take any model with an infinite number of elements.

However, let us assume that it holds a quite demanding metaphysical assumption: finitism. In such a case, there would be no model whose domain contains an infinite number of entities. Then $(Fin)$ would be true in all models and, according to Tarski's analysis, it would be logically true. But, observe, we would be forced to accept that a metaphysical feature of the world – viz. either having a finite number of entities or not having it – was relevant to establish if a sentence is logically true. Indeed, facts about the number of objects that can be in the universe are not *logical* facts. Moreover, Etchemendy argues that also assuming that the universe is necessarily infinite will not do. The assumption, equivalent to the assumption of the *Axiom of Infinity*, is not logical; so the account would be "influenced by extra logical facts" [8, p.116]. Once more, the account appears to be conceptually inadequate.

Etchemendy concludes that (Tarskian) model-theoretic validity of a sentence as $(Fin)$ has to do with an extra-logical fact, namely, the cardinality of the universe. And, in his view, this is a good argument for the extensional inadequacy of Tarski's semantics. But also for our purpose, it is straightforward that if a logic has to be *universally applicable*, then not even FOL can satisfy this requirement; indeed, its universal applicability seems to depend on the very general metaphysical structure of the world. One could reply[9] that if logic only depends on metaphysical facts about the world that hold in all domains, then it is universally applicable even if it makes ontological assumptions. Question: what do we mean with 'all domains'?

---

[9] We thank an anonymous referee for this suggestion.

Are they "all possible domains"? If so, which is the meaning of the *possibility* in question? We can say that they are all *logically possible domains*; in this case, we would have metaphysical features which are universally owned. But if the extension of these metaphysical features is the entire domain of logical possible worlds, how can we differentiate metaphysical and logical assumptions?

If there is a finite number of entities in the world, then (*Fin*) would be logically true. However the universal applicability is not grounded in a logical fact. On the contrary, it is *based* on a metaphysical fact. So *universal applicability*, according to which a logical theory can be applied to any realm of discourse, no matter what objects this discourse is concerned with, does not hold. But, even **FOL** cannot be applied to any realm of discourse or, at least, it is relevant, to establish what is logically true, a non logical feature of the domains, that is, their number of objects.

Of course one could conclude this analysis observing, as Dutilh Novaes has done, that: "Ontological neutrality is never possible anyway"[6, p. 86]. But one can reply that this is just a different argument against the *universal applicability* as a criterion of *logicality*.

## 5 Cognitive Primacy

What about *cogntive primacy*, i.e. the idea that the theory of plural quantification presupposes no extra-logical ideas in order to be understood?

Consider the following *prima facie* understanding of the requirement:

> If a theory does not presuppose extra-logical ideas, it means that it does not presupppose ideas, for example, from set theory or from other branches of mathematics.

So, for example, if a theory is logic it does not presuppose set theory or mathematics. If it is so, it is easy to observe that mathematical notions have a fundamental role in the epistemology of logic: part of the understanding of a logical system is constituted by the understanding of the synctactical structure of the language of the system and it is part of the syntactical structure to refer to mathematical structures. Take **FOL**: it implicitly appeals to the notion of denumerable infinity. So, to understand **FOL**, we have to understand for example the notion of an infinite vocabulary (say, with infinitely many individual variables).

In this respect, if the notion of denumerable infinity is mathematical, as it is, then according to Linnebo's requirement of cognitive primacy, **FOL** is not *pure* logic.

Moreover, consider that – since a language formation is a prerequisite of any logic, one can conclude – following Linnebo's request – that, not only **FOL** but also propositional logic is not logic!

Now, one may object – and Linnebo can do it – that independence of mathematical notions concerns just more sophisticated notions like the one in the following example: let $c$ be a function whose meaning is "the cardinality of..." and $\wp$ the powerset operator. So,

$$(C) \ \forall x(c(x) \neq c(\wp(x)))$$

($C$) presupposes the understanding of Cantor's theorem and other cognate notions. One can argue that in this example there are extra-logical ideas presupposed but they are *not essential* to understanding logic. But now the problem is: when extra-logical ideas are supposed to play an *essential* role in our understanding of logic?

An advocate of Linnebo's position could reply that the use of the notion of a denumerable infinity to characterize the syntax of FOL does not violate a criterion of cognitive primacy. Moreover, he can go on by pointing out that theorems of logic do not presuppose *extra-logical facts* and, for this reason, the way in which the language of the theory is framed does not matter.

To this reply, we answer that the content of theorems of a formalized theory should display the information contained in the axioms and in the relation of consequence connecting axioms with theorems. It is reasonable to claim that there is not a specific content presupposed by logical notions, at least for *propositional* and *first-order logic* (things are much more complex if we pass to higher-order logic). Instead, to understand the content of the theorems we need to understand the logical structure of the language in which they are formulated and the pattern of deduction from which they are proved. Otherwise, it would be suspect to claim that we understand a specific theorem in exam. But, to precisely intend this notion of consequence, we need a pack of mathematical concept as shown before.

So, we can concede to Linnebo that the logic is *cognitive prior* with respect to any specific content but – however – we think it presupposes a grasp of mathematical concepts to adequately capture the meaning of the logical consequences of axioms.

## 6 Ontological Innocence

Unlike *cognitive primacy* and *universal applicability*, Linnebo formulates the *ontological innocence* requirement in a *relational* way. Boolos' interpretation of MSOL is ontologically innocent *iff* it is not more committed than another theory assumed as ontologically innocent.

In this case, our basic theory is FOL. So, the criterion can be specify in the following way:

- a system $S*$ is ontologically innocent *iff* $S*$ is not more committed than $S$ and

- $S$ is ontologically innocent.

In the following, we shall try to characterize the relation between $S*$ and $S$; then, on the basis of this characterization, we shall cast some doubts on the *ontological innocence* requirement.

Suppose that our system $S*$ is not ontologically committed to any entities beyond those already accepted in the ordinary system $S$. Let us call the system $S*$ *ontologically conservative* on $S$. Ontological conservativity can be intended in, at least, two ways: as *numerical conservativity* and as *metaphysical conservativity*, i.e.:

$S*$ is *numerically conservative* (*NumCons*) on $S$ *iff* the intended model of $S*$ has the same size of the intended model of $S$.[10]

$S*$ is *metaphysically conservative* (*MetCons*) on $S$ *iff* in the intended model of $S*$ there are no different entities from those accepted in the domain of $S$.

The relations between these conditions are:

1. $S*$NumCons $S \Longrightarrow S*$ MetCons $S$
2. $S*$MetCons $S \Longrightarrow S*$ NumCons $S$

Ad 1). If $S*$ is numerically conservative on $S$, then it is not introducing anything new in respect to the entities introduced by $S$. But this is not unconditionally valid. Let us assume that the cardinality of the domains of $S$ and $S*$ is denumerable. Let us further assume that $S*$ introduces – via a *comprehension axiom* – a denumerable class of different entities from those already accepted in the domain of $S$. In this case, $S*$ is a numerically conservative extension of $S$ but it is not a metaphysical conservative extension of $S$. So, (1) holds provided that the cardinality of the domain is finite.

Ad 2). Let us assume that $S*$ is metaphysically conservative on $S$. Obviously, this does not say anything on the number of elements $S*$ is introducing in addition to the entities admitted by $S$. And this holds both for infinite and finite cardinalities. For instance, consider a system of FOL plus "there is a non denumerable quantity of individuals". This system is metaphysically conservative on FOL but it is not numerically conservative.

We have shown how to unpack Linnebo's third requirement: ontological innocence can be considered as the union of *numerical conservativity* and *metaphysical conservativity*. But, there is no guarantee that a system satisfying one of the two sub-criteria will be *ontological innocent*. Just the first one seems to fit Linnebo's intuitions, provided that the cardinality at play is finite. But again, we need to refer to an extra-logical fact to fully specify a criterion of logicality.

## 7 Skecth of a concluding proposal

From what previously said the three requirements of ontological innocence, cognitive primacy, and universal applicability seem to be too restrictive: not even FOL is able to satisfy them. One can also argue that they coincide with what FOL commits to. But, then, one can easily argue that they are *ad hoc*.

Are there any alternatives? One could abandon this path and follow, instead, one of well known route to logicality mentioned at the beginning of this paper. Otherwise, one could agree with Tarski, according to which:

---

[10] The reference to intended model is necessary since, by Loewenheim-Skolem results, any theory with a denumerable model has models of any cardinality.

[I]t would turn out to be necessary to treat such concepts as following logically, analytic sentence or tautology as relative concept which must be related to a definite but more or less arbitrary division of the terms of a language into logical and extra-logical. [21, p. 189].

Just to briefly sketch the train of thoughts of this Tarskian idea: one could keep fixed the intuition connected to the independence of logic from specific features of meaning and world but to *relativize*, at the same time, the very concept of *independence*. That is, once some general norms concerning, for instance, the general structure of the domain are fixed, it is possible to show that a system is a logical one *iff* its notions are universally applicable *modulo* these general norms previously stated.[11]

It is clear that any choice of the above mentioned norms should be justified and whatever the justification adopted is, this does not seem to affect the grounding idea according to which the nature of logic has to do with the *independence* from specific contents and domains.

## Aknowledgements

An early version of this paper have been read at the *SILFS 2014 – Triennial International Conference of the Italian Society for Logic and Philosophy of Sciences* held in Rome (Italy). We are indebted to the participants in the conference for stimulating discussion. Thanks also to Vittorio Morato and to a referee for their detailed comments and suggestions.

## BIBLIOGRAPHY

[1] BEALL, J. C., RESTALL, G., *Logical pluralism*. Oxford: Clarendon Press, 2006.
[2] BONNAY, D., 'Logicality and invariance', *Bulletin of Symbolic Logic*, (2008), 29–68.
[3] BOOLOS, G., 'To be is to be the Value of a Variable (or to be Some Values of Some Variables)', *Journal of Philosophy*, 81 (1984), 430–49.
[4] BOOLOS, G., 'Nominalist Platonism', *Philosophical Review*, 94 (1985), 327–44.
[5] CARRARA, M., MARTINO E., 'To be is to be the Value of a Possible Act of Choice', *Studia logica* 96(2010), 289-313.
[6] DUTILH NOVAES, C., 'The undergeneration of permutation invariance as a criterion for logicality' *Erkenntnis* 79(2014), 81-97.
[7] DUMMETT, M., *The Logical Basis of Metaphysics*, Harvard University Press, Harvard (Mass.), 1990.
[8] ETCHEMENDY, J., *The concept of logical consequence*, Harvard University Press, Harvard (Mass.),1990.
[9] GENTZEN, G., 'Untersuchungen über das logischen schliessen, *Math. Zeitschrift*, (1934), 405–31.
[10] GOMEZ-TORRENTE, M., 'The problem of logical constants', *The Bulletin of Symbolic Logic*, (2002), 1–37.
[11] KNEALE, W., 'The province of logic', *in* H. D. Lewis (ed.), *Contemporary British Philosophy*, George Allen & Unwin Ltd, London, 1956, 237–61.
[12] MACFARLANE, J., *What does it mean to say that logic is formal?*, University of Pittsburgh, 2000.
[13] LINNEBO, O., 'Plural Quantification Exposed', *Noûs*, 1 (2003), 71–92.
[14] LINNEBO, O., 'Plural Quantification', *The Stanford Encyclopedia of Philosophy (Fall 2014 Edition)*, Edward N. Zalta (ed.), URL = <http://plato.stanford.edu/archives/fall2014/entries/plural-quant/>.

---

[11] A discussion of this topic is out of the scope of this paper. On pluralism in logic see the milestone book written by Beall and Restall [1].

[15] PARSONS, C., *Mathematics in Philosophy: Selected Essays*, Cornell University Press, Ithaca, New York, 1983.
[16] PARSONS, C., 'The Structuralist View of Mathematical Objects', *Synthese*, 84 (1990), 303–346.
[17] POPPER, K., 'New foundations for logic', *Mind* 1947 (223), 193–235.
[18] QUINE, W.O., *Philosophy of Logic*, Harvard University Press, Harvard, 1970, 1986.
[19] RESNIK, M.D., 'Second-Order Logic Still Wild', *The Journal of Philosophy*, 85 (1988), 75–87.
[20] SHER, G., *The bounds of logic*, MIT Press, Cambridge, Mass, 1991.
[21] TARSKI, A., 'On the concept of following logically', *History and Philosophy of Logic*, 23 (2002) 155-196.
[22] TENNANT, N., *The Taming of the True*, Oxford University Press, Oxford, 1997.

# Cyclic Properties: from Linear Logic to Pregroups

Claudia Casadio and Mehrnoosh Sadrzadeh

ABSTRACT. We show that the algebra of pregroups, used in the type grammar recently introduced by the mathematician J. Lambek, exhibits a weak form of cyclic properties similar to those holding in Non-commutative Multiplicative Linear Logic (NMLL). We prove some algebraic inequalities for these notions and present them in a sequent calculus form. We motivate the advantages of this approach, both at the theoretical and at the descriptive level, applying it to the analysis of word order changes in certain natural languages.

**Keywords** Type grammar, linear logic, pregroup, cyclic rules, word order

## 1 Introduction

The calculus of pregroups is a type grammar introduced by the mathematician J. Lambek [20, 21], that has been applied to the logical analysis of many natural languages like English, German, French, Italian, and others [23]. Similarly to the Syntactic Calculus [19], the calculus of pregroups is free from structural rules (weakening, contraction, exchange) to the effect that sequences of formulas are not necessarily commutative. Non commutativity is a property of particular interest in the formal study of language, as proven by the extensive work on this subject [3, 11, 8, 17]. The syntax of natural languages admits, however, changes of word order like e.g. the patterns of *topicalisation* and *VP-preposing* studied by theoretical linguistics [15, 27]. We made this observation precise by developing a novel approach to pregroups based on the theoretical concept of *cyclic rule* [31, 1, 3] and the permutations that can be introduced by means of these rules. Whereas these notions have been studied for monoids and residuated monoids and associated sequent calculi, particularly in the framework of Non-commutative and Cyclic Linear Logic [18, 2, 4, 28, 29], no one has up to now studied their import in the calculus of pregroups.

In previous work [13], we showed how cyclic properties can be formulated as meta rules imposed on the lexicon of a pregroup type grammar, and used these to reason about word order alternation as a result of clitic movement in natural languages such as Farsi, Italian, and some examples from French. These rules were external to the system and did not relate to any internal property of pregroups. More recently in [14] we showed how pregroups admit certain weaker forms of cyclicity; we referred to these by *precyclicity* and presented some algebraic properties for them, further we applied these to develop precyclic transformations that allowed us to reason about word order alternations in Sanskrit.

In this paper, we build on the latter work. We first review the definition and properties of cyclicity in residuated monoids (which are the basis for the Syntactic Calculus). Using the known translation between residuated monoids and pregroups, we then develop definitions for the notions of cyclic and dualizing elements of a pregroup and prove some properties about them. For instance, we show how the inequalities of adjunction can be derived from those of precyclicity. We also review the cyclic properties from a sequent calculus point of view and go through the corresponding rules of Non-commutative/Cyclic Multiplicative Additive Linear Logics of Abrusci and Yetter [1, 2, 4, 31]. We show how the translations of these rules into the sequent calculus of pregroups, due to Buszkowski [6, 7], are sound. Finally, we apply these findings to reason about change of word order in natural language, offering some examples from different languages.

## 2  Pregroups: algebra and rules

A pregroup $P$ is a partially ordered monoid $(P, \cdot\, , 1, \leq, ()^l, ()^r)$ : $P$ a set of types, '$\cdot$' a non commutative multiplicative operation, 1 the unit of the monoid, and each element $p \in P$ has both a left adjoint $p^l$ and a right adjoint $p^r$:

$$p^l \cdot p \leq 1 \leq p \cdot p^l \qquad\qquad p \cdot p^r \leq 1 \leq p^r \cdot p$$

The two inequalities on the left side of 1 are referred to as *contractions*, while the two at the right side of 1 as *expansions*; the unit 1 and the multiplication are self dual [20, 6]:

$$1^l = 1 = 1^r \qquad (pq)^l = q^l p^l \qquad (pq)^r = q^r p^r \ .$$

where the (left, right) adjoint of multiplication is the multiplication of adjoints, but in the reverse order. Some other properties of pregroups are as follows:

- The adjoint operation is order reversing, that is:

$$p \leq q \implies q^r \leq p^r \qquad \text{and} \qquad p \leq q \implies q^l \leq p^l$$

- Composition of opposite adjoints is identity:

$$(p^l)^r = (p^r)^l = p$$

- Composition of the same adjoints is not identity:

$$p^{ll} = (p^l)^l \neq p, \qquad p^{rr} = (p^r)^r \neq p$$

This leads to the existence of iterated adjoints [20], so that each element of a pregroup can have countably many iterated adjoints:

$$\ldots, p^{ll}, p^l, p, p^r, p^{rr}, \ldots$$

A pregroup is *proper* whenever for any of its elements $p$ we have that $p^l \neq p^r$.

We may read $x \leq y$ as saying that everything of type $x$ is also of type $y$. Linguistic applications make particular use of the equation $a^{r\ell} = a = a^{\ell r}$, allowing the cancellation of double opposite adjoints and of the rules $a^{\ell\ell} a^\ell \to 1 \to a^\ell a^{\ell\ell}$, $a^r a^{rr} \to 1 \to a^{rr} a^r$, contracting and expanding identical *left* and *right* double adjoints respectively. Just *contractions* $a^\ell a \to 1$ and $a a^r \to 1$ are needed to determine constituent analysis and to show that a string of words is a sentence, while *expansions* $1 \to a a^\ell$, $1 \to a^r a$ are useful for expressing structural (syntactic and semantic) properties [23].

## 3 Algebraic Cyclicity

### 3.1 Residuated Monoids

A monoid $(M, \cdot, 1)$ is a set $M$ admitting an associative operation with unit 1. A monoid is partially ordered when $M$ is partially ordered and the order preserves the monoid operation, that is for every $a, b, e \in M$:

$$a \leq b \implies a \cdot e \leq b \cdot e \text{ and } e \cdot a \leq e \cdot b$$

We denote a partially ordered monoid $M$ by $(M, \cdot, 1, \leq)$. A residuated monoid, denoted by $(M, \cdot, 1, , \leq, /, \backslash)$, is a partially ordered monoid in which the monoid multiplication has a right $- \backslash -$ and a left $-/-$ adjoint, that is, for $a, b, e \in M$ [19, 23, 26, 4]:

$$b \leq a \backslash e \iff a \cdot b \leq e \iff a \leq e/b$$

An element $c$ of a partially ordered monoid $M$ is said to be *cyclic* whenever, for all $a, b \in M$:

$$a \cdot b \leq c \implies b \cdot a \leq c$$

Thus, one can define the notion of cyclicity for partially ordered monoids that are not necessarily residuated. Whenever a monoid is residuated, the cyclic condition becomes equivalent to:

$$c/a = a \backslash c$$

### 3.2 Cyclicity in residuated monoids

The notion of cyclicity does not depend on the residuated structure of a monoid, but only on its underlying partial order. We say that a partially ordered monoid (residuated or not) is cyclic whenever it has a cyclic element. In a residuated monoid, one can also define a new notion: that of *dualization*. An element $d$ of a residuated monoid is dualizing whenever for all $a \in M$ we have: $(d/a) \backslash d = a = d/(a \backslash d)$. If the dualizing element of $M$ is also cyclic, we obtain: $d/(d/a) = a = (a \backslash d) \backslash d$.

These notions were defined by Yetter [31] focusing on residuated lattice monoids $(M, \cdot, 1, /, \backslash, \vee, \wedge, \bot, \top)$. In such structures, the bottom element of the lattice $\bot$ is *dualizing*, and it can be used to define two notions of negation: $\neg^r a := a \backslash \bot$ and $\neg^l a := \bot/a$. If the two negations of a residuated lattice monoid coincide: $\neg^l a = \neg^r a$, then $\bot$ is also cyclic. Yetter used these notions to provide an algebraic semantics for linear logic and called *Girard Quantales* the structures in which the two negations coincide, i.e. in which $\bot$ is cyclic.

The passage from residuated monoids (on which Lambek's Syntaxtic Calculus is based [19]) to pregroups consists in replacing the two adjoints of the monoid multiplication with the two adjoints of the elements. If a residuated monoid has a dualizing object, i.e. an object $0 \in M$ satisfying $(0/p) \backslash 0 = p = 0/(p \backslash 0)$ for $p \in M$, then one can define for each element a left and a right negation as $p^0 := p \backslash 0$ and $^0p := 0/p$. It would then be tempting to think of these negations as the two pregroup adjoints, i.e. to define $p^0 = p^r$ and $^0p = p^l$. The problem with this definition is that the operation $\wp$ - the linear logic "par" - involved in $a \backslash b = a^\perp \wp b$, $b/a = b \wp a^\perp$, is different from the multiplicative operation (a . b) of pregroups. One can however translate Syntactic Calculus expressions into pregroups provided that both $\wp$ and $\otimes$ of CyMLL are identified with the pregroup unique operation: then all the $a \backslash b$ (or $b/a$) types will become $a^r b$ (or $b a^l$) [20].

## 3.3 Cyclicity in pregroups

By translating the terms and properties of residuated monoids into pregroups, we investigate whether and how the translations of the above notions may hold in a pregroup. In particular, we will have the translation of the unit object of a residuated monoid, which is again the unit object in a pregroup, and the translation of the definition of a dualizing object of a residuated monoid. Then we show that the translation of 1 satisfies the translation of the dualizing property, but not the translation of the cyclic property. By reusing the vocabulary to some extent, one can summarise this result and say that the unit 1 of a pregorup is a 'dualizing' element which is not necessarily 'cyclic'.

DEFINITION 1. Given an element $x$ of a residuated monoid $M$, we denote its translation into a pregroup by $t(x)$. For all $a, b \in M$, this translation is defined as follows:

$$t(1) = 1, \quad t(a \cdot b) = t(a) \cdot t(b), \quad t(a \backslash b) = t(a)^r \cdot t(b), \quad t(a/b) = t(a) \cdot t(b)^l$$

DEFINITION 2. A pregroup has a dualizing element, whenever the following equality holds in it:

$$t((d/a) \backslash d) = t(a) = t(d/(a \backslash d))$$

for $d$ and $a$ elements of a residuate monoid and $d$ dualizing. If that is the case, we call $t(d)$ the dualizing element of a pregroup.

DEFINITION 3. The dualizing element $t(d)$ of a pregroup is cyclic whenever we have

$$t(d/(d/a)) = t(a) = t((a \backslash d) \backslash d)$$

PROPOSITION 4. *The unit of a pregorup is dualizing.*

**Proof.** Recall that $t(1)$ is 1. So we have to show that for all $a \in P$, the translation of $(1/a) \backslash 1 = a = 1/(a \backslash 1)$ holds in a pregroup. That is we have to show:

$$t((1/a) \backslash 1) = t(a) = t(1/(a \backslash 1))$$

For the left hand side we have
$$t((1/a)\backslash 1) = (t(1/a))^r \cdot t(1) = (1 \cdot t(a)^l)^r \cdot 1 = ((t(a)^l)^r \cdot 1) \cdot 1 = (t(a)^l)^r = t(a)$$
For the right hand side we have
$$t(1/(a\backslash 1)) = t(1) \cdot t(a\backslash 1)^l = 1 \cdot (t(a)^r \cdot t(1))^l = 1 \cdot (1 \cdot (t(a)^r)^l) = (t(a)^l)^r = t(a)$$
∎

**PROPOSITION 5.** *The unit of a proper pregroup is not cyclic.*

**Proof.** We need to show that for all $a \in P$, the following is the case
$$t(1/(1/a)) \neq t(a) \neq t((a \backslash 1) \backslash 1)$$
For the left hand side we have:
$$t(1/(1/a)) = t(1) \cdot t(1/a)^l = t(1) \cdot (t(1) \cdot t(a)^l)^l = t(1) \cdot t(a)^{ll} \cdot t(1)^l = t(a)^{ll}$$
For the right hand side we have
$$t((a\backslash 1)\backslash 1) = t(a\backslash 1)^r \cdot t(1) = (t(a)^r \cdot t(1))^r \cdot t(1) = t(1)^r \cdot t(a)^{rr} \cdot t(1) = t(a)^{rr}$$
and since the pregroup is proper, it is the case that: $t(a)^{ll} \neq t(a)^{rr}$. ∎

However, as proved in [14], pregroups do admit a weak form of cyclicity, which we refer to by using the term *precyclicity*, described below:

**PROPOSITION 6.** *The following hold in any pregroup $P$, for $p, q, r \in P$*

(i) $pq \leq r \implies q \leq p^r r$ \qquad (ii) $q \leq rp \implies qp^r \leq r$

(iii) $qp \leq r \implies q \leq rp^l$ \qquad (iv) $q \leq pr \implies p^l q \leq r$

As a consequence we obtain:

**COROLLARY 7.** *The following hold in any pregroup $P$, for any $a, b \in P$:*

(1) $1 \leq ab \overset{(ll)}{\implies} 1 \leq ba^{ll}$ \qquad (2) $1 \leq ab \overset{(rr)}{\implies} 1 \leq b^{rr}a$

Informally, case (1) of the above corollary says that whenever a juxtaposition of types, e.g. $ab$, is above the monoidal unit, then so is a permuted version of it, where $a$ moves from the left of $b$ to the right of it, but as a result of this movement, $a$ gets marked with double adjoints $ll$ to register the fact that it came from the *left*. That is why this property is annotated with (and we thus refer to it by) $ll$. Case (2) is similar, except that in this case it is $b$ that moves from the right of $a$ to its left, hence it is marked with $rr$. This result can be expressed in different forms, what follows is one variant:

**PROPOSITION 8.** *For a pregroup $P$ and $p, q \in P$:*

(3) $1 \leq p^r q^l \implies 1 \leq q^l p^l$ \qquad (4) $1 \leq p^r q^l \implies 1 \leq q^r p^r$

(5) $pq \leq 1 \implies q^{ll} p \leq 1$ \qquad (6) $pq \leq 1 \implies qp^{rr} \leq 1$

**Proof.** The properties of the first line are obtained by taking $a = p^r$ and $b = q^l$ in the properties of Corollary 3; the properties of the second line are obtained by taking $a = q^l$ and $b = p^l$ in the properties of Corollary 3. ∎

Finally note that another interesting feature of the *precyclicity* property is the fact that it can be used to obtain the adjunction inequalities of a pregroup, perhaps indicating that the precyclic properties can be used a basis for the pregorup adjunctions properties:

PROPOSITION 9. *The pregroup adjunction inequalities follow from precyclicity.*

**Proof.** Consider the inequalities of the *left adjoint*: for all $p$ in a pregroup $P$, we have $p^l p \le 1 \le pp^l$. To prove the left hand side inequality, start from $p \le p$, from which it follows that $p \le p1$, then by inequality (iv) of Proposition 6, it follows that $p^l p \le 1$. To prove the right hand side inequality, start from $p \le p$, from which it follows that $1p \le p$, then by inequality (iii) of Proposition 6, it follows that $1 \le pp^l$. To prove the inequalities of the right adjoint, that is $pp^r \le 1 \le p^r p$, start from $p \le p$, from this it follows that $p \le 1p$ and that $p1 \le p$; from the former by inequality (ii) of Proposition 6 it follows that $pp^r \le 1$, and from the latter by inequality (i) of Proposition 6 it follows that $1 \le p^r p$. ∎

## 4 Cyclicity in Sequent Calculi

### 4.1 Linear Logic

The origins of the cyclic rules in sequent calculi go back to the following (restricted) form of exchange rule, first introduced by Girard [18]:

$$\frac{\vdash \Gamma, A}{\vdash A, \Gamma} \; CycExch$$

To remove the exchange rule (and hence all the structural rules) from Linear Logic, Abrusci generalised this rule in the following way, referring to its logic as Pure Non-Commutative Classical Linear Logic (**SPNCL'**)[1]:

$$\frac{\vdash \Gamma, A}{\vdash \neg^r \neg^r A, \Gamma} \; Cyc^{+2} \qquad \frac{\vdash A, \Gamma}{\vdash \Gamma, \neg^l \neg^l A} \; Cyc^{-2}$$

The semantics of this logic is a version of Girard's phase semantics where $\bot$ is dualizing and it is used to define negations as demonstrated in the previous section, that is $\neg^r A := A \backslash \bot$ and $\neg^l A := \bot / A$. Note that Girard's original Multiplicative Additive Linear Logic (**MALL**) [18], has the following exchange rule:

$$\frac{\vdash A_1, \cdots, A_n}{\vdash A_{\sigma(1)}, \cdots, A_{\sigma(n)}}$$

In this rule $\sigma$ is any permutation of $\{1, \cdots, n\}$. Later, Yetter restricted this rule so that $\sigma$ was not any permutation, but a cyclic permutation. In any case, in these logics the constant $\bot$ is both cyclic and dualizing, whereas in Abrusci's version of the logic (**SPNCL'**) [1], $\perp$ is dualizing but not cyclic. Abrusci also showed that this logic is equivalent to the system **SPNCL**, which does not include $Cyc^{+2}$ and $Cyc^{-2}$ but its *cut* and *multiplicative* rules have side conditions. These two logics are referred to by the umbrella terms **NMALL** or (**CyMALL**) for Non-commutative or Cyclic Multiplicative Additive Linear Logic [3].

### 4.2 Compact Bi-Linear Logic

The logic of pregroups is called a Compact Bi-Linear Logic [25], that is a Linear Logic which has two implications (hence the mention of the word bi-linear) and in which the *tensor* and *par* coincide (hence the mention of the word compact). The first sound and complete cut-free sequent calculus for such a logic is presented in [7]. The rules of this calculus are as follows, for $A, B$ single formulae and $\Delta, \Gamma$ finite sequences of formulae:

$$\frac{\vdash \Gamma, \Delta}{\vdash \Gamma, 1, \Delta} \; 1 \qquad \frac{\vdash \Gamma, A, B, \Delta}{\vdash \Gamma, A \cdot B, \Delta} \qquad \frac{\vdash \Gamma, \Delta}{\vdash \Gamma, A^r, A, \Delta} \; Adj^r \qquad \frac{\vdash \Gamma, \Delta}{\vdash \Gamma, A, A^l, \Delta} \; Adj^l$$

This calculus admits the following two cut rules:

$$\frac{\vdash A, \Delta \quad \vdash \Gamma, A^l}{\vdash \Gamma, \Delta} \; cut\; l \qquad \frac{\vdash A^r, \Delta \quad \vdash \Gamma, A}{\vdash \Gamma, \Delta} \; cut\; r$$

As elaborated on in detail in [11, 20], there is a correspondence between this logic and **NMALL**: the multiplication operation of a pregroup is the *tensor* product of **NMALL** [11, 20] and the two adjoints of a pregroup correspond to the two negations of **NMALL**; in particular we have:

$$A^{+2} := A^{rr} \qquad A^{-2} := A^{ll}$$

Given such a correspondence, the formulation of a calculus for pregroups can interestingly refer to the tensor fragment of the calculus **NMALL**, like the one introduced for instance in [1, 3]. For example the pregroup calculus presented in [7] and the tensor product fragment of **NMALL** presented in [1] share similar logical properties. In particular, one can show the following:

PROPOSITION 10. *The rules $Cyc^{+2}$ and $Cyc^{-2}$ of **NMALL** are sound in any pregroup $P$.*

**Proof.** Using definitions $A^{+2} := A^{ll}$ and $A^{-2} := A^{rr}$ from [11], the rules $Cyc^{+2}$ and $Cyc^{-2}$ become as follows in a pregroup setting:

$$\frac{\vdash A, \Gamma}{\vdash \Gamma, A^{ll}} \; (ll) \qquad \frac{\vdash \Gamma, A}{\vdash A^{rr}, \Gamma} \; (rr)$$

We use the truth-assignment map $h\colon \mathcal{L} \to P$ from the formulae of the logic $\mathcal{L}$ to a pregroup $P$, as given by [6]. A formula $A$ of the logic $\mathcal{L}$ is true under $h$ if $1 \leq h(A)$, for 1 the unit in the pregroup $P$. A sequent $\vdash A_1, A_2, \ldots, A_n$ of $\mathcal{L}$ is true if the formula $A_1 \cdot A_2 \cdot \ldots \cdot A_n$ is true. The soundness of the translations of the cyclic rules in a pregroup then follows from Corollary 3, by taking $h(A) = a$ and $h(\Gamma) = b$. ∎

## 5 Changes of word order in natural languages

Natural languages exhibit various kinds of word order changes with respect to the basic orders admitted by grammatical rules; these patterns are extensively studied in theoretical linguistics, where they are often referred to as *movements* of words or constituents [15, 27].

We obtain here the result of limiting word order changes in natural languages to the grammatical ones by introducing a set of *precyclic transformations* and *permutations* that specify the conditions for obtaining correct (language dependent) word order patterns. In the paper we consider some critical examples taken from a non inflectional language, like English, and two inflectional languages like Persian and Latin. But the procedure presented here can be easily extended to any language whatsoever. To analyse a natural language we use a *pregroup grammar*. Similar to other type-categorial grammars, a pregroup grammar is a free pregroup generated over a set of basic types together with the assignment of the pregroup types to the vocabulary of the language. For the purpose of this paper we assume the set of basic types $\{n, \pi, o, s\}$, representing four basic grammatical categories:

$n$: noun phrase  $\pi$: subject  $o$: object  $s$: sentence

The linguistic reading of a pregroup partial order $a \leq b$ is that a word of type $a$ is also of type $b$. We assume the partial orders $n \leq \pi$ and $n \leq o$, routinely used in pregroup grammars. The free pregroup generated over the above basic types includes *simple* types such as $n^l, n^r, \pi^l, \pi^r$, and *compound* types such as $(\pi^r s o^l)$. A sentence is defined to be grammatical whenever the multiplication (syntactic composition) of the types of its constituents is less than or equal to the type $s$. The computations that lead to deciding this matter are referred to as *grammatical reductions*. For example, the assignments of the words of the declarative sentence 'I saw him.' and its grammatical reduction are as in the Figure on page 150. The grammar of a wide range of natural languages have been analysed using pregroup grammars, see [23]. The computations that lead to type reductions are depicted by the under-link diagrams, reminiscent of the planar proof nets of non-commutative linear logic, as shown in the calculi developed in [3, 6, 21, 22].

There are grammatical regularities within languages that involve word order changes: e.g., certain language units within a sentence move from *after* the verb to *before* it, or from *before* the verb to *after* it, and the resulting juxtaposition of words is still a grammatical sentence. Pregroups were not able to reason about change of word order in a general way and we offer a solution here. We propose to enrich the pregroup grammar of a language with a set of *precyclic transformations* that allow for substituting certain type combinations with their *precyclic permutations*. These permutations differ from language to language and express different, language specific, movement patterns. Within each language, they are restricted to a specific set so that not all word orders become permissible. More formally, we define:

DEFINITION 11. In a pregroup $P$, whenever $1 \leq ab \implies 1 \leq ba^{ll}$ or $1 \leq b^{rr}a$, then we refer to $ba^{ll}$ and $b^{rr}a$ as precyclic permutations of $ab$ and denote this relationship by $ab \stackrel{\sigma}{\leadsto} ba^{ll}$ and $ab \stackrel{\sigma}{\leadsto} b^{rr}a$.

DEFINITION 12. In a pregroup $P$, for $ba^{ll}$ and $b^{rr}a$ precyclic permutations of $ab$, and any $A, B, C \in P$, we define the following precyclic transformations[1]:

$(ll)$-**transformation** $\quad A \leq B(ab)C \overset{(ll)}{\rightsquigarrow} A \leq B(ba^{ll})C$

$(rr)$-**transformation** $\quad A \leq B(ab)C \overset{(rr)}{\rightsquigarrow} A \leq B(b^{rr}a)C$

DEFINITION 13. A precyclic pregroup grammar is a pregroup grammar with a set of precyclic transformations.

**Reduction Procedure.** The idea is that the arguments of certain words and phrases with complex types, e.g. adjectives and verb phrases, can be moved before or after them, as an effect of stress or other semantic vs. pragmatic intentions. We will briefly describe how to extend the existing grammar of a language to include the word order changes resulting from these movements.

1. First you decide which words or phrases allow which forms of movement and encode this information about movement in the precyclic permutations of the type of each such word or phrase $w$, in the following way:

    (a) If $w$ is of type $p^r q$, i.e. it requires an argument of type $p$ *before* it, and $p$ can be moved after $w$, then allow for the cyclic permutation $p^r q \overset{\sigma}{\rightsquigarrow} qp^l$.

    (b) Else, if $w$ is of type $qp^l$, i.e. it requires an argument of type $p$ *after* it, and $p$ can be moved before $w$, then allow for the cyclic permutation $qp^l \overset{\sigma}{\rightsquigarrow} p^r q$.

2. Then you form a *precyclic pregroup grammar* from the pregroup grammar of a language by turning the above permutations into precyclic transformations in the following way:

    (a) If $w$ is from step 1(a), add an $(ll)$-transformation by taking $a = p^r$ and $b = q$ and computing $ba^{ll} = (qp^r)^{ll} = qp^l$.

    (b) Else, if $w$ is from step 1(b), add an $(rr)$-transformation by taking $a = q$ and $b = p^l$ and computing $b^{rr}a = (p^l)^{rr}q = p^r q$.

3. A string of words is grammatical, whenever either the types of its words, as assigned by the pregroup grammar, reduce to $s$, or their transformed versions do.

To exemplify, consider first English and its different word order patterns, as discussed in detail by [5]. The basic English word order is SVO (Subject-Verb-Object), but this order may change as a result of object topicalisation or VP-preposing. Topicalisation allows for the object to move from after the verb phrase to before it. VP-preposing allows for the infinitive verb phrase to move from after the auxiliary or modal verb to before it. These permissible movements are reflected by the following precyclic transformations:

| Moving Unit | Permutation | Transformation |
|---|---|---|
| Object | $so^l \overset{\sigma}{\rightsquigarrow} o^r s$ | $A \leq B(so^l)C \overset{(rr)}{\rightsquigarrow} A \leq B(o^r s)C$ |
| Infinitive | $si^l \overset{\sigma}{\rightsquigarrow} i^r s$ | $A \leq B(si^l)C \overset{(rr)}{\rightsquigarrow} A \leq B(i^r s)C$ |

[1] These transformations prevent us from making isolated assumptions such as $1 \leq so^l$ and stop generation of meaningless inequalities such as $1o \leq (so^l)o \leq s$.

As an example of topicalization, consider the simple transitive sentence 'I saw him', and its topicalized form 'Him I saw', which are typed as follows:

$$\text{I saw him.} \quad : \quad \pi(\pi^r s\, o^l) o \leq 1(s\, o^l) o \leq 1\, s\, 1 \leq s$$

$$\text{Him I saw.} \quad : \quad o\, \pi(\pi^r s\, o^l) \leq o\, (s\, o^l) \stackrel{(rr)}{\leadsto} o\, (o^r\, s) \leq s$$

We are not allowed to reduce the other possible four orderings (*him saw I, saw I him, saw him I, I him saw*) to $s$, since for obtaining similar permutations we need either the subject to move to after the verb, or subject and object invert their relative position; in both cases the consequence is that the subject and the verb occur in configurations like verb-subject (inversion) or subject-object-verb (separate) not admitted by the English grammar, as pointed out in [5]. Formally, to obtain similar ungrammatical word orders, we should need transformations based on the following unlawful permutations, which we have not included into the pregroup grammar:

$$(*)\ \pi^r s \stackrel{\sigma}{\leadsto} s\pi^l \quad (*)\ \pi^r s o^l \stackrel{\sigma}{\leadsto} s o^l \pi^l \quad (*)\ o^r s \pi^l \stackrel{\sigma}{\leadsto} s \pi^l o^l \quad (*)\ \pi^r s o^l \stackrel{\sigma}{\leadsto} o^r \pi^r s$$

As another example, consider the sentence 'He must love her': here we can have both topicalisation (case (1) below) and VP-preposing (case (2) below). The type assignments and derivations of these cases are as displayed on page 151.

Non-permissible combinations like 'must love her he' or 'must love he her' cannot be derived, because they require, as before, a transformation corresponding to the precyclic permutation $\pi^r s o^l \stackrel{\sigma}{\leadsto} s o^l \pi^l$, in which the subject is expected to occur after the verb, that has not been included into the pregroup grammar.

For an example of another language, consider Persian which has an SOV (Subject-Object-Verb) structure. A simple transitive sentence is 'Man u-ra didam', (I him saw) where the object 'u' is suffixed by the morpheme 'ra', hence this sentence has a more free word order than in English. As a result, either the subject and object and the subject-object cluster can move from before the verb to after it. The latter case will be a case of VP-movement, whereas the former two are cases of topicalisation, respectively, for subject and object. The permutations and transformations reflecting these movements are as follows:

| Moving Unit | Permutation | Transformation |
|---|---|---|
| Subject | $\pi^r s \stackrel{\sigma}{\leadsto} s\pi^l$ | $A \leq B(\pi^r s)C \stackrel{(ll)}{\leadsto} A \leq B(s\pi^l)C$ |
| Object | $o^r \pi^r s \stackrel{\sigma}{\leadsto} \pi^r s o^l$ | $A \leq B(o^r \pi^r s)C \stackrel{(ll)}{\leadsto} A \leq B(\pi^r s o^l)C$ |
| Subject-and-Object | $(\pi o)^r s \stackrel{\sigma}{\leadsto} s(\pi o)^l$ | $A \leq B((\pi o)^r s)C \stackrel{(ll)}{\leadsto} A \leq B(s(\pi o)^l)C$ |

On page 151 one can find the original sentence, its three permissible variations, their derivations. A fourth less common variation 'Didam u-ra man' has two stages of movement: first the object moves to after the verb, then the subject does the same. This is derivable by first applying the object permutation $o^r \pi^r s \stackrel{\sigma}{\leadsto} \pi^r s o^l$ and then the subject permutation $\pi^r s \stackrel{\sigma}{\leadsto} s\pi^l$. The non-permissible variation 'U-ra man didam' also needs two stages of permutation, but it is not derivable since the second stage needs the unlawful permutation $so^l \overset{\sigma}{\leadsto} o^r s$. This permutation is preceded by $o^r \pi^r s \overset{\sigma}{\leadsto} \pi^r so^l$ and is meant to place back the moved object to before the verb.

As a third example, consider Latin, in which word order is relatively free, and position is used to obtain the same effect that in English is secured by emphasis or stress: the role played by a word in the sentence is shown by its ending and not by its position. The basic word order, when no particular emphasis is expressed, is (SOV) like in Persian, but in Latin texts one finds that changes from the basic order are very frequent, due to the intention of putting emphasis upon some word or phrase [10]. The *first* position in the sentence is the most emphatic, and the position next in importance is the *last* one; since the subject generally plays the most important role, it is placed first in the sentence; then the verb is the next in importance, and is placed in the last position, with the direct object in the middle. Possessive pronouns and modifying adjectives normally occur *after* the noun, but when they are emphatic they are placed *before* it, or even at the beginning of the sentence, as in the example on page 152.

In case (1), the adjective *parvam* and its head noun *casam* have swapped order. In cases (2) and (3), *parvam* has moved to the beginning of the sentence, in (3) *casam* has moreover moved to the end of the sentence. The typings of these three word sequences, after the reduction of 'Filia' with 'mea' are as follows, where, for the shake of clarity, we underline the types to which the preciclic transformations are applied to:

(1) $\pi(\underline{o^r o})o(o^r \pi^r s) \overset{(ll)}{\leadsto} \pi(oo^l)o(o^r \pi^s) \leq \pi o(o^r \pi^r s) \leq s$

(2) $(\underline{o^r o})\pi o(o^r \pi^r s) \overset{(ll)}{\leadsto} (oo^l)\underline{\pi o}(o^r \pi^r s) \overset{(rr)}{\leadsto} (oo^l)o\pi^{rr}(o^r \pi^r s) \leq \underline{o\pi^{rr}}(o^r \pi^r s) \overset{(ll)}{\leadsto} \pi o(o^r \pi^r s) \leq s$

(3) $(\underline{o^r o})\pi(o^r \pi^r s)o \overset{(ll)}{\leadsto} (oo^l)\underline{\pi(o^r \pi^r s)}o \overset{(ll)}{\leadsto} (oo^l)o(\pi(o^r \pi^r s))^{ll} \leq \underline{o(\pi(o^r \pi^r s))^{ll}} \overset{(rr)}{\leadsto} \pi(\underline{o^r \pi^r s})o \overset{(ll)}{\leadsto} \pi(\pi^r so^l) \, o \leq s$

The permutations and transformations reflecting these movements in the Latin examples are as follows:

| Moving Unit | Permutation | Transformation |
|---|---|---|
| Subject modifier | $o^r o \overset{\sigma}{\leadsto} oo^l$ | $A \leq B(o^r o)C \overset{(ll)}{\leadsto} A \leq B(oo^l)C$ |
| Object | $o^r \pi^r s \overset{\sigma}{\leadsto} \pi^r so^l$ | $A \leq B(o^r \pi^r s)C \overset{(ll)}{\leadsto} A \leq B(\pi^r so^l)C$ |
| Subject-and-Object | $(\pi o) \overset{\sigma}{\leadsto} (\pi o)^{rr}$ | $A \leq B((\pi o)s)C \overset{(rr)}{\leadsto} A \leq B(s(\pi o)^{rr})C$ |

## 6 Conclusions

We have shown that precyclicity, a restricted form of cyclicity, holds in pregroups. With precyclicity we can reason about movement and word order change in natural languages. Over-generation is avoided by introducing transformations based on precyclic permutations of types of words that allow movement. We have provided witnesses for this phenomena with examples in English, Persian, and Latin. A decision procedure, using a cut-free sequent calculus and computing the complexity of parsing in this setting, constitutes the subject of future work.

I saw him. $\quad\quad\quad \pi(\pi^r s\, o^l)\, o \leq 1\, s\, 1 = s$
$\pi\ \ (\pi^r s\, o^l)$

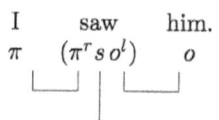

$\quad\quad\quad\quad\quad\quad\quad\quad\quad\quad\quad$ o

The subject is accessed by the verb by using the inequality $\pi\pi^r \leq 1$, similarly, the object is accessed by using the inequality $o^l o \leq 1$, and since 1 is the unit of juxtaposition the result is the type of the sentence.

He     must        love      her.
π      $(\pi^r s i^l)$   $(i o^l)$    o

(1) Her he must love.   $o \, \pi (\pi^r s i^l)(i o^l) \leq o \, (s o^l) \overset{(rr)}{\rightsquigarrow} o \, (o^r s) \leq s$

(2) Love her he must.   $(i o^l) \, o \, \pi \, (\pi^r s \, i^l) \leq i \, (s \, i^l) \overset{(rr)}{\rightsquigarrow} i \, (i^r \, s) \leq s$

I      him       saw
man    u-ra      didam.    (1) U-ra didam man. $o(o^r \pi^r s)\pi \leq (\pi^r s)\pi \overset{(ll)}{\rightsquigarrow} (s\pi^l)\pi \leq s$
π      o         $(o^r \pi^r s)$   (2) Man didam u-ra. $\pi(o^r \pi^r s)o \overset{(ll)}{\rightsquigarrow} \pi(\pi^r s o^l)o \leq s$
                                    (3) Didam man u-ra. $(o^r \pi^r s)\pi o = ((\pi o)^r s)\pi o \overset{(ll)}{\rightsquigarrow}$
                                    $(s(\pi o)^l)\pi o = (s o^l \pi^l)\pi o \leq s$

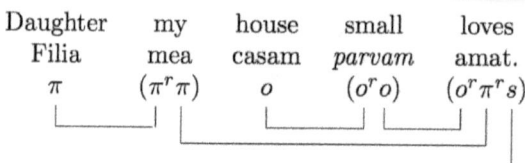

# BIBLIOGRAPHY

[1] Abrusci, M.: Phase Semantics and Sequent Calculus for Pure Noncommutative Classical Linear Propositional Logic. J Symb Logic 56, 1403–1451 (1991)

[2] Abrusci, V. M., Ruet, P.: Non-commutative Logic I: the Multiplicative Fragment. Ann. Pure Appl. Logic 101(1): 29-64 (1999)

[3] Abrusci, V. M.: Classical Conservative Extensions of Lambek Calculus. Studia Logica, 71, 277–314 (2002)

[4] Abrusci, V. M.: On Residuation. C. Casadio, B. Coecke, M. Moortgat, P. Scott (eds.), Categories ans Types in Logic, Language, and Physics: 14-27, 2014

[5] Ades, A. E., Steedman, M. J.: On the Oreder of Words. Linguistics and Philosophy 4 (4): 517-558 (1982)

[6] Buszkowski, W.: Lambek Grammars Based on Pregroups. In: De Groote, P. , Morrill, G., Retoré, C. (eds.) Logical Aspects of Computational Linguistics. LACL 2099, pp. 95–109. Springer (2001)

[7] Buszkowski, W.: Cut elimination for Lambek calculus of adjoints. In Abrusci V. M. and C. Casadio (Eds.), New perspectives in logic and formal linguistics, proceedings of the 5th Roma Workshop (pp. 85-93). Rome (2002)

[8] Buszkowski, W.: Type Logics and Pregroups. Studia Logica, 87(2/3), 145–169 (2007)

[9] Casadio, C., Lambek, J.: An Algebraic Analysis of Clitic Pronouns in Italian. In: De Groote, P., Morrill, G., Retoré, C. (eds.) Logical Aspects of Computational Linguistics. LACL 2099. Springer, Berlin 110–124 (2001)

[10] Casadio, C., Lambek, J.: A Computational Algebraic Approach to Latin Grammar, Research on Language and Computation, Volume 3, Pages 45–60, 2005.

[11] Casadio, C., Lambek, J.: A Tale of Four Grammars. Studia Logica, 71, 315-329 (2002)

[12] Casadio, C., Lambek, J. (eds.): Recent Computational Algebraic Approaches to Morphology and Syntax, Polimetrica, Milan, 2008.

[13] Casadio, C. and Sadrzadeh, M.: Clitic Movement in Pregroup Grammar: A Cross-Linguistic Approach. Lecture Notes in Computer Science, 197–214 (2011)

[14] Casadio, C. and Sadrzadeh, M.: Word Order Alternations in Sanskrit via Precyclicity in Pregroup Grammars. Horizons of the Mind. A Tribute to Prakash Panangaden: Essays Dedicated to Prakash Panangaden on the Occasion of His 60th Birthday, van Breugel, F., Kashefi, E., Palamidessi, C., Rutten, J. (eds.), Springer International Publishing, 229–249(2014).
[15] Chomsky, N.: The Minimalist Program. The MIT Press, Cambridge, Mass. (1995)
[16] Foret, A.: A modular and parameterized presentation of pregroup calculus. Journal of Information and Computation 208(5): 510-520 (2010).
[17] Francez, N., Kaminski, M: Commutation-Augmented Pregroup Grammars and Mildly Context-Sensitive Languages. Studia Logica 87(2/3), 295-321 (2007)
[18] Girard, J.Y.: Linear Logic. *Theoretical Computer Science*, 50, 1-102 (1987)
[19] Lambek, J.: The Mathematics of Sentence Structure. American Math Monthly 65, 154–169 (1958)
[20] Lambek, J.: Type Grammar Revisited. Logical Aspects of Computational Linguistics, LNAI 1582, 1–27 (1999)
[21] Lambek, J.: Type Grammars as Pregroups. Grammars 4(1), 21–39 (2001)
[22] Lambek, J.: A computational Algebraic Approach to English Grammar. Syntax 7(2), 128–147 (2004)
[23] Lambek, J.: From Word to Sentence. A Computational Algebraic Approach to Grammar. Polimetrica, Monza (MI) (2008)
[24] Lambek, J.: Exploring Feature Agreement in French with Parallel Pregroup Computations, *Journal of Logic, Language and Information* 19, 75–88 (2010)
[25] Lambek, J.: From Word to Sentence. A Computational Algebraic Approach to Grammar. Polimetrica, Monza (MI) (2008)
[26] Moortgat, M.: Categorical Type Logics. In: van Benthem, J., ter Meulen, A. (eds.) Handbook of Logic and Language, 93–177. Elsevier, Amsterdam (1997)
[27] Morrill, G.: Categorial Grammar. Logical Syntax, Semantics, and Processing. Oxford University Press, Oxford (2010)
[28] Retore, C.: Pomset Logic. A Non-commutative Extension of Classical Linear Logic. TLCA 1997: 300-318.
[29] Gillbert, C. and Retore C.: Category Theory, Logic and Formal Linguistics: some connections old and new. Applied Logic 12(1): 1–13 (2014).
[30] Sadrzadeh, M.: Pregroup Analysis of Persian Sentences. in [12].
[31] Yetter, D. N.: Quantales and (non-Commutative) Linear Logic. J Symb Logic, 55 (1990)

# Another way out of the Preface Paradox?

Gustavo Cevolani

## 1 Introduction

The so called Preface Paradox runs as follows [11]. Suppose you write a book, in which you advance a great number of claims $b_1, b_2, \ldots, b_m$. Since you can adequately defend each one of them, it seems rational for you to accept their conjunction, call it $b$. Even so, you admit in the preface that your book will contain at least a few errors. This apparently amounts to say that at least one of the claims in your book is false, i.e., that you accept the disjunctive statement $\neg b_1 \vee \neg b_2 \vee \cdots \vee \neg b_m$. But this statement is logically equivalent to $\neg b$; thus, it seems that you are entitled to rationally accept both $b$ and its negation. "Rationality, plus modesty, thus forces [you] to a contradiction" [16, p. 162].

In this note, I explore a possible way out of the Preface Paradox based on the notion, to be introduced below, of "approximate" belief: i.e., on the idea that, in some circumstances, you may assert $b$ while believing, in fact, a different statement $h$ which is "close" to $b$ (in a suitably defined sense). This idea is inspired by a solution to the Preface Paradox recently put forward by Hannes Leitgeb ("A way out of the preface paradox?", *Analysis*, 2014) which is presented in section 2. Another relevant suggestion comes from a paper by Sven Ove Hansson [9], who highlights an interesting link between the Preface Paradox and the logic of belief change. I discuss this suggestion in section 3, where an account of approximate belief is proposed. I conclude, in section 4, by briefly discussing the main differences between the present account and Leitgeb's solution.

## 2 What does the author really believe?

According to a well-known definition [16, p. 1], a paradox is a "an apparently unacceptable conclusion derived by apparently acceptable reasoning from apparently acceptable premises". Thus, solving or dissolving a paradox amounts to showing that "either the conclusion is not really unacceptable, or else the starting point, or the reasoning, has some non-obvious flaw" (*ibidem*). In our case, the line of reasoning leading to the Preface Paradox is quite clear. First, some general, background assumptions are more or less explicitly stated. They are labeled A0–A2 below:

**A0.** (Rationality) The author of the book is (ideally) rational.

**A1.** (Conjunctive closure) The beliefs of a rational author are closed under conjunction; i.e., if the author accepts $b_1, b_2, \ldots, b_m$ then he accepts $b$.

**A2.** (Consistency) The beliefs of a rational author are (logically) consistent.

Secondly, the premises of the paradox are presented:

**P1.** The author accepts $b_1, b_2, \ldots, b_m$.

**P2.** The author accepts $\neg b_1 \vee \neg b_2 \vee \cdots \vee \neg b_m$.

Then the paradox is easily derived. On the one hand, given A0, it follows from A1 and P1 that the author accepts $b$. On the other hand, P2 implies, by logic alone, that the author accepts $\neg b$. It follows, against A2, that the author accepts both $b$ and $\neg b$ (and hence their conjunction, again by A0 and A1).

In the attempt to find a way out of the Preface Paradox, most commentators have questioned either A1 or A2 as the most problematic assumptions [7, 5]. In his analysis, Leitgeb [10] focuses instead on P1, and challenges the assumption that the author of the book actually believes $b$. His idea is that, by publishing the book, the author doesn't really accept all the claims $b_1, \ldots, b_m$ in the book. Thus, he doesn't believe their conjunction $b$, but a strictly weaker claim: namely, that "the vast majority" of these claims are true. This provides a straightforward way out of the paradox, since this weaker claim is logically compatible with $\neg b$, i.e., with what the author states in the preface. More generally, Leitgeb argues that when someone makes a great number $m$ of assertions, as opposed to one or few claims, what he really believe is just that most of them are true.

More formally, let $k$ be a natural number not greater than $m$, but "sufficiently close" to $m$.[1] According to Leitgeb [10, p. 12], what the author accepts by publishing the book is not $b$, but its "statistical weakening" $S_k(b)$, defined as the disjunction of all the conjunctions of $k$ different sentences among $b_1, \ldots, b_m$.

EXAMPLE 1. In the following, I'll repeatedly make use of the toy (and, as such unrealistic) example where $m = 3$ and $k = 2$ [10, p. 12]. In this case, the statistical weakening of $b = b_1 \,\&\, b_2 \,\&\, b_3$ is

$$S_2(b_1 \,\&\, b_2 \,\&\, b_3) = (b_1 \,\&\, b_2) \vee (b_1 \,\&\, b_3) \vee (b_2 \,\&\, b_3).$$

Note that the precise value of $k$ is highly context-dependent and does not need to be explicitly stated, not even by the author of the book [10, pp. 12, 14]. In any case, as far as $k$ is smaller than $m$, $S_k(b)$ is strictly weaker than $b$ in the sense that $b$ entails $S_k(b)$, but not vice versa. Hence, $S_k(b)$ is compatible with $\neg b$, so that the author could accept both of them and still maintain the consistency of what he believes.

Leitgeb's solution is interesting also because it naturally suggests a different, more general account of the Preface Paradox. From a purely logical point of view, it is clear that any statement $h$ which, like $S_k(b)$, is compatible with $\neg b$ (and hence doesn't entail $b$) provides a way out of the paradox, if $h$ is taken to represent the "real" content of the author's beliefs. In this connection, a recent paper by Hansson

---

[1]Leitgeb [10, p. 12] assumes $1 \leq k \leq m$ but, given the intended interpretation, it seems safe to say that $k$ should be not smaller than $\frac{m}{2}$.

[9] provides a potentially fruitful suggestion. Hansson notes, in passing, that the author in the Preface Paradox apparently faces a problem of "belief contraction" as studied in the AGM theory of belief revision [9, pp. 1024–1025].[2] This means that our author initially accepts $b$ but has reasons to believe that $\neg b$ is the case; accordingly, he should give up his belief in $b$ or, in the AGM jargon, he should perform a contraction of $b$ by $b$ itself, denoted $(b-b)$. This would lead him to accept a new statement $h = (b-b)$ that is strictly weaker than $b$ and hence compatible with $\neg b$. As in the case of Leitgeb's solution, this would provide a way out of the paradox.

Hansson's suggestion, however, adds an important idea to Leitgeb's strategy of weakening $b$ in order to solve the paradox. In belief revision theory, in fact, a relevant *caveat* applies: belief contraction, and belief change in general, has to be "conservative" [8, sec. 3.5 and pp. 91 ff.]. This means that, after the change, the beliefs of the author should be as close as possible to his previous beliefs; in other words, belief change should be "minimal", in that it preserves as much as possible of the content of the original belief state.

This idea of minimal change leads us to the following proposal, inspired by both Leitgeb's solution and Hansson's suggestion. Let say that someone *approximately believes* $b$—or has an *approximate belief* in $b$—when, while asserting that $b$ is the case, he actually accept some other statement $h$ which is "close" to $b$ in some adequately defined sense (to be clarified in the next section). If $h$ is compatible with $\neg b$, but still close to $b$, this offers a solution to the Preface Paradox in line with Leitgeb's strategy. Both Leitgeb's and Hansson's proposals can then be recovered as the special cases where $h$ is, respectively, the statistical weakening $S_k(b)$ of $b$ or the contraction $(b-b)$. In the former case, "approximation" to $b$ is construed as $k$ being close to $m$, i.e., the "vast majority" of the claims $b_1, \ldots, b_m$ being true. In my proposal, what matters is not the number $k$ of purportedly true claims, but the overall closeness or similarity of $h$ to $b$. The following section shows how this notion of approximate belief can be made precise.[3]

## 3  Approximate belief

In this section, we will consider a couple of different ways of formally reconstructing the notion of approximate belief in the context of the Preface Paradox.

**Preliminaries**  To keep things simple, let's consider a propositional language $\mathcal{L}_n$ with a finite number $n$ of atomic sentences $a_1, \ldots, a_n$.[4] The constituents of $\mathcal{L}_n$ are

---

[2]The AGM account of belief revision [8] has been developed in the eighties by Carlos Alchourrón, Peter Gärdenfors, and David Makinson, and is named after them. Note that I'm not suggesting that Hansson would underwrite the proposal advanced below. Hansson is not proposing a solution to the Preface Paradox; he just highlights that what is paradoxical in this situation is exactly that the author "has reasons to contract by [$b$] but refrains from doing so since such a contraction would be cognitively unmanageable", and hence retains his belief in $b$.

[3]Philosophers of science are familiar with various notions of approximation in different contexts [12]; the need for such notions is increasingly acknowledged also in traditional and formal epistemology (see, respectively, [1, pp. 327 ff.] and [6]).

[4]All definitions in this section can be easily generalized to more complex languages, including monadic and "nomic" languages [12, 13, 4].

the $q = 2^n$ maximally informative conjunctions $c_1, \ldots, c_q$ of $\mathcal{L}_n$. Each constituent has the form $\pm a_1 \& \ldots \& \pm a_n$, where $\pm$ can be $\neg$ or nothing, and can be thought of as the most complete description of a possible world given the expressive resources of $\mathcal{L}_n$.

It is well-known that any statement $x$ of $\mathcal{L}_n$ can be expressed, in normal form, as the disjunction of all constituents in its "range", which is the set of constituents entailing $x$: $x \equiv \bigvee_{c_i \vDash x} c_i$. (For this reason, we abuse notation by letting $x$ denote both a statement and its range.) Equivalently, one may think of $x$ as the set of possible worlds in which $x$ is true. It is often instructive to consider what may be called the "conjunctive statements" of $\mathcal{L}_n$ [3]. These are finite, consistent conjunctions of "basic" statements $\pm a_i$, i.e., of atomic sentences or their negations. Constituents are a special case of conjunctive statements, containing exactly $n$ conjuncts. I will often refer to conjunctive statements in the examples below.

To make sense of the notion of approximation, one needs to introduce a distance measure $\Delta(c_i, c_j)$ defined on any pair of constituents $c_i$ and $c_j$ of $\mathcal{L}_n$, expressing the similarity or closeness between the two corresponding possible worlds. In the following, I will assume that $\Delta(c_i, c_j)$ is the normalized Hamming (also known as Dalal) distance between $c_i$ and $c_j$, i.e., the plain number of atomic sentences on which $c_i$ and $c_j$ disagree, divided by $n$.[5] There are various ways to define, on the basis of $\Delta$, the distance between a statement $x$ and a constituent $c_i$. For instance, the minimum distance $\Delta_{min}(x, c_i)$ between $x$ and $c_i$ is defined as $\min_{c_j \in x} \Delta(c_j, c_i)$, i.e., as the distance from $c_i$ of the closest constituents of $x$.

**Approximation by minimal belief contraction** For our purposes, the following notion will prove useful [13, p. 171]. Given two statements $x$ and $y$ in normal forms, the set $D_x(y)$ of the $y$-worlds closest to $x$ is defined as follows:

$$D_x(y) = \{ c_i \in y : \Delta_{min}(c_i, x) \leq \Delta_{min}(c_j, x) \text{ for all } c_j \in y \}.$$

In words, $D_x(y)$ contains all constituents in (the range of) $y$ at minimum distance from $x$. In the context of belief revision theory, this immediately provides a definition of the contraction $(x - y)$ of $x$ by $y$, as follows (*ibidem*):

$$(x - y) = \bigvee D_x(\neg y) \vee x$$

The contraction of $x$ by $y$ thus enlarges the set of possibilities admitted by $x$ with the set of the $\neg y$-worlds closest to $x$ (see below for examples).

Now, following Hansson, suppose that the author in the Preface Paradox initially believes $b$ but decides to give up his belief in $b$. In this case, the contraction $(b - b)$ will contain all the possibilities within $b$, along with all the closest possibilities "around" $b$ (see figure 1). To see the above definition at work, it is instructive to consider the special case where the claims $b_1, \ldots, b_m$ in the book are basic statements in the sense defined before, and hence $b$ is a conjunctive statement. In such case, one can check that, by giving up $b$, the author comes to believe that at least $m - 1$ of his $m$

---

[5]This assumption will significantly simplify the definitions and the examples below, but it is not essential. See [12, 13] for a more general treatment.

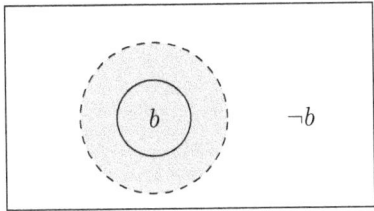

Figure 1. Each point of the rectangular surface represents a constituent or possible world. The solid circle represents the range of statement $b$. The dashed circle includes the worlds at minimum distance from $b$; the shadowed area is the contraction $(b - b)$.

beliefs are true:[6]

$$(b - b) = S_{m-1}(b) \qquad (1)$$

Thus, when $b$ is a conjunctive statement, belief contraction leads to a special case of Leitgeb's solution, where the author accepts the strongest possible statistical weakening of $b$ (with $k = m - 1$).

EXAMPLE 2. Suppose that $b$ is the conjunctive statement $b_1 \& b_2 \& b_3$. Then:

$$(b - b) = (b_1 \& b_2) \vee (b_1 \& b_3) \vee (b_2 \& b_3) = S_2(b)$$

(compare example 1 in section 2).

All other cases admitted by Leitgeb's solution, and corresponding to values of $k$ smaller than $m - 1$, are excluded here since they would result in non-conservative contractions of $b$, i.e., statements too distant from $b$. (With reference to Figure 1, such statements would be represented, for decreasing $k$, by increasingly larger circles around $b$.)

**Approximation by distance minimization** Up to this point, I followed Hansson's suggestion of reconstructing the Preface Paradox as a problem of belief change. According to this idea, and in agreement with a special case of Leitgeb's solution, the author escapes the paradox by accepting a statement $h$ which is close to $b$ in the sense that $h$ coincides with a conservative contraction of $b$. The idea of approximate belief introduced in section 2 is however more general than this, since $h$ can be close to $b$ without being a contraction of $b$. To make sense of this notion in full generality, one needs to define a measure for the distance between two arbitrary statements $x$

---

[6]Proof. Suppose that $b$ is a conjunctive statement. An arbitrary constituent $c$ belongs to $D_b(\neg b)$ iff $c$ disagrees with $b$ (otherwise it would be in the range of $b$) exactly on one of the conjuncts of $b$ (otherwise $\Delta_{min}(c, b)$ wouldn't be minimal). It follows that $b \cup D_b(\neg b)$ contains all constituents which disagree with $b$ at most on one claim of $b$. This is the range of the contraction $(b - b)$, which then says that at least $m - 1$ claims of $b$ are true.

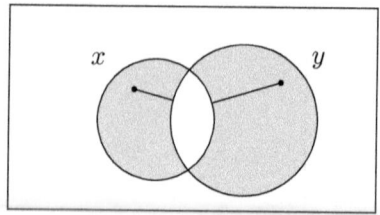

Figure 2. The symmetric difference (shadowed) between (the ranges of) $x$ and $y$. The minimum distance of an arbitrary constituent of each theory from the other theory is displayed.

and $y$ of $\mathcal{L}_n$.[7] Niiniluoto [12, p. 248] proposes the following normalized measure:

$$\delta(x,y) = \frac{\alpha}{q} \sum_{c_i \in y \setminus x} \Delta_{min}(c_i, x) + \frac{\alpha'}{q} \sum_{c_j \in x \setminus y} \Delta_{min}(c_j, y) \qquad (2)$$

where $0 < \alpha, \alpha' \leq 1$. The distance between $x$ and $y$ is thus based on the symmetric difference $(x \setminus y) \cup (y \setminus x)$ between (the ranges of) $x$ and $y$ (see section 2). If, e.g., $y$ is construed as the "target" which $x$ has to approximate, then the worlds in the symmetric difference between $x$ and $y$ reflect two kind of "errors" of $x$. The members of $y \setminus x$ can be construed as the mistaken exclusions of $x$, i.e., possibilities admitted by $y$ and wrongly excluded by $x$; while the elements of $x \setminus y$ are the mistaken inclusions of $x$, i.e., possibilities excluded by $y$ and wrongly admitted by $x$ (see also [4]). The minimum distances of all errors are then summed up, with weights $\alpha$ and $\alpha'$ reflecting the relative seriousness of the two kinds of errors. Note that $\delta(x,y)$ takes is minimal value just in case $x$ and $y$ are the same statement.

The above distance measure can be employed to define a notion of approximate belief as applied to the case of the Preface Paradox, as follows. Let us say that the author approximately believes $b$ when he accepts a statement $h$ such that:

$$h \nvDash b \text{ and } \delta(h, b) \text{ is minimal}$$

This guarantees that the author's beliefs are both close to $b$ and compatible with $\neg b$. Thus, any statement $h$ meeting the condition above provides a possible way out of the paradox.

Note that, in order to be a good approximation of $b$, $h$ has to include possibilities which are close to $b$ and exclude possibilities which are far from $b$. This is guaranteed when $h$ is chosen as a subset of $b \cup D_b(\neg b)$, since $D_b(\neg b)$ contains the closest possibilities to $b$ among those excluded by $b$ itself (see again figure 1). Thus, belief contraction turns out to be the special case where $h$ is chosen as $b \cup D_b(\neg b)$ itself. In

---

[7]Different measures of this kind have been studied in the philosophy of science literature concerning verisimilitude or truthlikeness [12, 14]. In fact, note that when $x$ is an arbitrary statement and $y$ is the true constituent of $\mathcal{L}_n$ (describing the actual world, i.e., "the whole truth" about the domain), the verisimilitude of $x$ can be defined as a decreasing function of the distance between $x$ and $y$.

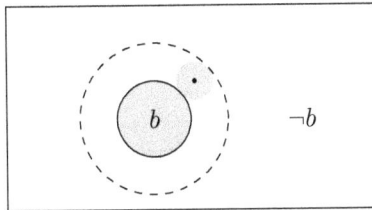

Figure 3. A statement $h$ (shadowed) at minimum distance from $b$.

general, however, $(b-b)$ will be too weak a statement to be a good approximation of $b$; accordingly, $h$ will typically be stronger than $(b-b)$. More precisely, one can check that $\delta(h,b)$ is minimized when:[8]

$$h = b \cup \{c_i\} \text{ where } c_i \in D_b(\neg b),$$

i.e., when $h$ includes all worlds in $b$ and exactly one of the worlds at minimum distance from $b$ (see figure 3).

EXAMPLE 3. Let $b$ be the conjunctive statement $b_1 \& b_2 \& b_3$ in $\mathcal{L}_4$. Then the following statements are at minimum distance from $b$:

$$b_1 \& b_2 \& (b_3 \vee b_4),$$
$$b_1 \& b_2 \& (b_3 \vee \neg b_4),$$
$$b_1 \& b_3 \& (b_2 \vee b_4),$$
$$b_1 \& b_3 \& (b_2 \vee \neg b_4),$$
$$b_2 \& b_3 \& (b_1 \vee b_4),$$
$$b_2 \& b_3 \& (b_1 \vee \neg b_4).$$

In any case, the author will keep believing two of his original claims and will suspend the judgment on the remaining one. By taking the disjunction of all the statements above, one finds again the contraction $(b-b)$ of example 2.

The example above shows that there are in general many different statements $h$ at minimum distance from $b$. In specific cases, one may think that pragmatic factors will guide the choice in favor of one or the other of these different approximations of $b$. In this connection, contracting $b$ by $b$ can be construed as the safe strategy of choosing *all* the best candidate approximations to $b$. This avoids the problem posed by their multiplicity and guarantees an unique result, $(b-b)$, which, however, is not maximally close to $b$. While sub-optimal in this sense, such a strategy may be rational, if one recalls that $(b-b)$ is after all a good approximation to $b$ if compared to other solutions, like $S_k(b)$ for low values of $k$ (cf. figure 1).

---

[8] Proof. Distance $\delta(h,b)$ is minimal when both addenda in equation 2 are minimal. For fixed values of $\alpha$ and $\alpha'$ (and a given choice of $\mathcal{L}_n$), this is guaranteed when both $\sum_{c_i \in b \setminus h} \Delta_{min}(c_i, h)$ and $\sum_{c_j \in h \setminus b} \Delta_{min}(c_j, b)$ are minimized. The former sum is minimized, and equals 0, if $h$ is chosen such that $b \vDash h$, since in that case $b \setminus h = \emptyset$. On the other hand, if, as required, $h \nvDash b$ then the latter sum cannot be zero, since $h \setminus b$ has to include at least a constituent "outside" $b$. Thus, $\delta(h,b)$ is minimized if $h$ is chosen to include just one of the closest constituents to $b$, besides those of $b$ itself.

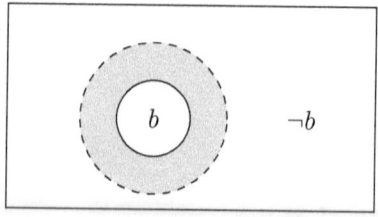

Figure 4. The revision $(b * \neg b)$ of $b$ by $\neg b$ (shadowed).

**Approximation by minimal belief revision** Before concluding, it may be instructive to consider still another strategy of determining an unique approximation $h$ of $b$. The two solutions considered above share a common feature: both the statistical weakening $S_k(b)$ and the contraction $(b-b)$ are entailed by $b$. Indeed, it is easy to check that, in order to minimize the distance from $b$, $h$ has to be a consequence of $b$ (since in this case the second addendum in equation 2 is 0). However, also statements not entailed by $b$ can be quite (although not maximally) close to $b$. In particular, it may be the case that $h$ is logically incompatible with $b$ while being close to $b$. In this connection, an interesting special case is when $h$ is the revision of $b$ by $\neg b$, i.e., the result of accepting $\neg b$ when one believes $b$. This is also a way of reconstructing the situation of an author who, having published $b$ in the book, asserts in the preface that $\neg b$ is actually the case (in line with premise P2 of the Preface Paradox).

The revision of $x$ by $y$ is defined in general as follows [13, p. 171]:

$$(x * y) = \bigvee D_x(\neg y),$$

i.e., as the set of possibilities admitted by $\neg y$ which are closest to $x$. In the present case, the revision of $b$ by $\neg b$ reduces to the worlds "around" $b$, i.e., at minimum distance from $b$ (see figure 4).

EXAMPLE 4. If $b$ is the conjunctive statement $b_1 \& b_2 \& b_3$ then:

$$(b * \neg b) = (b_1 \& b_2 \& \neg b_3) \vee (b_1 \& \neg b_2 \& b_3) \vee (\neg b_1 \& b_2 \& b_3)$$

In short, the author believes that exactly one of the claims in the book is false, the others being true.

As in the case of contraction, revision turns out to be a special case of distance minimization, where $h$ is chosen as $D_b(\neg b)$. In general, however, $h$ will differ from both the contraction and the revision of $b$ (as the foregoing examples show). Still, contraction and revision provide two instructive illustrations of approximate belief, especially as far as the Preface Paradox is concerned. These correspond to two alternative ways of of approximating $b$ through minimal belief changes, which lead in turn to two alternative ways out of the paradox. The first, contraction-based solution amounts to choose $h$ such that $h$ entails neither $b$ nor $\neg b$; this amounts to questioning both premises P1 and P2 of the paradox, since in this case the author believes neither the conjunction of the claims in the book nor its negation. The second, revision-based solution is to say that $h$ entails $\neg b$, so that the author indeed accepts $\neg b$ and rejects $b$; in turn, this amounts to rejecting P1 while fully endorsing P2. In this connection, both Leitgeb's proposal from section 2 and the one based on distance minimization favor the former, contraction-based solution over the latter.

## 4 Concluding remarks

In this paper, I followed Leitgeb's idea that when someone makes a great number of different claims $b_1, \ldots, b_m$, he doesn't actually accept their conjunction $b$ but some weaker statement $h$. I also argued that $h$ should be construed as a good approximation of $b$, or that $h$ should be close to $b$. These notions of approximation and closeness can be made precise once a distance measure among the possibilities in the logical space is defined. In the case of the Preface Paradox, the author of the book approximately believes $b$ in the sense that he accepts a statement $h$ which doesn't entail $b$ but still is close as possible to $b$.

As shown in section 3, $h$ doesn't coincide, in general, with Leitgeb's statistical weakening $S_k(b)$ of $b$, or with the minimal changes of $b$ obtained through contraction and revision. In fact, $h$ will be closer to $b$ than each of the three statements $S_k(b)$, $(b-b)$, and $(b*\neg b)$, which are all too weak to be good approximations of $b$. Still, these weaker statements, and especially the latter two, may be plausible approximations of $b$ in some contexts, since they sometimes uniquely determine the actual beliefs of the author (as in the simple examples from section 3). On the contrary, as already observed, there are in general many statements $h$ which are maximally close to $b$; in this sense, the notion of approximate belief as defined here may be "cognitively unmanageable" [9, p. 1024] and hence less psychologically plausible than those alternatives. In other words, the author may be unable to specify the statement $h$ which he really believes; and this may be the reason why, in the book, he actually asserts $b$ [10, p. 14].

In any case, the main conceptual difference between the account proposed here and Leitgeb's solution has to do with the notion of belief itself. As Leitgeb [10, p. 14] notes, his solution of the paradox has the advantage of allowing the author to accept $S_k(b)$ with high confidence, in the sense that the probability of $S_k(b)$ can be high even if the probability of $b$ is very low. This depends on the fact that $S_k(b)$ is a much weaker statement than $b$, and probability is inversely related to logical strength (in the sense that if $x$ entails $y$ then $x$ cannot be more probable than $y$). On the contrary, approximation as defined here is positively correlated with logical strength at least in the following sense. In order to be a good approximation of $b$, $h$ has to hold in roughly the same set of possible worlds (cf. figure 2); this means that $h$ will entail most of the consequences of $b$ (recall the principle of conservatism of the AGM theory). Accordingly, as compared to $S_k(b)$, $h$ is much stronger, and hence a less probable statement.

As a consequence, if belief requires high probability, $h$ (as well as $b$ itself) cannot be really believed by the author of the book. On the other hand, it is well-known that the "high probability" view of belief is problematic, and the Preface Paradox is often used exactly to show that it is untenable [7]. For this reason, it is useful to consider other notions of belief or acceptance, compatible with the possibility of believing also propositions which are not highly probable. One such notion is adopted within the fallibilist tradition in epistemology and philosophy of science, according to which even our best beliefs (e.g., scientific hypotheses) are typically false or highly improbable [15, 12]. The idea of approximate belief defended here apparently provides an account of the Preface Paradox in line with this tradition (for a related but different treatment see [2]). While publishing $b_1, \ldots, b_m$ in his book, the author is conscious of his own fallibility, and, accordingly, doesn't believe that $b$ is actually the case. Still, he remains committed to the claim that $b$ is, so to speak, roughly the case. This situation may be understood by saying that what the author really believes is a statement $h$, which, while not highly probable, is a good approximation of what the author in fact asserts (i.e., $b$). In this way, this notion of approximate belief provides another possible solution to the Preface Paradox, alternative to Leitgeb's one.

## Acknowledgements

I thank an anonymous reviewer for useful comments on the previous version of this paper; financial support from the priority program *New Frameworks of Rationality*, SPP 1516 (Deutsche Forshungsgemeinshaft, grant CR 409/1-2), andthe FIRB project *Structures and Dynamics of Knowledge and Cognition* (Italian Ministry of Scientific Research, Turin unit, D11J12000470001) is gratefully acknowledged.

## BIBLIOGRAPHY

[1] R. Audi. *Epistemology: a contemporary introduction to the theory of knowledge*. Routledge, New York, 3$^{rd}$ edition, 2011.
[2] G. Cevolani. Fallibilism, verisimilitude, and the Preface Paradox. *Erkenntnis*, 2016, forthcoming.
[3] G. Cevolani, V. Crupi, and R. Festa. Verisimilitude and belief change for conjunctive theories. *Erkenntnis*, 75(2):183–202, 2011.
[4] G. Cevolani, R. Festa, and T. A. F. Kuipers. Verisimilitude and belief change for nomic conjunctive theories. *Synthese*, 190(16):3307–3324, 2013.
[5] D. Christensen. *Putting Logic in its Place: Formal Constraints on Rational Belief*. Oxford University Press, 2004.
[6] L. Decock, I. Douven, C. Kelp, and S. Wenmackers. Knowledge and Approximate Knowledge. *Erkenntnis*, 79(6):1129–1150, 2014.
[7] R. Foley. Beliefs, degrees of belief, and the lockean thesis. In F. Huber and C. Schmidt-Petri, editors, *Degrees of Belief*, pages 37–47. Springer, Dordrecht, 2009.
[8] P. Gärdenfors. *Knowledge in Flux: Modeling the Dynamics of Epistemic States*. MIT Press, Cambridge (Massachusetts), 1988.
[9] Sven Ove Hansson. Bootstrap contraction. *Studia Logica*, 101(5):1013–1029, 2013.
[10] H. Leitgeb. A way out of the preface paradox? *Analysis*, 74(1):11–15, 2014.
[11] D. C. Makinson. The paradox of the preface. *Analysis*, 25(6):205–207, June 1965.
[12] I. Niiniluoto. *Truthlikeness*. Reidel, Dordrecht, 1987.
[13] I. Niiniluoto. Revising beliefs towards the truth. *Erkenntnis*, 75(2):165–181, 2011.
[14] G. Oddie. Truthlikeness. In Edward N. Zalta, editor, *The Stanford Encyclopedia of Philosophy*. Summer 2014 edition, 2014.
[15] K. R. Popper. *Conjectures and Refutations: the Growth of Scientific Knowledge*. Routledge and Kegan Paul, London, 3$^{rd}$ edition, 1963.
[16] R. M. Sainsbury. *Paradoxes*. Cambridge University Press, 3$^{rd}$ edition, 2009.

# Characterizing Logical Consequence in Paraconsistent Weak Kleene

Roberto Ciuni and Massimiliano Carrara

ABSTRACT. In this paper we present Parconsistent Weak Kleene (PWK), a logic that first appeared in the works of Sören Halldén and Arthur Prior, and we establish a characterization result for PWK-consequence, thus providing necessary and sufficient conditions for $B$ to be a consequence of $\Gamma$ in PWK.

## 1 Introduction

In [7] and [15], Sören Halldén and Arthur Prior independently discuss a logic based on the following three tenets: (a) there are cases where classical truth value assignment is not possible, (b) in such cases, the presence of a third, non-classical, truth value propagates from one sentence to any compound sentence including it, and finally, (c) valid inferences go from non-false premises to non-false conclusions. The so-called Weak Kleene Logic (or Bochvar Logic) is built in accordance with tenets (a)–(b), but it assumes that classical truth is the only value to be preserved by valid inference.[1] If we endorse (c) and include the non-classical value among the designated values, we get a *paraconsistent counterpart* of Bochvar Logic, that we call *Paraconsistent Weak Kleene* or PWK for short.[2]

In this paper, we give a characterization result of the relation of logical consequence in propositional PWK, that is, we provide necessary and sufficient conditions for a formula $B$ to be the logical consequence of a set $\Gamma$ of formulas.

There are two main rationales for this result. First, our result has a general mathematical interest in the areas of three-valued logics. Indeed, few results have been provided on PWK, but an exploration of the formalism reveals interesting connections with Relevant Logic. Second, our result generalizes a result by Paoli [12], that considers syntactical restrictions that obtain by imposing the First-Degree-Entailment (FDE for short) requirements to PWK. It is thus of interest in relation

---

[1] For Bochvar Logic, see [4], [10] and [16].

[2] In this paper, we are using the label 'paraconsistent Kleene logic' as short for 'paraconsistent counterpart of a Kleene Logic'. This use is suggested by the fact that paraconsistent logics as Priest's Logic of Paradox LP and the present PWK obtain by keeping the 'strong matrices' introduced by [8] and the 'weak matrices' by [4] and [8], respectively, and extending the set of designated elements as to include the non-classical value. Our choice does not presuppose anything more, since paraconsistency does not belong to the range of applications for which Kleene Logics have been designed (which included phenomena of *underdetermination*, by contrast). We use the label PWK accordingly.

to existing background.

The paper proceeds as follows: in section 2 we introduce PWK and its relation of logical consequence. In section 3 we prove the characterization result and in section 4 we discuss its relevance against existing scientific background. In section 5, we discuss some features of PWK and we briefly look at its connection with Relevant Logic.

## 2   Paraconsistent Weak Kleene

The logic we discuss here dates back to [7] by Sören Halldén, where it is proposed as a *logic of non-sense* (an umbrella-term that, in Halldén's usage, included logical paradoxes, vagueness, ambiguity).[3] Prior used PWK as the propositional fragment of the modal logic Q (see [15]), that he proposed as a quantified modal logic for *contingently non-existing entities*.

Here we will not discuss the cogency of the readings by Halldén and Prior, since this lies beyond the aim of this paper. Let us just clarify two points, though.

Halldén and Prior do not use the name PWK, and they do not explore much the formal properties of the apparatus they introduce. However, two points make it crystal-clear that they are using PWK as their propositional logic. First, they use the language and the semantics we are going to use to interpret the propositional connectives, (though they also extend the language with further operators). Second, they accept (a)–(c). Remarkably, Halldén and Prior do not seem to notice that the apparatus they are using is *paraconsistent*, but as we know, in every many-valued Kleene logic that endorses (c), contradictions are satisfiable. With this said, we can go to the logic PWK.

The language $\mathcal{L}$ of PWK consists of the set $\{\neg, \vee, \wedge\}$ of connectives (negation, disjunction and conjunction) and the set Atom of atomic sentences $\{p, q, r \ldots\}$. The *arbitrary formulas* $A, B, C, D, \ldots$ of PWK are defined by the usual recursive definition. We denote the set of such formulas by Form and use Greek upper-case letters $\Gamma, \Phi, \Psi, \Sigma, \ldots$ to denote sets of arbitrary formulas. Given a formula $A$, we define the set $\mathsf{Atom}(A) := \{p \mid p \in \mathsf{Atom} \text{ and } p \text{ occurs in } A\}$ of the atomic sentences (occurring) in $A$. We also follow the standard definition of the set $\mathsf{Sub}(A) := \{B \mid B \in \mathsf{Form} \text{ and } B \text{ occurs in } A\}$ of the subformulas of $A$, the set $\mathsf{Atom}(\Gamma) := \{p \mid p \in \mathsf{Atom}(A) \text{ for some } A \in \Gamma\}$ of the atoms of formulas in $\Gamma$, and the set $\mathsf{Sub}(\Gamma) := \{B \mid B \in \mathsf{Sub}(A) \text{ for some } A \in \Gamma\}$ of the subformulas of formulas in $\Gamma$. Clearly, $\mathsf{Atom}(\Gamma) \subseteq \mathsf{Sub}(\Gamma)$.

The semantics of PWK comprises a non-classical value **n** beside the two values **t** and **f** of classical logic CL, as all Kleene logics or paraconsistent counterparts of them—the label '**n**' here is indeed short for 'non-classical'. Formulas of $\mathcal{L}$ are assigned a truth value by the evaluation function $V : \mathsf{Atom} \longmapsto \{\mathbf{t}, \mathbf{n}, \mathbf{f}\}$ from atomic sentences to truth-values. We generalize truth-assignments to arbitrary formulas as follows:

---

[3]See also [16] for the proposal by Halldén.

DEFINITION 1. *A valuation $V$ : Form $\longmapsto \{\mathbf{t},\mathbf{n},\mathbf{f}\}$ is the unique extension of a mapping $V$ : Atom $\longmapsto \{\mathbf{t},\mathbf{n},\mathbf{f}\}$ that is induced by the truth tables in Table 1.*

Table 1. Truth Tables for PWK

| | $\neg A$ | $A \vee B$ | t | n | f | $A \wedge B$ | t | n | f |
|---|---|---|---|---|---|---|---|---|---|
| t | f | t | t | n | t | t | t | n | f |
| n | n | n | n | n | n | n | n | n | n |
| f | t | f | t | n | f | f | f | n | f |

Table 1 encodes the typical feature of Bochvar's logic: for every truth function $f$ corresponding to a connective in the language, if any input of $f$ is the non-classical value, so is the output. In a nutshell, $\mathbf{n}$ transmits from any component $B$ of a formula to the entire formula $A$, regardless of the connectives appearing in $A$. Table 1 also reveals that we could have introduced $\wedge$ as a derived symbol: the definition $A \wedge B := \neg(\neg A \vee \neg B)$ is adequate, since definiens and definiendum have exactly the same truth tables (we leave this easy exercise to the reader). A striking feature of the language is that no conditional is present. We will define a possible candidate below. Whether such a candidate can fit minimal criteria for the conditional or not, we will briefly discuss in Remark 6. In any case, we will feel free to take the derived operator as a notational convenience.

We let $\mathcal{V}_{\mathsf{PWK}} = \{V \mid V : \text{Form} \longmapsto \{\mathbf{t},\mathbf{n},\mathbf{f}\}\}$ be the set of valuations of PWK. The following fact will be helpful in what follows:

FACT 2. *For all formulas $A$ in $\mathcal{L}$ and valuation $V \in \mathcal{V}_{\mathsf{PWK}}$, $V(A) = \mathbf{n}$ iff $V(B) = \mathbf{n}$ for some $B \in \mathsf{Sub}(B)$ iff $V(p) = \mathbf{n}$ for some $p \in \mathsf{Atom}(A)$.*

**Proof.** The left-to-right (LTR) direction is trivial: as in every three-valued Kleene logic (or paraconsistent counterpart), if a formula $A$ whatever has the non-classical value, at least one of its components has it. By applying this line of reasoning, we reach a smallest possible component, namely an atomic sentence, having the non-classical value. The right-to-left (RTL) direction immediately follows from the fact that $\mathbf{n}$ transmits from smaller components to entire formulas no matter what connectives are involved. Also, this feature implies that, if $V(p) = \mathbf{n}$ for some $p \in \mathsf{Atom}(A)$, then $V(A) = \mathbf{n}$. The interesting point is that this holds *no matter of* what $V(q)$ is for any $q \in \mathsf{Atom}(B)/\{p\}$. This will be relevant in the proof of Theorem 14. ∎

We let $\mathcal{D} = \{\mathbf{t},\mathbf{n}\}$ be the set of *designated values* of PWK. With this at hand, we define *satisfaction*, *dissatisfaction* and *satisfiability*, together with the notion of the class of valuations that satisfy all formulas in a given set:

DEFINITION 3.

1. *An evaluation $V \in \mathcal{V}_{\mathsf{PWK}}$ satisfies a formula $A$ iff $V(A) \in \mathcal{D}$.*

2. An evaluation $V \in \mathcal{V}_{\mathsf{PWK}}$ *dissatisfies a formula $A$ iff $V(A) = \mathbf{f}$.*

3. *A formula $A$ is satisfiable iff there exists an evaluation $V \in \mathcal{V}_{\mathsf{PWK}}$ such that $V(A) \in \mathcal{D}$.*

4. $\mathcal{V}(\Gamma) := \{V \mid V \in \mathcal{V}_{\mathsf{PWK}} \text{ and } V(A) \in \mathcal{D} \text{ for all } A \in \Gamma\}$

These definitions are standard, but they will prove convenient in what follows. Logical consequence is defined as usual:

DEFINITION 4 (Logical Consequence).
$\Gamma \models_{\mathsf{PWK}} B$ *iff every valuation $V \in \mathcal{V}_{\mathsf{PWK}}$ that satisfies all formulas $A \in \Gamma$ also satisfies $B$.*

We write $A, B \models_{\mathsf{PWK}} C$ for $\{A, B\} \models_{\mathsf{PWK}} C$, and '$A$ is valid' is defined as $\varnothing \models_{\mathsf{PWK}} B$.

***Notable failures.*** As expected by a many-valued logic that designates more than one value, the relation of logical consequence for PWK (hereafter, PWK-consequence) does not coincide with that of Classical Logic (since now on, CL-consequence).[4] In particular, PWK shares some failures of cases of CL-consequence together with the famous Logic of Paradox LP—see [14]—which also designates a non-classical value and is based on the so-called Strong Kleene Matrix. Let us define $A \to B := \neg A \vee B$. We will discuss below whether this connective can really count as a conditional, but for the time being let us just use it as a convenient device. Here are some notable failures:

| | | | |
|---|---|---|---|
| 1 | $A, \neg A \vee B \not\models B$ | $A, A \to B \not\models B$ | MP |
| 2 | $\neg B, \neg A \vee B \not\models \neg A$ | $\neg B, A \to B \not\models \neg A$ | MT |
| 3 | $\neg A \vee B, \neg B \vee C \not\models \neg A \vee C$ | $A \to B, B \to C \not\models A \to C$ | TR $\to$ |
| 4 | $\neg A \vee (B \wedge \neg B) \not\models \neg A$ | $A \to (B \wedge \neg B) \not\models \neg A$ | RAA |
| 5 | $A \wedge \neg A \not\models B$ | | ECQ |

As for 1, suppose $V(A) = \mathbf{n}$ and $V(B) = \mathbf{f}$. This suffices to have the premises designated, but the conclusion undesignated. By switching those two values between $A$ and $B$, we get a countermodel for 2. The versions with $\to$ make it crystal-clear that the rules failing are Modus Ponens (MP) and Modus Tollens (MT), respectively. As for 3, suppose $V(A) = \mathbf{t}$, $V(B) = \mathbf{n}$ and $V(C) = \mathbf{f}$: we will have the premises designated and the conclusion false. This is failure of the Transitivity of $\to$, as is clear by trading $\to$ for the appropriate combinations of $\neg$ and $\vee$. Coming to 4, $V(A) = \mathbf{t}$ and $V(B) = \mathbf{n}$ falsifies the rule, which is nothing but Reductio ad Absurdum (RAA). Finally, any valuation $V$ such that $V(A) = \mathbf{n}$ and $V(B) = \mathbf{f}$ will falsify Ex Contradictione Quodlibet, thus making PWK a paraconsistent logic.

A little familiarity with Priest's Logic of Paradox LP suffices to see that PWK share failures 1–5 with LP (see [2] and [14]). A distinctive feature of PWK, however, is failure of Conjunction Simplification (CS):

---

[4]See [10, 66].

$$6 \quad A \wedge B \not\models B \qquad \text{CS}$$

$V(A) = V(A \wedge B) = \mathbf{n}$ and $V(B) = \mathbf{f}$ is enough to falsify CS. This marks a crucial difference with LP, where CS is a valid rule.

**Notable Validities.** It is easy to check that the following formulas are valid in PWK:

$$
\begin{array}{ll}
7 & (A \wedge (A \to B)) \to B \\
8 & (\neg B \wedge (A \to B)) \to \neg A \\
9 & ((A \to B) \wedge (B \to C)) \to (A \to C) \\
10 & (A \to (B \wedge \neg B)) \to \neg A \\
11 & (A \wedge \neg A) \to B \\
12 & (A \wedge B) \to B
\end{array}
$$

These formulas are verified by every valuation $V \in \mathcal{V}_{\text{PWK}}$ that assign classical values ($\mathbf{t}$ or $\mathbf{f}$) to their antecedents—this equates with no subformula in the antecedent having value $\mathbf{n}$, as clear by Table 1. If any subformula whatever in the antecedent is assigned $\mathbf{n}$, the antecedent itself is assigned $\mathbf{n}$ (once again, Table 1 suffices to check this). Due to the definition of $\to$ and the truth table of disjunction, this suffices for the entire conditional to have value $\mathbf{n}$ and be designated. But the two cases above are the only possible in PWK.

The above helps establish that the Deduction Theorem does not hold for PWK:

**FACT 5.** *It is not the case that* $\models_{\text{PWK}} A \to B$ *iff* $A \models_{\text{PWK}} B$

**Proof.** Clearly, $\models_{\text{PWK}} A \to B$ can hold and yet $A \models_{\text{PWK}} B$ can fail, as is clear from validities 7–12 and failures 1–6. Of course, one direction of the Deduction Theorem holds: if $A \models_{\text{PWK}} B$, then $\models_{\text{PWK}} A \to B$. ∎

REMARK 6 (Conditional in PWK). *Whether $\to$ can really play the role of a conditional depends on the features we want a conditional to have. Validation of MP is usually included in the pack, and so failure 1 above, in its $\to$-version, would answer for the negative. However, some researchers from the many-valued tradition have recently argued that MP is not meaning-constitutive for the conditional (see [3]). We will not survey the debate here. Suffice it to say that, no matter what stance of the two above one takes, lack of a detachable conditional is not fatal to PWK: as for its kin LP, such a conditional can be added. One way (among many others) to do that, for example, is to extend the connectives of PWK with the detachable conditional from $\text{RM}_3$, that is a formalism related to Relevant Logic[5] and for which*

---

[5] The acronym RM in the name of the logic points at the result of adding the *M*ingle Axiom $(A \to (A \to A))$ to (a system of) *R*elevant Logic. The reason why $\text{RM}_3$ cannot be considered a system of Relevant Logic is that from its *mingle axiom* the formula $\neg(A \to A) \to (B \to B)$ is derivable, which does not satisfy the variable sharing properties that in turn defines Relevant Logic.

a three-valued semantics is also designed.[6] Notice that, by contrast, the reading of $\neg$ and $\vee$ as negation and disjunction, respectively, is unproblematic: $\neg$ inverts classical values and keeps the non-classical one fixed, as every negation does in Kleene logics and paraconsistent cognates; $\vee$ restitutes a designated formula any time one of its disjunct is designated. Whether $\wedge$ can be read as a conjunction is of course a natural question, in light of failure 7 (and of Fact 9 below). In order to keep the presentation compact, we defer the issue to section 5.

## 3 Characterizing PWK-Consequence

We prove some facts about PWK-consequence before going to the characterization result. Let us first go through the relations between the tautologies of PWK and classical tautologies:

FACT 7. $\models_{\mathsf{PWK}} A$ iff $\models_{\mathsf{CL}} A$

**Proof.** Take the class $\mathcal{V}_{\mathsf{CL}}$ of valuations of CL (or 'classical valuations'). It is clear by Table 1 that $\mathcal{V}_{\mathsf{CL}} \subset \mathcal{V}_{\mathsf{PWK}}$: in particular, those valuations $V \in \mathcal{V}_{\mathsf{PWK}}$ where no atomic sentence $p$ is assigned value $\mathbf{n}$ will be classical valuations. $\mathcal{V}_{\mathsf{CL}} \subseteq \mathcal{V}_{\mathsf{PWK}}$ proves the LTR direction. As for the RTL direction: take a formula $A$ that is valid in CL and suppose that it is not valid in PWK. This means that there is a PWK-valuation $V$ is such that $V(A) = \mathbf{f}$. We can easily construct a corresponding $V' \in \mathcal{V}_{\mathsf{CL}}$ retaining the value of $A$ from $V$. But this implies that some classical valuation falsifies $A$, thus contradiction the initial hypothesis. ∎

Let us now explore monotonicity. On the one hand, PWK-consequence is *monotonic*:

FACT 8. If $\Gamma \models_{\mathsf{PWK}} B$, then $\Gamma, A \models_{\mathsf{PWK}} B$

**Proof.** Due to $\mathcal{V}(\Gamma \cup \{A\}) \subseteq \mathcal{V}(\Gamma)$. ∎

On the other hand, we have that

FACT 9. It can be the case that $A_1, \ldots, A_n \models_{\mathsf{PWK}} B$ and $A_1 \wedge \cdots \wedge A_n \not\models B$

**Proof.** For instance, suppose $B$ is $A_n$. All valuations $V$ such that, for all $i \in \{1, \ldots, n\}$, $V(A_i) \in \mathcal{D}$ suffices to verify $A_1, \ldots, A_n \models_{\mathsf{PWK}} B$, while a valuation where $V(A_n) = \mathbf{f}$ and $V(A_j) = \mathbf{n}$ for all $j \in \{1, \ldots, n-1\}$ suffices to have $V(A_1 \wedge \cdots \wedge A_n) \in \mathcal{D}$ and $V(B) = \mathbf{f}$, thus implying that $A_1 \wedge \cdots \wedge A_n \not\models B$. This is possible because the set $\mathcal{V}(\{A_1, \ldots, A_n\})$ of valuations satisfying *all* formulas in $\{A_1, \ldots, A_n\}$ may not coincide with the set $\mathcal{V}(A_1 \wedge \cdots \wedge A_n)$—the latter may also include valuations where *some* of $A_1, \ldots, A_n$ is undesignated, on condition that at least one of them has value $\mathbf{n}$. ∎

---

[6]The troubles with adding a detachable conditional to paraconsistent Kleene logics arises only in the context of paraconsistent truth theory: many of the proposed conditionals are detachable but also validate absorption, which opens the way for Curry Paradox. But as far as the truth predicate does not enter the language, many different detachable conditionals will do.

Fact 9 tells us that conjunction appearing in the language does not behave as the comma appearing in the metalanguage: the latter *releases* all the premises, while the former may not release all the conjuncts. Thus, in the premises of an inference in PWK, we cannot trade the comma for the conjunction. In sum, the comma proves stronger than PWK's conjunction.

## 3.1 Characterization Result

We now provide necessary and sufficient conditions for a formula $B$ to be a PWK-consequence of a set $\Gamma$ of formulas. We call this a *characterization* of PWK-consequence and the relative result we call a *characterization result*. In order to prove the characterization result, we first go through some preliminary issues and results. First, we individuate two *necessary conditions* for $B$ to be a PWK-consequence of $\Gamma$. Let $\models_{\mathsf{CL}}$ be the standard relation of classical consequence:

FACT 10. *If $\models_{\mathsf{PWK}} B$, then $\Gamma \models_{\mathsf{PWK}} B$*

By Fact 8.

FACT 11. *If $\Gamma \models_{\mathsf{PWK}} B$, then $\Gamma \models_{\mathsf{CL}} B$*

**Proof.** Suppose it were not so: there would be a classical valuation $V \in \mathcal{V}_{\mathsf{CL}}$ such that $V(B) = \mathbf{f}$ and $V(A) = \mathbf{t}$ for every $A \in \Gamma$. But since $\mathcal{V}_{\mathsf{CL}} \subset \mathcal{V}_{\mathsf{PWK}}$, this would contradict $\Gamma \models_{\mathsf{PWK}} B$. ∎

PROPOSITION 12.
*If $\Gamma \models_{\mathsf{PWK}} B$ and $\not\models_{\mathsf{PWK}} B$, then there is at least a non-empty set $\Gamma' \subseteq \Gamma$ of formulas such that $\mathsf{Atom}(\Gamma') \subseteq \mathsf{Atom}(B)$.*

**Proof.** By contraposition. Assume $\not\models_{\mathsf{PWK}} B$ and $\mathsf{Atom}(\Gamma') \not\subseteq \mathsf{Atom}(B)$ for all non-empty sets $\Gamma' \subseteq \Gamma$. The latter implies $\mathsf{Atom}(\Gamma) \not\subseteq \mathsf{Atom}(B)$. We have three possible cases here:

1. $\mathsf{Atom}(\Gamma) \cap \mathsf{Atom}(B) = \emptyset$

2. $\mathsf{Atom}(\Gamma) \supset \mathsf{Atom}(B)$

3. $\mathsf{Atom}(\Gamma) \cap \mathsf{Atom}(B) \neq \emptyset$ and $\mathsf{Atom}(\Gamma)/\mathsf{Atom}(B) \neq \emptyset$

In the first case, there is a valuation $V \in \mathcal{V}_{\mathsf{PWK}}$ such that $V(A) \in \mathcal{D}$ for all $A \in \Gamma$ and $V(B) = \mathbf{f}$. As for the other two cases, take the set $\mathsf{Atom}(\Gamma)/\mathsf{Atom}(B)$—which is non-empty in both cases. For every $A \in \Gamma$ we have that $\mathsf{Atom}(A) \cap \mathsf{Atom}(\Gamma)/\mathsf{Atom}(B) \neq \emptyset$. Indeed, $\{A\} \in \Gamma$, and from this and the initial hypothesis, $\mathsf{Atom}(A) \not\subseteq \mathsf{Atom}(B)$; but since $\mathsf{Atom}(\Gamma) = \mathsf{Atom}(B) \cup \mathsf{Atom}(\Gamma)/\mathsf{Atom}(B)$, $\mathsf{Atom}(A) \not\subseteq \mathsf{Atom}(B)$ implies $\mathsf{Atom}(A) \subseteq \mathsf{Atom}(\Gamma)/\mathsf{Atom}(B)$. Since the valuation of the atoms in $\mathsf{Atom}(A) \cap \mathsf{Atom}(\Gamma)/\mathsf{Atom}(B)$ is independent from that of the atoms in $\mathsf{Atom}(B)$, there is a valuation $V \in \mathcal{V}_{\mathsf{PWK}}$ such that $V(p) = \mathbf{n}$ for all $p \in \mathsf{Atom}(A) \cap \mathsf{Atom}(\Gamma)/\mathsf{Atom}(B)$ and all $A \in \Gamma$, and such that $V(B) = \mathbf{f}$. By Fact 2, this valuation is such that $V(A) = \mathbf{n}$ for all $A \in \Gamma$, while $V(B) \notin \mathcal{D}$. This proves $\Gamma \not\models_{\mathsf{PWK}} B$. As a consequence, if $\Gamma \models_{\mathsf{PWK}} B$ and $\not\models_{\mathsf{PWK}} B$, then $\mathsf{Atom}(\Gamma') \subseteq \mathsf{Atom}(B)$ for at least a set $\Gamma' \subseteq \Gamma$ of formulas. ∎

**PROPOSITION 13.** If $\Gamma \models_{\mathsf{PWK}} B$ and $\not\models_{\mathsf{PWK}} B$, then $\Gamma' \models_{\mathsf{CL}} B$ for some non-empty $\Gamma' \subseteq \Gamma$ such that $\mathsf{Atom}(\Gamma') \subseteq \mathsf{Atom}(B)$.

**Proof.** By contraposition. Take the set $\Gamma^+ = \{\Gamma' \mid \Gamma' \subseteq \Gamma \text{ and } \mathsf{Atom}(\Gamma') \subseteq \mathsf{Atom}(B)\}$, whose existence and non-emptyness are guaranteed by Proposition 12. Suppose $\Gamma' \not\models_{\mathsf{CL}} B$ for all $\Gamma' \in \Gamma^+$. This implies $\Gamma' \not\models_{\mathsf{PWK}} B$ for all $\Gamma' \in \Gamma^+$. Take now the set $\Gamma^- = \{\Gamma'' \mid \Gamma'' \subseteq \Gamma \text{ and } \mathsf{Atom}(\Gamma'') \not\subseteq \mathsf{Atom}(B)\}$. Clearly, there is a valuation $V \in \mathcal{V}_{\mathsf{PWK}}$ such that $V(B) = \mathbf{f}$ and for all $\Gamma'' \in \Gamma^-$, $V(p) = \mathbf{n}$ for some $p \in \mathsf{Atom}(\Gamma'')/\mathsf{Atom}(B)$. As a consequence, $\Gamma'' \not\models_{\mathsf{PWK}} B$. But of course, there will also be a valuation $V' \in \mathcal{V}_{\mathsf{PWK}}$ such that $V'(B) = \mathbf{f}$ and for all $\Gamma''' \in \Gamma^+ \cup \Gamma^-$, $V(q) = \mathbf{n}$ for some $q \in \mathsf{Atom}(\Gamma''')$. But since $\Gamma^+ \cup \Gamma^- = \Gamma$, this implies $\Gamma \not\models_{\mathsf{PWK}} B$. ∎

With this at hand, we are ready to prove our *characterization result*:

**THEOREM 14.**

$$\Gamma \models_{\mathsf{PWK}} B \text{ iff } \Gamma \models_{\mathsf{CL}} B \text{ and } \begin{cases} \models_{\mathsf{PWK}} B, & \text{or} \\ \mathsf{Atom}(\Gamma') \subseteq \mathsf{Atom}(B) & \text{for at least a non-empty} \\ & \Gamma' \subseteq \Gamma \text{ s.t. } \Gamma' \models_{\mathsf{CL}} B. \end{cases}$$

**Proof.** The LTR direction immediately follows from Fact 11 and Proposition 13. As for the RTL direction, we prove it in two steps. Let us first assume that $\models_{\mathsf{PWK}} B$ holds—notice that this suffices to have $\Gamma \models_{\mathsf{CL}} B$, by Fact 10 and Fact 11. Given the assumption, we have $\Gamma \models_{\mathsf{PWK}} B$ by Fact 10. Let us now assume $\not\models_{\mathsf{PWK}} B$, $\Gamma \models_{\mathsf{CL}} B$ and $\mathsf{Atom}(\Gamma') \subseteq \mathsf{Atom}(B)$ for at least a $\Gamma' \subseteq \Gamma$ s.t. $\Gamma' \models_{\mathsf{CL}} B$. Then we have two possible cases: either all the assignments to the premises are classical, or at least some atom in them has value $\mathbf{n}$. From $\Gamma \models_{\mathsf{CL}} B$ and $\mathcal{V}_{\mathsf{CL}} \subseteq \mathcal{V}_{\mathsf{PWK}}$, for all $\Gamma' \subseteq \Gamma$ such that $\Gamma' \models_{\mathsf{CL}} B$ and valuation $V$ such that $V(A) = \mathbf{t}$ for all $A \in \Gamma'$, we will have $V(B) \in \mathcal{D}$. Suppose now that $V(A) = \mathbf{n}$ for at least a $A \in \Gamma'$, where $\Gamma' \subseteq \Gamma$, $\mathsf{Atom}(\Gamma') \subseteq \mathsf{Atom}(B)$ and $\Gamma' \models_{\mathsf{CL}} B$. By Fact 2, $V(A) = \mathbf{n}$ implies $V(p) = \mathbf{n}$ for at least one $p \in \mathsf{Atom}(A)$, and by this, $\mathsf{Atom}(\Gamma') \subseteq \mathsf{Atom}(B)$ and again Fact 2, we have that $V(B) = \mathbf{n}$. Thus, we have $\Gamma' \models_{\mathsf{PWK}} B$. But this implies that $\Gamma \models_{\mathsf{PWK}} B$ also holds by monotonicity of PWK-consequence (Fact 8). ∎

Theorem 14 explains all the failures 1–6: those inferences do not satisfy the necessary and sufficient criteria by the theorem. The paradigmatic case is the failure of MP: clearly, in such a rule the atomic sentences in the premises are a superset of the atomic sentences in the conclusion.

Notice that Theorem 14 provides an adequate characterization of logical consequence even in case the set of premises is empty, though in this situation the characterization will be trivial. Indeed, when $\Gamma = \emptyset$, the condition stated by theorem will reduce to

$$\Gamma \models_{\mathsf{PWK}} B \text{ iff } \Gamma \models_{\mathsf{CL}} B \text{ and } \models_{\mathsf{PWK}} B.$$

which is guaranteed by Fact 7—the fact characterizing tautologies in PWK. On the one hand, the collapse holding when $\Gamma = \emptyset$ can make our characterization from Theorem 14 look odd, but on the other hand, this does not conflict with the adequacy of the characterization, which has also the virtue of being the most general possible.

Before closing, we focus on the inclusion requirement specified in Theorem 14 for a (nonempty) subset $\Gamma'$ of the set of premises $\Gamma$. An interesting feature here is that the proviso that $\Gamma' \models_{CL} \Gamma$ cannot be dropped without compromising the result. For instance, take the classically valid inference $C \wedge \neg C, A \vee B \models_{CL} (A \vee B) \wedge D$. The inference is *not* valid in PWK—as is easy to check—even though a subset (namely, $A \vee B$) of the set of premises satisfies the inclusion requirement. Theorem 14 implies that, in order to have a case of a PWK-consequence, at least a subset satisfies the inclusion requirement *and* is in the relation of classical consequence with the conclusion. None of the possible subsets of $\{C \wedge \neg C, A \vee B\}$ satisfy both conditions w.r.t. $(A \vee B) \wedge D$. Thus, the proviso that $\Gamma' \models_{CL} \Gamma$ is essential for the characterization.

This marks an important difference with the characterization proposed by [16] for logical consequence in Bochvar's logic (see below).

## 4 Discussion

Theorem 14 generalizes the result proved by [12]. There, Paoli considers FDE-formalisms connected to a variety of logics, including PWK. In particular, he introduces the logic H, which is PWK augmented with the (standardly defined) entailment connective $\Rightarrow$ from FDE-formalisms. The logics PWK and H are related by the fact: $(\star)$ $\models_H A \Rightarrow B$ iff $A \models_{PWK} B$,[7] where $A \Rightarrow B$ is a standardly defined FDE-entailment.[8] Paoli proves:

$$\models_H A \Rightarrow B \text{ iff } A \models_{CL} B \text{ and either } \models_{CL} B \text{ or } \mathsf{Atom}(A) \subseteq \mathsf{Atom}(B).$$

Due to $(\star)$, Paoli's result[9] turns to be a special case of our one. In particular, our result generalizes Paoli's in two respects. First, it shows that the same characterization can be given if we consider the full language $\mathcal{L}$. Indeed, a straightforward corollary of Theorem 14 is:

COROLLARY 15.
$A \models_{PWK} B$ iff $A \models_{CL} B$ and *either* (i) $\models_{PWK} B$ or (ii) $\mathsf{Atom}(A) \subseteq \mathsf{Atom}(B)$.

Second, Theorem 14 has a wider generality, since it establishes a characterization for multiple-premise consequence, while the result in [12] is illuminating just if we confine ourselves to single-premise consequence. In particular, Paoli's result highlights the role of the atom inclusion requirement in PWK-consequence, but it cannot show how exactly this role is played when we have more than one premise. Indeed, if we extend our consideration to multiple-premise consequence, then the

---
[7] See [12].
[8] There are many different ways to characterize FDE-logics and -fragments. Here, we find it natural to follow the one adopted in [12].
[9] See Theorem 1 of [12].

simple atom-inclusion condition presented in Corollary 15 and Paoli's result does not suffice for a characterization: $C, A \models_{\mathsf{PWK}} A \vee B$ holds, but of course it may be that $\mathsf{Atom}(\{C, A\}) \nsubseteq \mathsf{Atom}(A \vee B)$—suppose $C$ is $p$, $A$ is $q$ and $B$ is $r$. The condition $\mathsf{Atom}(\Gamma) \subseteq \mathsf{Atom}(B)$ alone is not the right one for the multiple-premise case. And yet atom-inclusion still plays a decisive role in PWK-consequence, as is proved by $A, A \to B \nvDash_{\mathsf{PWK}} B$. The methodology underlying our more general Theorem 14 is indeed to check for a subset $\Gamma'$ of the premises that satisfies the atom inclusion appearing in Corollary 15. Since by Fact 8 consequentiality transmits to $\Gamma$, the theorem allows us to capture the multiple-premise cases of PWK-consequence. Thus, Theorem 14 offers a full understanding of the atom-inclusion condition and its impact in determining the class of sets of formulas/formula pairs that are in the relation of PWK-consequence.

Our theorem proves interesting also in light of established results in Kleene logics and related systems. The characterization of consequence in Bochvar Logic ($\models_{\mathsf{B}}$) by [16, Theorem 2.3.1] also includes an inclusion condition: $\Gamma \models_A \phi$ iff $\Gamma \models_{\mathsf{CL}} A$ and every atom in $A$ occurs in some formula from $\Gamma$. Thus, the characterization *reverses* the atom inclusion condition presented in our Theorem 14. Notice that no counterpart of the 'subset condition' from Theorem 14 is needed for Bochvar Logic, and thus the two multiple-premise consequence relations are not exact duals.

Finally, *containment logics* ([13, 6]) also impose a condition of variable inclusion on logical consequence. The direction of the inclusion is usually the same as in Bochvar's logics usually, but a recent paper in this tradition also investigates the reverse direction, which characterizes PWK-consequences (see [5]).

## 5 Open Problems and Directions

We close this paper with a look at open problems and directions on the topic of PWK-conjunction $\wedge$. The connective shares a crucial feature of *compatibility operators*, which tell them apart from standard conjunction: compatibility operators do not simplify, exactly as PWK-conjunction. More precisely, PWK-conjunction displays some similarity with the *fusion* operator ∘ from Relevant Logic.[10]

The interesting point is that, semantics of choice aside, the behavior of $\wedge$ does not entirely reduce to that of fusion.[11] One the one hand, ∘ shares with PWK-conjunction the failure of CS, and it is easy to proof the both can be introduced when each of the conjuncts is proved separately. However, there are also notable differences between ∘ and our $\wedge$. The first is not *idempotent*, while the second is (that is, $A \circ A \nvDash A$ and $A \wedge A \models_{\mathsf{PWK}} A$)—see [11, 168].[12] Also, Fact 9 does not hold for ∘ (see again [11, 167]): the connective is indeed introduced to guarantee an equivalence with the comma of multiple premises (which is lost for standard conjunction in Relevant Logic). Finally, *fusion* is intended as a dual of *implication*

---

[10]The provenance of the operator ∘ can be traced out of the tradition of Relevant Logic, and precisely in [9], where it is explicitly proposed as a *compatibility operator*.

[11]Our comparison with fusion is based on [11, 166–168].

[12]Notice, however, that in the system RM and similar systems closely related to Relevant Logic, fusion is actually idempotent (see [1]).

in Relevant Logic, that is $A \circ B := \neg(A \to \neg B)$ where $\neg$ and $\to$ are (some) relevant negation and conditional, respectively.[13] It is easy to see that, in PWK, we could set $A \wedge B := \neg(A \to \neg B)$ for the conditional introduced in section 2; however, the latter (should it qualify as an acceptable conditional) falsifies the Deduction Theorem (see Fact 5). It is then questionable that the definition above characterizes a compatibility/implication pair.

At the same time, $A \wedge B \models_{\mathsf{PWK}} A \vee B$—we leave this to the reader—and together with CS, this points at the compatibility of $A$ and $B$ *and* the actual availability of one of them. The corresponding reading of $\wedge$ would be '$A$ and $B$ are compatible and one of them actually holds', which also fit with the idempotence of $\wedge$.

Whether PWK-conjunction can really be read as a compatibility operator depends on the elaboration of an *intensional semantics* that captures the behavior of $\wedge$ as defined by the three-valued semantics above, while at the same time providing truth conditions for $\wedge$ that prove conceptually insightful.[14] We believe that this semantics can be obtained by elaborating on the Routley-Meyer semantics for Relevant Logic, and we plan to explore this issue in some future research.

## Acknoweldgements
The authors wish to thank an anonymous referee for their helpful comments.

## BIBLIOGRAPHY
[1] Anderson Alan R. and Belnap Nuel (1975) Entailment. The Logic of Relevance and Necessity, Princeton, Princeton University Press.
[2] Beall JC (2011) 'Multiple-conclusion LP and Default Classicality', Review of Symbolic Logic, 4/2: 326–336.
[3] Beall JC (2013) 'Free of Detachment: logic, rationality and gluts', Nous, article first published online.
[4] Bochvar Dmitri A. (1938) 'On a Three-Valued Calculus and its Application in the Analysis of the Paradoxes of the Extended Functional Calculus', Matamaticheskii Sbornik, 4: 287–308.
[5] Ferguson Thomas S. (2015) Logic of Nonsense and Parry Systems, Journal of Philosophical Logic, 44/1: 65–80.
[6] Fine Kit (1986) Analytic Implication, Notre Dame Journal of Formal Logic, 27/2: 169–179.
[7] Halldén Sören (1949) The Logic of Nonsense. Uppsala, Uppsala University.
[8] Kleene Stephen C. (1952) Metamathematics. Amsterdam, North Holland.
[9] Lewis Clarence I. (1918) A Survey of Symbolic Logic. Berkeley, CA, University of California Press.
[10] Malinowski Grzegorz (2007) 'Many-Valued Logic and its Philosophy', in Gabbay Dov and Woods John (eds.) Handbook of the History of Logic, volume 8, Amsterdam, North-Holland, pp. 13–94.
[11] Mares Edwin (2004) Relevant Logic. A Philosophical Interpretation, Cambridge, Cambridge University Press.
[12] Paoli Francesco (2007) 'Tautological Entailments and their Rivals, in Bezieau Jean Yves, Carnielli Walter, Gabbay Dov (eds.) Handbook of Paraconsistency, London, College Publications, pp. 153–175.
[13] Parry William T. (1933) Ein Axiomensystem für eine neue Art von Implikation (analytische Implikation). In *Ergebnisse eines mathematischen Kolloquiums*, 4: 5–6.
[14] Priest Graham (2006) In Contradiction, Oxford, Oxford University Press (2nd edition).
[15] Prior Arthur (1967) Past, Present and Future, Oxford, Oxford University Press.

---

[13]Notice that this definition is adequate just in some relevant systems. However, these systems capture the original rationale for the introduction of $\circ$.

[14]Notice that this does not equate to turn PWK into a system of Relevant Logic: to this purpose, it is necessary to add informational incompleteness and avoid the paradoxes of implication.

[16] Urquhart Alasdair (2002) 'Basic Many-Valued Logic', in Gabbay Dov and Guenthner Friederich (eds.) Handbook of Philosophical Logic, volume 2, Dordrecht, Kluwer, pp. 249–296.

# Logic of Implicit and Explicit Justifiers

Alessandro Giordani

ABSTRACT. The aim of this paper is to provide an intuitive semantics for systems of justification logic which allows us to cope with the distinction between implicit and explicit justifiers. The paper is subdivided into three sections. In the first one, the distinction between implicit and explicit justifiers is presented and connected with a proof-theoretic distinction between two ways of interpreting sequences of sentences; that is, as sequences of axioms in a certain set and as proofs constructed from that set of axioms. In the second section, a basic system of justification logic for implicit and explicit justifiers is analyzed and some significant facts about it are proved. In the final section, an adequate semantics is proposed, and the system is proved to be sound and complete whit respect to it.

*Keywords:* justification logic; epistemic logic; implicit justification; explicit justification; Fitting semantics.

## 1 Introduction

Justification logic is one of the most interesting developments of epistemic logic[1]. It extends the expressive power of the language of standard epistemic logic by introducing sentences like $t : \varphi$, to be intended as $\varphi$ is justified in virtue of $t$, or $t$ is a justifier for $\varphi$. Axioms for systems of justification logic can be introduced from different points of view. A first approach is to rest on our basic intuitions concerning how justifiers are related with both propositions and other justifiers. A slightly different approach is to focus on principles that characterize well-known systems of logic which are strictly connected with the structure of justification, such as systems of provability logic[2]. In fact, in standard systems of provability logic, a sentence like $\Box \varphi$ is interpreted as stating that $\varphi$ is provable in some mathematical base theory, so that there is a proof of $\varphi$ in that theory. Thus, a sentence stating that $t$ is a justifier for $\varphi$ is intuitively interpreted as stating that $t$ refers to a proof of $\varphi$. However, this is not the sole interpretation of a sentence like that. In particular, if $\mathcal{A}$ is a set of logical and non-logical axioms, then two options concerning the way of interpreting that $t$ is a justifier for $\varphi$ are available.[3]

---

[1][1], [3], and [10] are excellent introductions to this topic. In these works, a number of applications of systems of justification logic for the study of the notions of evidence and justification in epistemology are also provided.

[2]See [7] for an extensive introduction to systems of provability logic and their representation in modal logic.

[3]In what follows, I assume that proofs are constructed in Hilbert style systems where *modus ponens* is the only primitive rule.

**Option 1.** A sentence stating that $t$ is a justifier for $\varphi$ says that $t$ refers to a proof of $\varphi$ from $\mathcal{A}$. In this case, $t$ refers to a finite sequence of sentences, with final sentence $\varphi$, where every sentence either is some axiom in $\mathcal{A}$ or is obtained from previous sentences by applying *modus ponens*.

**Option 2.** A sentence stating that $t$ is a justifier for $\varphi$ says that a proof of $\varphi$ is obtainable from a sequence $t$ of theorems. In this case, $t$ refers to a finite sequence of sentences, where every sentence is derivable from axioms in $\mathcal{A}$, from which a proof of $\varphi$ can be constructed.

Hence, while in the first case $t$ refers to an explicit proof of $\varphi$, in the second case it refers to the basic sentences from which such a proof can be constructed and, in particular, to the basic axioms that can be used to prove it.

The first interpretation gives rise to a general notion of *explicit* justifier, which is extremely intuitive, since it is based on the idea that it is possible to identify what sentences are justified by $t$ by just considering the structure of $t$. In fact, since $t$ refers to a proof, all the sentences that are involved in $t$ are certainly justified by $t$. In the light of this, I will use the standard notation $t : \varphi$ to say that $t$ refers to a proof of $\varphi$. By contrast, the second interpretation gives rise to a general notion of *implicit*, or potential, justifier, according to which $t$ is a justifier for all the sentences that are contained in the logical closure of the axioms contained in $t$. I will use the notation $[t]\varphi$ to say that $t$ refers to a sequence of sentences from which $\varphi$ is provable.

*Remark 1:* The notion of potential justification is to be distinguished from the notion of possible justification. Indeed, every sentence that is provable from $\mathcal{A}$ has a proof exploiting a certain set of axioms in $\mathcal{A}$, but it is not true that every sentence that is so provable has a proof exploiting *the same set* of axioms in $\mathcal{A}$. Thus, the notion of potential justification is more fine-grained than the notion of possible justification.

*Remark 2:* The notion of explicit justification is distinguished from the notion of potential justification. Indeed, every sentence that is provable from a sequence of theorems from $\mathcal{A}$ is the final sentence in a proof consisting in a sequence of theorems from $\mathcal{A}$, but it is not true that every sequence of theorems from $\mathcal{A}$ gives rise to the proof of a unique sentence. In general, while $t : \varphi$ implies $[t]\varphi$, for every $t$, it is not true that $[t]\varphi$ implies $t : \varphi$, for every $t$.

I find both the first and the second interpretations worth of investigation, even if only the first one has received a systematic treatment in the current research on justification logic[4]. In the following sections, I will develop a system of logic where both assertions of explicit justification, like $t : \varphi$, and assertions of implicit justification, like $[t]\varphi$, are treated in a unified framework[5]. In particular, my two

---

[4] See [2], [4], and [9], for a survey of different directions in which the logic of justification can be developed.

[5] In [5], an interesting analysis of the distinction between implicit and explicit justifiers is proposed, but the notion of implicit justification is not distinguished from the notion of possible justification. As a consequence, there is no way of articulating the state described by a sentence like $[t]\varphi$. In [11], the distinction between $t : \varphi$ and $[t]\varphi$ is present, but the semantic analysis of $[t]\varphi$, as we will see, is not completely satisfactory.

main aims are to provide an axiomatization of the previous notions and to introduce a suitable semantics for them. Accordingly, in the next section, a basic system of justification logic for implicit and explicit justifiers is offered and some significant facts about it are proved, while in the final section a suitable semantics is proposed, and the system is proved to be sound and complete with respect to it.

## 2 Axiomatic characterization

Let us start with introducing an adequate axiomatic system for capturing both the notion of explicit justification and the notion of implicit justification. Let us call the basic system **IEJ**. The standard language of a system of justification logic is characterized by two set of rules, specifying the set of terms and the set of formulas of the language.[6] The language of **IEJ** is characterized in the same way. The set of terms and formulas are defined according to the following grammar.

$t := j \mid c \mid t+s \mid t \times s \mid !t$, where $j$ is a variable and $c$ is a constant for justifiers

$\varphi := p \mid \neg\varphi \mid \varphi \wedge \psi \mid [t]\varphi \mid t : \varphi$, where $p$ is a variable for propositions

The operators $+$, $\times$, and $!$ are used to construct new justifiers from basic ones. As usual, $t+s$ is interpreted as a justifier providing justification for all the sentences that can be justified either by $t$ or by $s$, while $t \times s$ is interpreted as a justifier providing justification for all the sentences that can be justified by applying *modus ponens* to premises justified by $t$ and by $s$. In addition, $!$ is a justification checker that returns a justifier $!t$ for the sentence stating that $t$ is a justifier for $\varphi$, provided that $t$ is indeed such a justifier. Finally, a justification sentence like $[t]\varphi$ is interpreted as $t$ is an implicit justifier for $\varphi$, whereas a justification proposition like $t : \varphi$ is interpreted as $t$ is an explicit justifier for $\varphi$.

### 2.1 Axioms

The basic system **IEJ** is constituted by three groups of axioms: the first group is a standard system for classical propositional logic, while the two other groups are introduced in order to characterize explicit and implicit justifications. It is worth noting that axioms are considered as a priori justified, so that any epistemic agent accepts logical axioms, including the ones concerning justification, as immediately evident. This intuition is made precise by introducing a *constant specification*, which can be construed according to the following definition.

**Definition 1:** *Constant specification.*

Let $CS!$ be the set of $c : \varphi$, such that $c$ is a constant for justifiers and $\varphi$ is an axiom of **IEJ**. Then, a *constant specification* $CS$ is a subset of $CS!$ and an *axiomatically appropriate constant specification* is a constant specification where, for all the axioms $\varphi$ of **IEJ**, there is a constant $c$ such that $c : \varphi \in CS$.

In particular, we will only work with axiomatically appropriate constant specifications. In this way, every logical axiom is associated with a justification constant, witnessing that the axiom is accepted as justified. Thus, let $CS$ be an axiomatically

---

[6]See [1], [9], and [10], for a detailed exposition, and [3] for the connection between operators on justifiers and operators on proofs within the context of the logic of provability.

appropriate constant specification. Then, **IEJ** is characterized, relative to $CS$, by the following axioms.

**Group 1:** classical tautologies and *modus ponens*.

For the notion of explicit justification, let us use the standard axioms provided in [8] and [10].

**Group 2:** axioms concerning explicit justification and *internalization rule*.

**EJ1:** $t : (\varphi \to \psi) \to (s : \varphi \to t \times s : \psi)$

**EJ2:** $t : \varphi \vee s : \varphi \to t+s : \varphi$

**EJ3:** $t : \varphi \to\, !t : (t : \varphi)$

**RJ:** $c : \varphi$, where $\varphi$ is an axiom in **IEJ** such that $c : \varphi \in CS$.

Group 2 includes the axioms which characterize the standard notion of explicit justification. **EJ1** states that, given two justifiers, $t$ and $s$, a justifier like $t \times s$ provides justification to any sentence that can be derived from implications justified by $t$ and sentences justified by $s$ by applying *modus ponens*. The idea is that *modus ponens* is the basic deduction rule and that propositional deduction is accepted by the epistemic agent as providing justification. **EJ2** states that given two justifiers, $t$ and $s$, a justifier like $t + s$ provides justification to any proposition justified by either $t$ or $s$. **EJ3** states that justification is internally accessible, so that all justified propositions can be acknowledged as such. Finally, **RJ** allows us to have axioms justified by basic justifiers.

For the notion of implicit justification, I will use the set of axioms provided in [11].

**Group 3:** axioms and rules concerning implicit justification.

**IJ1:** $[t](\varphi \to \psi) \to ([s]\varphi \to [t \times s]\psi)$

**IJ2:** $[t]\varphi \vee [s]\varphi \to [t+s]\varphi$

**IJ3:** $[t]\varphi \to [!t][t]\varphi$

**IJ4:** $t : \varphi \to [t]\varphi$

**IJ5:** $[c]\varphi \to [t]\varphi$, where $c$ is a constant

**IJ6:** $[t \times t]\varphi \leftrightarrow [t+t]\varphi \leftrightarrow [!t]\varphi \leftrightarrow [t]\varphi$

Group 3 includes the axioms which characterize an intuitive notion of implicit justification. The first three axioms state that the notion of implicit justification is similar to the notion of explicit justification, as far as the basic operations are concerned. In particular, **IJ2** captures the idea that the logical closure of a certain set of sentences is included in the logical closure of any set of sentences that includes that first set. **IJ4** states that what is explicitly justified by $t$ is implicitly justified by the same justifier. Indeed, if $t$ refers to a proof of a certain sentence, then that sentence is certainly contained in the logical closure of the set of sentences in $t$. Hence, the idea that any set of sentences is included in its logical closure is respected. **IJ5** states that the axioms, which are a priori justified, are always implicitly justified by any justifier, since they are contained in the logical closure of any set of sentences. Finally, **IJ6** says that $t \times t$, $t + t$, $!t$, and $t$ provide implicit justification to the same propositions. This axiom captures the idea that what is implicitly justified by $t$ is precisely what can be inferred from sentences in $t$, so that nothing new is implicitly justified when inferences are performed from sentences in $t$. Hence, the idea that the logical closure of the logical closure of a set of sentences is included in the logical closure of that set is respected. In conclusion, the crucial properties of a logical closure operator $Cn$

1. $X \subseteq Cn(X)$
2. $Cn(Cn(X)) \subseteq Cn(X)$
3. $X \subseteq Y \Rightarrow Cn(X) \subseteq Cn(Y)$

are incorporated in the treatment of any implicit justification operator.[7]

## 2.2 Theorems

In **IEJ**, some fundamental theorems are derivable, which concern rules for explicit and implicit justification. In particular, we get the following crucial rules.

**REJ:** $\vdash_{\mathbf{IEJ}} \varphi \Rightarrow \vdash_{\mathbf{IEJ}} t : \varphi$, for *some* term $t$.

The proof is by induction on the length of the derivation.
 Suppose $\varphi$ is an axiom. Then, $\vdash_{\mathbf{IEJ}} c : \varphi$, for some constant $c$, by **RJ**. Suppose $\varphi$ is obtained by an application of **RJ**. Then, $\varphi = c : \psi$, for some $c$ and some axiom $\psi$. Hence, $\vdash_{\mathbf{IEJ}} !c : (c : \psi)$, by **EJ3**, and so $\vdash_{\mathbf{IEJ}} !c : \varphi$. Suppose $\varphi$ is obtained by an application of *modus ponens* to $\psi \to \varphi$ and $\psi$. Then, by induction hypothesis, $\vdash_{\mathbf{IEJ}} t : (\psi \to \varphi)$ and $\vdash_{\mathbf{IEJ}} s : \psi$, for some $t$ and $s$. Hence, $\vdash_{\mathbf{IEJ}} t \times s : \varphi$, by **EJ1**.

**REJ** is a rule of explicit justification, stating that every theorem of **IEL** is justified by some justifier. **REJ** is a version of a *non-standard* rule of necessitation, since not every theorem is justified by the same term $t$. Hence, a modality like $t :$ is not a standard modality.

**RIJ:** $\vdash_{\mathbf{IEJ}} \varphi \Rightarrow \vdash_{\mathbf{IEJ}} [t]\varphi$, for *every* term $t$.

The proof is again by induction on the length of the derivation.
 Suppose $\varphi$ is an axiom. Then, $\vdash_{\mathbf{IEJ}} [t]\varphi$, for every term $t$, by **RJ**, **IJ4** and **IJ5**. Suppose $\varphi$ is obtained by an application of **RJ**. Then, $\varphi = c : \psi$, for some

---

[7] See Tarski [12], chapters V and XII. To be more precise, while property 1 is reflected by axiom **IJ4** and property 2 is reflected by axiom **IJ2**, property 3 is reflected by $[t]\varphi \to [t][t]\varphi$, which is a consequence of axioms **IJ3** and **IJ6**.

$c$ and some axiom $\psi$. Hence, $\vdash_{\mathbf{IEJ}} !c : (c : \psi)$, by **EJ3**, $\vdash_{\mathbf{IEJ}} [!c](c : \psi)$, by **IJ4**, $\vdash_{\mathbf{IEJ}} [c](c : \psi)$, by **IJ6**, and so $\vdash_{\mathbf{IEJ}} [t](c : \psi)$, by **IJ5**. Suppose $\varphi$ is obtained by an application of *modus ponens* to $\psi \to \varphi$ and $\psi$. Then, by induction hypothesis, $\vdash_{\mathbf{IEJ}} [t](\psi \to \varphi)$ and $\vdash_{\mathbf{IEJ}} [t]\psi$, for every $t$. Hence, $\vdash_{\mathbf{IEJ}} [t]\varphi$, by **IJ1**.

**RIJ** is a rule of implicit justification, stating that every theorem of **IEL** is justified by every justifier. **RIJ** is a version of a *standard* rule of necessitation, since every theorem is justified by the same term $t$. Hence, a modality like $[t]$ might be a standard modality. In fact, the next proposition shows that $[t]$ actually is a standard modality.

**KIJ:** $\vdash_{\mathbf{IEJ}} [t](\varphi \to \psi) \to ([t]\varphi \to [t]\psi)$, for *every* term $t$.

$\vdash_{\mathbf{IEJ}} [t](\varphi \to \psi) \to ([t]\varphi \to [t \times t]\psi)$ by **IJ1**

$\vdash_{\mathbf{IEJ}} [t](\varphi \to \psi) \to ([t]\varphi \to [t]\psi)$ by **IJ6**

Finally, we are also able to obtain the following propositions.

**IJ7:** $\vdash_{\mathbf{IEJ}} [s]\varphi \to [t \times s]\varphi$, for *every* term $s$.

$\vdash_{\mathbf{IEJ}} \varphi \to \varphi$ axiom in group 1

$\vdash_{\mathbf{IEJ}} c : (\varphi \to \varphi)$ by **RJ**

$\vdash_{\mathbf{IEJ}} [c](\varphi \to \varphi)$ by **IJ4**

$\vdash_{\mathbf{IEJ}} [t](\varphi \to \varphi)$ by **IJ5**

$\vdash_{\mathbf{IEJ}} [t](\varphi \to \varphi) \to ([s]\varphi \to [ttimess]\varphi)$ by **IJ1**

$\vdash_{\mathbf{IEJ}} [s]\varphi \to [t \times s]\varphi$ by logic

**IJ8:** $\vdash_{\mathbf{IEJ}} [t]\varphi \to [t \times s]\varphi$, for *every* term $s$.

$\vdash_{\mathbf{IEJ}} \varphi \to ((\varphi \to \varphi) \to \varphi)$ axiom in group 1

$\vdash_{\mathbf{IEJ}} [t](\varphi \to ((\varphi \to \varphi) \to \varphi))$ by **IJ1**

$\vdash_{\mathbf{IEJ}} [t]\varphi \to [t]((\varphi \to \varphi) \to \varphi)$ by **KIJ**

$\vdash_{\mathbf{IEJ}} [t]((\varphi \to \varphi) \to \varphi) \to ([s](\varphi \to \varphi) \to [t \times s]\varphi)$ by **IJ1**

$\vdash_{\mathbf{IEJ}} [t]\varphi \to ([s](\varphi \to \varphi) \to [t \times s]\varphi)$ by logic

$\vdash_{\mathbf{IEJ}} [t]\varphi \to [t \times s]\varphi$ by group 1, **IJ4**, **IJ5**, and logic

Hence, by **IJ7** and **IJ8**, a modality like $[t \times s]$ is both stronger than $t$ and stronger than $s$, in accordance with its intended interpretation.

Now, it is worth noting that, in the light of **RIJ** and **KIJ**, every $[t]$ is a standard modality. This suggests a new semantics for the basic system of logic for explicit and implicit justification, which is more insightful than the semantics proposed in [11]. To be sure, the new semantics fits the intuition that, while explicit operators can be modeled by means of syntactic assignments, implicit operators are to be modeled by means of conditions on the set of possible epistemic states.

## 3  Semantic characterization

The semantic framework for standard systems of justification logic is due to Fitting [9]. In Fitting semantics, a frame is a tuple $\langle W, \mathcal{R}, \mathcal{E} \rangle$, where $W$ is a non-empty set of states, $\mathcal{R}$ is a transitive relation on $W$, and $\mathcal{E}$ is a function from states and justifiers to sets of formulas. Within this framework, explicit justification is modeled by introducing a syntactic function that, given a justifier $t$ and a possible world $w$, selects the set of all formulas for which $t$ provides explicit justification at $w$. In particular, $\varphi \in \mathcal{E}(w,t)$ states that, at $w$, $t$ is a justifier that can serve as possible evidence for $\varphi$. In a similar way, we might model implicit justification by introducing a function that, given a justifier $t$ and a possible world $w$, selects the set of all formulas for which $t$ provides implicit justification at $w$.[8] Hence, a frame is a tuple $\langle W, \mathcal{R}, \mathcal{E}, \mathcal{E}^* \rangle$, where

- $W$ is a non-empty set of states
- $\mathcal{R} \subseteq W \times W$ is transitive
- $\mathcal{E}$ is such that $\mathcal{E}(w,t)$ is a set of formulas, for every $w$ and $t$
- $\mathcal{E}^*$ is such that $\mathcal{E}^*(w,t)$ is a set of formulas, for every $w$ and $t$

In addition, $\mathcal{E}$ and $\mathcal{E}^*$ must satisfy the following constraints.

1. Conditions on $\mathcal{E}$.

   $\varphi \to \psi \in \mathcal{E}(w,t)$ and $\varphi \in \mathcal{E}(w,s) \Rightarrow \psi \in \mathcal{E}(w, t \times s)$

   $\mathcal{E}(w,t) \cup \mathcal{E}(w,s) = \mathcal{E}(w, t+s)$

   $\varphi \in \mathcal{E}(w,t) \Rightarrow t:\varphi \in \mathcal{E}(w, !t)$

   $\mathcal{R}(w,v) \Rightarrow \mathcal{E}(w,t) \subseteq \mathcal{E}(v,t)$

2. Conditions on $\mathcal{E}^*$.

   $\varphi \to \psi \in \mathcal{E}^*(w,t)$ and $\varphi \in \mathcal{E}^*(w,s) \Rightarrow \psi \in \mathcal{E}^*(w, t \times s)$

   $\mathcal{E}^*(w,t) \cup \mathcal{E}^*(w,s) = \mathcal{E}^*(w, t+s)$

   $\varphi \in \mathcal{E}^*(w,t) \Rightarrow [t]\varphi \in \mathcal{E}^*(w, !t)$

   $\mathcal{E}(w,t) \subseteq \mathcal{E}^*(w,t)$

   $\mathcal{E}^*(w,c) \subseteq \mathcal{E}^*(w,t)$, for every $c$

   $\mathcal{E}(w, t \times t) = \mathcal{E}^*(w, t+t) = \mathcal{E}^*(w, !t) = \mathcal{E}^*(w,t)$

   $\mathcal{R}(w,v) \Rightarrow \mathcal{E}^*(w,t) \subseteq \mathcal{E}^*(v,t)$

Once these conditions are posed, one can prove a completeness theorem for **IEJ**.[9] To be sure, the conditions on $\mathcal{E}$ and $\mathcal{E}^*$ are introduced precisely for ensuring the soundness of the axioms in group 2 and 3.

---
[8]This is the strategy pursued in [11].
[9]See [11], section 4.2.

## 3.1 A new semantics for implicit and explicit justification

An apparent limitation of the previously introduced semantic framework is that implicit and explicit justifications are modeled in the same way. In particular, while it is normal to model the set of explicitly justified propositions by means of a selection function like $\mathcal{E}$, since we do not expect such a set to be closed with respect to any logical rule, it is not intuitive to model the set of implicitly justified propositions by means of a selection function like $\mathcal{E}^*$, since, in this case, we do expect such a set to be closed with respect to the logical rules, and indeed **RIJ** and **KIJ** confirm our expectation. Hence, it should be more appropriate to develop the logic of implicit justification by means of conditions linking epistemic states, which are the standard tools for treating implicit epistemic modalities. The rest of this section is thus dedicated to develop this kind of semantics.

**Definition 2:** Basic frame for **IEJ**.

A basic frame for **IEJ** is a tuple $\langle W, \mathcal{S}, \mathcal{E} \rangle$, where $W$ is a set of epistemic states, $\mathcal{S}$ is a function that assigns to every $w \in W$ and every term $t$ a set of states $\mathcal{S}(w,t)$, and $\mathcal{E}$ is a function that assigns to every $w \in W$ and every term $t$ a set of formulas $\mathcal{E}(w,t)$. In addition, $\mathcal{S}$ and $\mathcal{E}$ must satisfy the following conditions.

1. Conditions on $\mathcal{S}$

    *S1*: $\mathcal{S}(w, t \times s) \subseteq \mathcal{S}(w,t) \cap \mathcal{S}(w,s)$

    *S2*: $\mathcal{S}(w, t + s) \subseteq \mathcal{S}(w,t) \cap \mathcal{S}(w,s)$

    *S3*: $\mathcal{S}(w,t) \subseteq \mathcal{S}(w,c)$, for all $c$

    *S4*: $\mathcal{S}(w, t \times t) = \mathcal{S}(w, t+t) = \mathcal{S}(w, !t) = \mathcal{S}(w,t)$

    *S5*: $v \in \mathcal{S}(w,t) \Rightarrow \mathcal{S}(v,t) \subseteq \mathcal{S}(w,t)$

2. Conditions on $\mathcal{E}$.

    *E1*: $\varphi, \varphi \to \psi \in \mathcal{E}(w,t) \Rightarrow \psi \in \mathcal{E}(w,t)$

    *E2*: $\mathcal{E}(w,t) \cup \mathcal{E}(w,s) = \mathcal{E}(w, t+s)$

    *E3*: $\varphi \in \mathcal{E}(w,t) \Rightarrow [t]\varphi \in \mathcal{E}(w, !t)$

    *E4*: $v \in \mathcal{S}(w,t) \Rightarrow \mathcal{E}(w,t) \subseteq \mathcal{E}(v,t)$

**Definition 3:** Basic model for **IEJ**.

A model for **IEJ** is a tuple $M = \langle W, \mathcal{S}, \mathcal{E}, V \rangle$, where

- $\langle W, \mathcal{S}, \mathcal{E} \rangle$ is a frame for **IEJ**

- $V$ is such that $V(p) \subseteq W$ for any propositional variable $p$

As usual, a valuation function for propositional variables is introduced as a function that assigns to each propositional variable a set of epistemic states, which are the states where the proposition denoted by the variable is true.

**Definition 4:** Truth at a world in a model for **IEJ**.
The notion of truth of a formula is defined as follows:

$M, w \models p \Leftrightarrow w \in V(p)$
$M, w \models \neg \varphi \Leftrightarrow M, w \not\models \varphi$
$M, w \models \varphi \wedge \psi \Leftrightarrow M, w \models \varphi$ and $M, w \models \psi$
$M, w \models [t]\varphi \Leftrightarrow M, w \models \varphi$, for all $v$ such that $v \in S(w, t)$
$M, w \models t : \varphi \Leftrightarrow M, w \models \varphi$, for all $v$ such that $v \in S(w, t)$, and $\varphi \in \mathcal{E}(w, t)$

The notions of logical consequence and logical validity are defined as usual.

## 3.2 Characterization

Let us now show that the previous system can be completely characterized by the class of basic frames. It is not difficult to show that the axioms in groups 2 are valid with respect to the class of all frames and that *modus ponens* preserves validity.[10] Let us then focus on the axioms of group 3 and prove their validity.

**IJ1:** ⊩ $[t](\varphi \to \psi) \to ([s]\varphi \to [t \times s]\psi)$

Suppose $M, w \models [t](\varphi \to \psi)$ and $M, w \models [s]\varphi$. Suppose, in addition, that $u \in S(w, t \times s)$. Since $S(w, t \times s) \subseteq S(w, t) \cap S(w, s)$, by conditions $S1$, $u \in S(w, t)$ and $u \in S(w, s)$. Since $M, v \models \varphi \to \psi$, for all $v$ such that $v \in S(w, t)$, and $M, v \models \varphi$, for all $v$ such that $v \in S(w, s)$, by the definition of truth, $M, u \models \varphi \to \psi$ and $M, u \models \varphi$, and so $M, u \models \psi$. Thus, $M, u \models \psi$, for all $u$ such that $u \in S(w, t \times s)$, and so $M, w \models [t \times s]\psi$.

**IJ2:** ⊩ $[t]\varphi \vee [s]\varphi \to [t+s]\varphi$

Suppose either $M, w \models [t]\varphi$ or $M, w \models [s]\varphi$. Then, either $M, v \models \varphi$, for all $v$ such that $v \in S(w, t)$, or $M, v \models \varphi$, for all $v$ such that $v \in S(w, s)$. In both cases, $M, v \models \varphi$, for all $v$ such that $v \in S(w, t+s)$, since $S(w, t+s) \subseteq S(w, t) \cap S(w, s)$, by condition $S2$, and so $M, v \models \varphi$, for all $v$ such that $v \in S(w, t+s)$. Hence, $M, w \models [t+s]\varphi$.

**IJ3:** ⊩ $[t]\varphi \to [!t][t]\varphi$

Suppose $M, w \models [t]\varphi$, so that $M, v \models \varphi$, for all $v$ such that $v \in S(w, t)$, and $u \in S(v, t)$. Then, $S(u, t) \subseteq S(v, t)$ and $S(v, t) \subseteq S(w, t)$, by condition $S5$, and so $S(u, t) \subseteq S(w, t)$. Hence, $M, u \models \varphi$, for all $u$ such that $u \in S(v, t)$, and so $M, v \models [t]\varphi$. Since this is so for all $v$ such that $v \in S(w, t)$, and $S(w, !t) = S(w, t)$ by condition $S4$, $M, w \models [!t][t]\varphi$.

**IJ4:** ⊩ $t : \varphi \to [t]\varphi$

Straightforward, by the definition of $M, w \models t : \varphi$ and $M, w \models [t]\varphi$.

**IJ5:** ⊩ $[c]\varphi \to [t]\varphi$, where $c$ is a constant

Straightforward, by the definition of $M, w \models [t]\varphi$ and condition $S3$.

---

[10] The proof is a straightforward adaptation of the proof proposed in [9], section 3.

**IJ6:** ⊩ $[t \times t]\varphi \leftrightarrow [t+t]\varphi \leftrightarrow [!t]\varphi \leftrightarrow [t]\varphi$

Straightforward, by the definition of $M, w \models [t]\varphi$ and condition *S4*.

Thus we obtain the following

**Theorem 1: IEJ** *is sound with respect to the class of all basic frames for* **IEJ**. (relative to a specific constant specification)

The proof of the completeness theorem is more involved. As usual, the proof is based on a canonicity argument.[11] Therefore, let us start by defining the canonical model for **IEJ**. Let $w/[t] = \{\varphi \mid [t]\varphi \in w\}$ and $w/t = \{\varphi \mid t : \varphi \in w\}$, for all terms $t$. Then, the canonical model is the tuple $M = \langle W, \mathcal{S}, \mathcal{E}, V \rangle$, where

- $W$ is the set of maximally **IEJ**-consistent sets of formulas
- $\mathcal{S}$ is such that $v \in \mathcal{S}(w,t) \Leftrightarrow w/[t] \subseteq v$
- $\mathcal{E}$ is such that $\mathcal{E}(w,t) = w/t$

**Corollary 1:** $v \in \mathcal{S}(w,t) \cap \mathcal{S}(w,s) \Leftrightarrow w/[t] \cup w/[s] \subseteq v$.
Straightforward:
$v \in \mathcal{S}(w,t) \cap \mathcal{S}(w,s) \Leftrightarrow v \in \mathcal{S}(w,t)$ and $v \in \mathcal{S}(w,s)$
$v \in \mathcal{S}(w,t) \cap \mathcal{S}(w,s) \Leftrightarrow w/[t] \subseteq v$ and $w/[s] \subseteq v$
$v \in \mathcal{S}(w,t) \cap \mathcal{S}(w,s) \Leftrightarrow w/[t] \cup w/[s] \subseteq v$

**Lemma 1:** M is a model for **IEJ**.
We have to show that the conditions on $\mathcal{S}$ and $\mathcal{E}$ are satisfied.

**Part 1:** the conditions on $\mathcal{S}$ are satisfied.

- *S1*: $\mathcal{S}(w, t \times s) \subseteq \mathcal{S}(w,t) \cap \mathcal{S}(w,s)$.

Suppose $v \in \mathcal{S}(w, t \times s)$, so that $w/[t \times s] \subseteq v$, by the definition of $\mathcal{S}$. Since $w$ is maximal, $[t]\varphi \in w \Rightarrow [t \times s]\varphi \in w$, by **IJ8**, and $[s]\varphi \in w \Rightarrow [t \times s]\varphi \in w$, by **IJ7**. Thus, $\varphi \in w/[t] \Rightarrow \varphi \in w/[t \times s]$ and $\varphi \in w/[s] \Rightarrow \varphi \in w/[t \times s]$. Therefore, $w/[t] \cup w/[s] \subseteq w/[t \times s]$, and so $w/[t] \cup w/[s] \subseteq v$. Hence, $\mathcal{S}(w, t \times s) \subseteq \mathcal{S}(w,t) \cap \mathcal{S}(w,s)$, by corollary 1.

- *S2*: $\mathcal{S}(w, t + s) \subseteq \mathcal{S}(w,t) \cap \mathcal{S}(w,s)$.

Suppose $v \in \mathcal{S}(w, t+s)$, so that $w/[t+s] \subseteq v$, by the definition of $\mathcal{S}$. Since $w$ is maximal, $[t]\varphi \vee [s]\varphi \in w \Rightarrow [t+s]\varphi \in w$, by **IJ2**, and so $[t]\varphi \in w \Rightarrow [t+s]\varphi \in w$ and $[s]\varphi \in w \Rightarrow [t+s]\varphi \in w$. Thus, $\varphi \in w/[t] \Rightarrow \varphi \in w/[t+s]$ and $\varphi \in w/[s] \Rightarrow \varphi \in w/[t+s]$. Therefore, $w/[t] \cup w/[s] \subseteq w/[t+s]$, and so $w/[t] \cup w/[s] \subseteq v$. Hence, $\mathcal{S}(w, t+s) \subseteq \mathcal{S}(w,t) \cap \mathcal{S}(w,s)$, by corollary 1.

---

[11] See [6], chapter 4, for an introduction to modal completeness and, in particular, completeness by canonicity. In what follows I will omit the standard parts and definitions, and focus on the new parts of the proofs.

- *S3*: $\mathcal{S}(w,t) \subseteq \mathcal{S}(w,c)$.

It is to prove that $w/[c] \subseteq v/[t]$, which follows from **IJ5**.

- *S4*: $\mathcal{S}(w, t \times t) = \mathcal{S}(w, t+t) = \mathcal{S}(w, !t) = \mathcal{S}(w,t)$.

It is to prove that $w/[t \times t] = w/[t+t] = w/[!t] = w/[t]$, which follows from **IJ6**.

- *S5*: $v \in \mathcal{S}(w,t) \Rightarrow \mathcal{S}(v,t) \subseteq \mathcal{S}(w,t)$.

Since $w/[!t] = w/[t]$, by **IJ6**, it suffices to prove that, if $w/[!t] \subseteq v$, then $w/[t] \subseteq v/[t]$. Suppose $w/[!t] \subseteq v$ and $\varphi \in w/[t]$. Then, $[t]\varphi \in w$, so that $[!t][t]\varphi \in w$, by **IJ3** and $w \in W$. Therefore, $[t]\varphi \in v$, and so $w/[t] \subseteq v/[t]$.

**Part 2:** the conditions on $\mathcal{E}$ are satisfied.

The proof of conditions $E1$, $E2$, and $E3$ is well-known.[12] We only check $E4$.

- *E4*: $v \in \mathcal{S}(w,t) \Rightarrow \mathcal{E}(w,t) \subseteq \mathcal{E}(v,t)$.

Suppose $v \in \mathcal{S}(w,t)$, so that $w/[t] \subseteq v$, by the definition of $\mathcal{S}$. Since $w$ is maximal, $t : \varphi \in w \Rightarrow !t : (t : \varphi) \in w$, by **EJ3**. By **IJ4**, $!t : (t : \varphi) \in w \Rightarrow [!t](t : \varphi) \in w$. By **IJ6**, $[!t](t : \varphi) \in w \Rightarrow [t](t : \varphi) \in w$. Hence, $t : \varphi \in w \Rightarrow [t](t : \varphi) \in w$, and so $t : \varphi \in w \Rightarrow t : \varphi \in w/[t] \subseteq v$. Therefore, $w/t \subseteq v/t$, from which the conclusion follows.

**Lemma 2 (Truth Lemma):** $M, w \models \varphi \Leftrightarrow \varphi \in w$.
The interesting cases are the modal ones.

1. $M, w \models [t]\varphi \Leftrightarrow [t]\varphi \in w$.

    $M, w \models [t]\varphi \Leftrightarrow M, w \models \varphi$, for all $v$ such that $v \in \mathcal{S}(w,t)$

    $M, w \models [t]\varphi \Leftrightarrow \varphi \in w$, for all $v$ such that $w/[t] \subseteq v$, by I.H.

    $M, w \models [t]\varphi \Leftrightarrow \varphi \in w/[t]$, since $w/[t]$ is a closed set

    $M, w \models [t]\varphi \Leftrightarrow [t]\varphi \in w$, by the definition of $w/[t]$

2. $M, w \models t : \varphi \Leftrightarrow t : \varphi \in w$.

Suppose $M, w \models t : \varphi$. Then $\varphi \in \mathcal{E}(w,t)$, by the definition of truth. Thus, $t : \varphi \in w$, by the definition of $\mathcal{E}$. Suppose now $t : \varphi \in w$. Then $[t]\varphi \in w$, by **IJ4**. Thus, $M, w \models [t]\varphi$, by I.H., and $\varphi \in \mathcal{E}(w,t)$, by the definition of $\mathcal{E}$.

This concludes the proof. We then obtain the following

**Theorem 2:** *IEJ is complete with respect to the class of all basic frames for IEJ.* (relative to a specific constant specification)

---

[12] See, for instance, [9], section 8.

## 3.3 Developments

In this paper, I have presented a complete basic system of logic of implicit and and explicit justification. This work can be extended in at least three different directions. A first possibility is to introduce a hierarchy of systems of increasing power based on **IEJ**. In effect, it is not difficult to see that systems dealing with consistent and correct justifiers can be obtained by introducing axioms like

**EJD:** $t : \varphi \to \neg(t : \neg\varphi)$      **EJT:** $t : \varphi \to \varphi$
**IJD:** $[t]\varphi \to \neg[t]\neg\varphi$      **IJT:** $[t]\varphi \to \varphi$

and modifying the conditions on $\mathcal{S}$ so to account for their validity. Along similar lines, more powerful systems might be developed. A second possibility is to make the system dynamic, by looking at the connections with recent intuitions proposed in [13] and [14]. The idea in this case is to interpret $t : \varphi$ as saying that a proof $t$ of $\varphi$ has been announced, i.e. discovered and published, and to adapt the semantics of the logic of announcement to the present framework. A final possibility is to connect the idea of implicit justification involved in modalities like $[t]$ with the more usual idea of implicit knowledge provided in [5], and to look for an integrated system, where notions like conclusive evidence and default evidence are also accounted for.

## BIBLIOGRAPHY

[1] Artemov, S. (2001). Explicit provability and constructive semantics. The Bulletin for Symbolic Logic, 7: 1–36.
[2] Artemov, S. (2008). The logic of justification. The Review of Symbolic Logic, 1: 477–513.
[3] Artemov. S. and Nogina, E. (2004). Logic of knowledge with justifications from the provability perspective. Technical Report, City University of new York, 2004.
[4] Artemov. S. and Nogina, E. (2005). Introducing justification into epistemic logic. Journal of Logic and Computation, 15: 1059–1073.
[5] Baltag, A., Renne, B. and Smets, S. (2014). The logic of justified belief, explicit knowledge, and conclusive evidence. Annals of Pure and Applied Logic, 165: 49–81.
[6] Blackburn, B., de Rijke, M., and Venema, Y. (2001). Modal Logic. Cambridge University Press.
[7] Boolos, G. (1993). Reasoning about Knowledge. Cambridge University Press.
[8] Fitting, M. (2005). The logic of proofs, semantically. Annals of Pure and Applied Logic, 132: 1–25.
[9] Fitting, M. (2008). Justification logics, logics of knowledge, and conservativity. Annals of Mathematics and Artificial Intelligence, 53: 153-167.
[10] Fitting, M. (2009). Reasoning with justifications. In Makinson, D., Malinowski, J. and Wansing, H. editors, Towards Mathematical Philosophy, Trends in Logic 28, Springer, 2009. pp. 107–123.
[11] Giordani, A. (2013). A Logic of Justification and Truthmaking. The Review of Symbolic Logic, 20: 323–342.
[12] Tarski, A. (1956). Logic, Semantics, Metamathematics. Oxford: Oxford University Press.
[13] van Benthem, J. and Pacuit, E. (2011). Dynamic logics of evidence-based beliefs. Studia Logica, 99: 61–92.
[14] van Benthem, J. and Velazquez-Quesada, F. (2010). The dynamics of awareness. Synthese, 177: 5–27.

# A System of Proof for Lewis Counterfactual

Sara Negri and Giorgio Sbardolini

ABSTRACT. A deductive system for Lewis counterfactual is presented, based directly on Lewis' influential generalisation of relational semantics with ternary similarity relations. This deductive system builds on a method for enriching the syntax of sequent calculus by labels for possible worlds. The resulting labelled sequent calculus is shown to be equivalent to the axiomatic system **VC** of Lewis. It is further shown to have the structural properties that are needed for an analytic proof system that supports root-first proof search. Completeness of the calculus is proved in a direct way, such that for any given sequent either a formal derivation or a countermodel is provided; it is also shown how finite countermodels for unprovable sequents can be extracted from failed proof search, by which the completeness proof turns into a proof of decidability.

## 1 Introduction

Kripke's relational semantics was a decisive turning point for modal logic: earlier axiomatic studies were replaced by a semantic method that displayed the connections between modal axioms and conditions on the accessibility relation between possible worlds. Based on a development of Kripke's semantic framework, David Lewis put forward a study of conditionals in the classic work *Counterfactuals* (1973). Counterfactual conditionals have long been of interest in Philosophy, for they play a crucial role in our understanding of scientific laws, causation, metaphysics and epistemology.

The success of the semantic methods has not been followed by equally powerful syntactic theories of modal and conditional reasoning: Concerning the former, the situation was so depicted by Melvin Fitting in his article (2007) in the Handbook of Modal Logic: "No proof procedure suffices for every normal modal logic determined by a class of frames"; Concerning the latter, as stated by Graham Priest "there are presently no known tableau systems" for Lewis' logic for counterfactuals (2008, p. 93).

In Negri (2005) it was shown how Kripke semantics can be exploited to enrich the syntax of systems of proof. In particular, a more expressive language turned out to be crucial, with a formal notation of labels representing possible worlds. The approach has been extended to wider frame classes in later work (Negri 2016), and in Dyckhoff and Negri (2015) it was shown how the method can capture any nonclassical logic characterized by arbitrary first-order frame conditions in their relational semantics. Notably, in these calculi, all the rules are invertible and a strong form of completeness holds for them, with a simultaneous construction of formal proofs for derivable sequents, or countermodels for underivable ones (Negri 2014a).

The semantics of Lewis' conditional is interestingly different from standard modal logics in that counterfactuals are analyzed in terms of a similarity relation among worlds. Ternary relations of comparative similiarity were proposed by Lewis himself as a formal account of the topological truth conditions for counterfactuals, in the setting of a sphere semantics, a special form of neighbourhood semantics. Interestingly, this gives an $\exists\forall$-nesting of quantifiers in the truth conditions for the counterfactual conditional, which makes the determination of the rules of the calculus a challenging task. The solution presented here makes use of indexed modalities, which allow to split the semantic clause in two separate parts; correspondingly, the rules for the counterfactual conditional depend on rules for the indexed modality, which are standard modal labelled rules. The result is a sequent system, called **G3LC** below, which is a sound and complete Gentzen-style calculus for Lewis' original counterfactual. The system has all the structural rules (weakening, contraction, and cut) admissible, and all its rules are invertible. Furthermore, we establish decidability of the calculus by means of a finitary root-first proof search procedure that for every sequent yields either a derivation or a countermodel.

We introduce **G3LC** in the next section. In Section 3, some interesting structural properties of **G3LC** are presented, in particular a cut elimination theorem. For lack of sufficient space some proofs are omitted, others are just sketched.[1] In Section 4, it is shown that Lewis' axioms and rules are, respectively, admissible and derivable, which allows to show that the calculus is complete (by soundness and by Lewis' own proof of completeness). Finally, Section 5 contains direct completeness and decidability results. Related literature and further work are discussed in the concluding section.

## 2 A sequent calculus for Lewis conditional

We follow precise steps for moving from the meaning of logical constants to sequent calculus rules; the method is fully general, and it allows us to internalize the semantics into the syntax of a good sequent calculus.[2] To begin with, the language is extended by labelled formulas of the form $x : A$, and by expressions of the form $xRy$. Labelled formulas $x : A$ correspond to the statement that $A$ is true at node/possible world $x$; expressions of the form $xRy$ correspond to relations between nodes/possible worlds in a frame. Then the compositional clauses that define the truth of a formula at a world are translated into natural deduction inference rules for labelled expressions; third, such rules are appropriately converted into sequent calculus rules; fourth, the characteristic frame properties are converted into rules for the relational part of the calculus following the method of translation of axioms

---

[1] Cf. Negri and Sbardolini (2016) for complete proofs and in-depth discussion.

[2] The details of the procedure are presented for intuitionistic and standard modal logic in Negri and von Plato (2014).

into sequent calculus rules introduced and developed in Negri and von Plato (1998, 2001, 2011). In this way, the frame properties are carried over to the calculus by the addition of rules for binary accessibility relations regarded as binary atomic predicates with the labels as arguments. In this section, this method is applied to the case of Lewis' counterfactual conditional.

The truth conditions for Lewis' conditional are spelled out in terms of a three-place similarity relation $\preceq$ among worlds, the intuitive meaning of "$x \preceq_w y$" being "$x$ is at least as similar to $w$ as $y$ is" (Lewis 1973a, 1973b). The following properties are generally assumed:

1. *Transitivity*: If $x \preceq_w y$ and $y \preceq_w z$ then $x \preceq_w z$,

2. *Strong connectedness*: Either $x \preceq_w y$ or $y \preceq_w x$,

3. *L-Minimality*: If $x \preceq_w w$ then $x = w$.

Through the conversion method outlined above these turn into the following sequent calculus rules:

$$\frac{x \preceq_w z, x \preceq_w y, y \preceq_w z, \Gamma \Rightarrow \Delta}{x \preceq_w y, y \preceq_w z, \Gamma \Rightarrow \Delta} \; Trans \qquad \frac{x \preceq_w y,, \Gamma \Rightarrow \Delta \quad y \preceq_w x, \Gamma \Rightarrow \Delta}{\Gamma \Rightarrow \Delta} \; SConn$$

$$\frac{x = w, \Gamma \Rightarrow \Delta}{x \preceq_w w, \Gamma \Rightarrow \Delta} \; LMin$$

Lewis' conditional is symbolized by $A \mathbin{\Box\!\!\rightarrow} B$, which intuitively reads "If it had been the case that $A$, it would be the case that $B$". The truth conditions are as follows:

$w \Vdash A \mathbin{\Box\!\!\rightarrow} B$ iff either

1. There is no $z$ such that $z \Vdash A$, or
2. there is $x$ such that $x \Vdash A$ and for all $y$, if $y \preceq_w x$ then $y \Vdash A \supset B$.

As previously anticipated, the truth condition for $A \mathbin{\Box\!\!\rightarrow} B$ has a universal quantification in the scope of an existential one, and thus it is not of a form that can be directly translated into rules following the method of generation of labelled sequent rules for intensional operators (as expounded in Negri 2005); a more complex formalism in the line of the method of *systems of rules* (Negri 2016) would have to be invoked to maintain the primitive language.

The rules for the labelled calculus for Lewis' conditional can be presented following the general method of embedding neighbourhood semantics for non-normal modal logics into the standard relational semantics for normal modal systems through the use of *indexed modalities*.[3] Specifically, the relation of similarity is used to define a ternary accessibility relation

$$xR_w y \equiv y \preceq_w x$$

---

[3] The method is formulated in general terms in Gasquet and Herzig (1996) for classical modal logics and used in Giordano et al. (2008) for a tableau calculus for preference-based conditional logics.

In turn, this relation defines an indexed necessity as follows:

$$x \Vdash \Box_w A \equiv \forall y. x R_w y \to y \Vdash A$$

Then the truth condition for the conditional may be replaced by the following

$w \Vdash A \mathbin{\Box\!\!\rightarrow} B$ iff either

1. There is no $z$ such that $z \Vdash A$, or
2. there is $x$ such that $x \Vdash A$ and $x \Vdash \Box_w(A \supset B)$.

Observe that the presentation of a calculus formulated in terms of indexed modalities is faithful to Lewis' original idea of conditional implication as a variably strict conditional.

The rules for Lewis conditional and for the indexed modality are obtained from their respective truth conditions following the general method of Negri (2005) for turning the truth conditions of standard modalities into rules of a labelled sequent calculus: quantification over worlds is replaced by the condition that certain variables in the rules (eigenvariables) should be fresh; the right to left direction in the truth conditions gives the right rule and the other direction gives the left rule. Since the truth condition for Lewis' conditional is a disjunction, there are two right rules (one for each disjunct) and accordingly one left rule with two premisses.

$$\frac{xR_w y, \Gamma \Rightarrow \Delta, y : A}{\Gamma \Rightarrow \Delta, x : \Box_w A} R\Box_w \; (y \text{ fresh}) \qquad \frac{xR_w y, x : \Box_w A, y : A, \Gamma \Rightarrow \Delta}{xR_w y, x : \Box_w A, \Gamma \Rightarrow \Delta} L\Box_w$$

$$\frac{z : A, \Gamma \Rightarrow \Delta, w : A \mathbin{\Box\!\!\rightarrow} B}{\Gamma \Rightarrow \Delta, w : A \mathbin{\Box\!\!\rightarrow} B} R\mathbin{\Box\!\!\rightarrow}_1 \; (z \text{ fresh})$$

$$\frac{\Gamma \Rightarrow \Delta, w : A \mathbin{\Box\!\!\rightarrow} B, x : A \quad \Gamma \Rightarrow \Delta, w : A \mathbin{\Box\!\!\rightarrow} B, x : \Box_w(A \supset B)}{\Gamma \Rightarrow \Delta, w : A \mathbin{\Box\!\!\rightarrow} B} R\mathbin{\Box\!\!\rightarrow}_2$$

$$\frac{w : A \mathbin{\Box\!\!\rightarrow} B, \Gamma \Rightarrow \Delta, z : A \quad x : A, x : \Box_w(A \supset B), \Gamma \Rightarrow \Delta}{w : A \mathbin{\Box\!\!\rightarrow} B, \Gamma \Rightarrow \Delta} L\mathbin{\Box\!\!\rightarrow} \; (x \text{ fresh})$$

The complete system is presented in Table 1. The system is thus obtained as an extension of the propositional part of the contraction- and cut-free sequent calculus G3K for basic modal logic introduced in Negri (2005). In addition there are rules for the similarity and the equality relation. For the latter, there are just two rules, reflexivity and the scheme of *replacement*, $Repl_{At}$, where $At(x)$ stands for an atomic labelled formula $x : P$ or a relation of the form $y = z$, $yR_w z$, with $x$ one of $y, w, z$. Symmetry of equality follows as a special case of $Repl_{At}$ as well as *Euclidean transitivity* which, together with symmetry, gives the usual *transitivity*.[4]

Before proceeding to the results, we give a definition of weight of formulas:

---

[4] The general reasons for the architecture behind the rules of equality are discussed in Negri and von Plato (2001, S6.5) for extensions of first-order systems, and the equality rules for labelled systems are given in Negri (2005) and Negri and von Plato (2011).

**Table1 : Lewis basic counterfactual conditional sequent system (G3LC)**
*based on ternary similarity*

**Initial sequents:**

$x : P, \Gamma \Rightarrow \Delta, x : P$

**Propositional rules:**

$$\frac{x : A, x : B, \Gamma \Rightarrow \Delta}{x : A\&B, \Gamma \Rightarrow \Delta} \, L\& \qquad \frac{\Gamma \Rightarrow \Delta, x : A \quad \Gamma \Rightarrow \Delta, x : B}{\Gamma \Rightarrow \Delta, x : A\&B} \, R\&$$

$$\frac{x : A, \Gamma \Rightarrow \Delta \quad x : B, \Gamma \Rightarrow \Delta}{x : A \vee B, \Gamma \Rightarrow \Delta} \, L\vee \qquad \frac{\Gamma \Rightarrow \Delta, x : A, x : B}{\Gamma \Rightarrow \Delta, x : A \vee B} \, R\vee$$

$$\frac{\Gamma \Rightarrow \Delta, x : A \quad x : B, \Gamma \Rightarrow \Delta}{x : A \supset B, \Gamma \Rightarrow \Delta} \, L\supset \qquad \frac{x : A, \Gamma \Rightarrow \Delta, x : B}{\Gamma \Rightarrow \Delta, x : A \supset B} \, R\supset$$

$$\frac{}{x : \bot, \Gamma \Rightarrow \Delta} \, L\bot$$

**Similarity rules:**

$$\frac{xR_w z, xR_w y, yR_w z, \Gamma \Rightarrow \Delta}{xR_w y, yR_w z, \Gamma \Rightarrow \Delta} \, Trans \qquad \frac{xR_w y, \Gamma \Rightarrow \Delta \quad yR_w x, \Gamma \Rightarrow \Delta}{\Gamma \Rightarrow \Delta} \, SConn \ (x, y, w \ in \ \Gamma, \Delta)$$

$$\frac{x = x, \Gamma \Rightarrow \Delta}{\Gamma \Rightarrow \Delta} \, Ref \ (x \ in \ \Gamma, \Delta) \qquad \frac{x = y, At(x), At(y), \Gamma \Rightarrow \Delta}{x = y, At(x), \Gamma \Rightarrow \Delta} \, Repl_{At}$$

$$\frac{x = w, wR_w x, \Gamma \Rightarrow \Delta}{wR_w x, \Gamma \Rightarrow \Delta} \, LMin$$

**Conditional rules:**

$$\frac{xR_w y, \Gamma \Rightarrow \Delta, y : A}{\Gamma \Rightarrow \Delta, x : \Box_w A} \, R\Box_w \ (y \ fresh) \qquad \frac{xR_w y, x : \Box_w A, y : A, \Gamma \Rightarrow \Delta}{xR_w y, x : \Box_w A, \Gamma \Rightarrow \Delta} \, L\Box_w$$

$$\frac{z : A, \Gamma \Rightarrow \Delta, w : A \mathbin{\Box\!\!\rightarrow} B}{\Gamma \Rightarrow \Delta, w : A \mathbin{\Box\!\!\rightarrow} B} \, R{\mathbin{\Box\!\!\rightarrow}}_1 \ (z \ fresh)$$

$$\frac{\Gamma \Rightarrow \Delta, w : A \mathbin{\Box\!\!\rightarrow} B, x : A \quad \Gamma \Rightarrow \Delta, w : A \mathbin{\Box\!\!\rightarrow} B, x : \Box_w(A \supset B)}{\Gamma \Rightarrow \Delta, w : A \mathbin{\Box\!\!\rightarrow} B} \, R{\mathbin{\Box\!\!\rightarrow}}_2$$

$$\frac{w : A \mathbin{\Box\!\!\rightarrow} B, \Gamma \Rightarrow \Delta, z : A \quad x : A, x : \Box_w(A \supset B), \Gamma \Rightarrow \Delta}{w : A \mathbin{\Box\!\!\rightarrow} B, \Gamma \Rightarrow \Delta} \, L{\mathbin{\Box\!\!\rightarrow}} \ (x \ fresh)$$

DEFINITION 1. The weight $\mathtt{w}(A)$ of a formula $A$ is defined inductively by the following:

$\mathtt{w}(\gamma) = 1$ for $\gamma$ the constant $\bot$, an atomic formula, or a relational atom,
$\mathtt{w}(A \circ B) = \mathtt{w}(A) + \mathtt{w}(B) + 1$ for $\circ$ conjunction, disjunction, or implication,
$\mathtt{w}(\Box_x A) = \mathtt{w}(A) + 1$,
$\mathtt{w}(A \boxbox B) = \mathtt{w}(A) + \mathtt{w}(B) + 3$.

Notice that since $\neg A$ is defined by $A \supset \bot$, $\mathtt{w}(\neg A) \equiv \mathtt{w}(A) + 2$. Notice also that $\mathtt{w}(\Box_x(A \supset B)) < \mathtt{w}(A \boxbox B)$.

The following lemma is proved by induction on the weight of $A$:

LEMMA 2. *All the sequents of the form* $x : A, \Gamma \Rightarrow \Delta, x : A$ *are derivable in* **G3LC**.

## 3 Structural properties

The proof of admissibility of the structural rules in **G3LC** follows the pattern presented in Negri and von Plato (2011, 11.4). Likewise, some preliminary results are needed, namely height-preserving admissibility of substitution (in short, hp-substitution) and height-preserving invertibility (in short, hp-invertibility) of the rules. Recall that the *height* of a derivation is its height as a tree, i.e. the length of its longest branch, and that $\vdash_n$ denotes derivability with derivation height bounded by $n$ in a given system. In what follows, the results are all referred to **G3LC**. The following is proved by induction on the height of the derivation:

PROPOSITION 3. *If* $\vdash_n \Gamma \Rightarrow \Delta$, *then* $\vdash_n \Gamma(y/x) \Rightarrow \Delta(y/x)$.

With a straightforward induction, it follows that:

PROPOSITION 4. *The rules of left and right weakening are hp-admissible.*

In a way similar to the proof of Lemma 11.7 in Negri and von Plato (2011), a result of *hp-invertibility* of the rules of **G3LC** can be proved next, i.e. for every rule of the form $\frac{\Gamma' \Rightarrow \Delta'}{\Gamma \Rightarrow \Delta}$, if $\vdash_n \Gamma \Rightarrow \Delta$ then $\vdash_n \Gamma' \Rightarrow \Delta'$, and for every rule of the form $\frac{\Gamma' \Rightarrow \Delta' \quad \Gamma'' \Rightarrow \Delta''}{\Gamma \Rightarrow \Delta}$ if $\vdash_n \Gamma \Rightarrow \Delta$ then $\vdash_n \Gamma' \Rightarrow \Delta'$ and $\vdash_n \Gamma'' \Rightarrow \Delta''$.

LEMMA 5. *All the propositional rules are hp-invertible.*

As for invertibility of the rules for the conditional, we have

LEMMA 6. *The following hold:*
  (i) *If* $\vdash_n \Gamma \Rightarrow \Delta, x : \Box_w A$, *then* $\vdash_n xR_w y, \Gamma \Rightarrow \Delta, y : A$,
  (ii) *If* $\vdash_n w : A \boxbox B, \Gamma \Rightarrow \Delta$, *then* $\vdash_n x : A, x : \Box_w(A \supset B), \Gamma \Rightarrow \Delta$.

Observe that Lemma 6(ii) states hp-invertibility of $L \boxbox$ with respect to the second premiss; its hp-invertibility with respect to the first premiss is a special case of Proposition 4. Therefore, as a general result we have:

COROLLARY 7. *All the rules are hp-invertible.*

The rules of contraction of **G3LC** have the following form, where $\phi$ is either a relational atom of the form $xR_w y$ or a labelled formula $x : A$:

$$\frac{\phi, \phi, \Gamma \Rightarrow \Delta}{\phi, \Gamma \Rightarrow \Delta} LC \qquad \frac{\Gamma \Rightarrow \Delta, \phi, \phi}{\Gamma \Rightarrow \Delta, \phi} RC$$

By simultaneous induction on the height of derivation for left and right contraction, it follows that:

THEOREM 8. *The rules of left and right contraction are hp-admissible.*

And finally:

THEOREM 9. *Cut is admissible.*

**Proof.** The proof is by induction on the weight of the cut formula and subinduction on the sum of the heights of derivations of the premisses (cut-height). The cases pertaining initial sequents and the propositional rules of the calculus are dealt with as in Theorem 11.9 of Negri and von Plato (2011) and therefore omitted here. Also the cases with cut formula not principal in both premisses of cut are dealt in the usual way by permutation of cut, with possibly an application of hp-substitution to avoid a clash with the fresh variable in rules with variable condition. So, the only cases to focus on are those with cut formula of the form $\square_w A$ or $A \,\square\!\!\rightarrow\, B$ which is principal in both premisses of cut. The former case presents, apart from the indexing on the accessibility relation, no difference with respect to the case of a plain modality, so we proceed to analyse the latter. This case splits into two subcases, depending on whether the left premiss is derived by $R\square\!\!\rightarrow_1$ or $R\square\!\!\rightarrow_2$.

In the first case there is a derivation of the form

$$\frac{\dfrac{\mathcal{D}_1}{y : A, \Gamma \Rightarrow \Delta, w : A \,\square\!\!\rightarrow\, B}}{\Gamma \Rightarrow \Delta, w : A \,\square\!\!\rightarrow\, B} R\square\!\!\rightarrow_1 \quad \dfrac{\dfrac{\mathcal{D}_2}{w : A \,\square\!\!\rightarrow\, B, \Gamma' \Rightarrow \Delta', z : A} \quad \dfrac{\mathcal{D}_3}{y : A, y : \square_w(A \supset B), \Gamma' \Rightarrow \Delta'}}{w : A \,\square\!\!\rightarrow\, B, \Gamma' \Rightarrow \Delta'} L\square\!\!\rightarrow \quad Cut$$
$$\Gamma, \Gamma' \Rightarrow \Delta, \Delta'$$

This is converted into a derivation with three cuts of reduced height as follows (we have to split the result of the conversion to fit it in the page): First, a derivation $\mathcal{D}_4$

$$\frac{\mathcal{D}_1 \qquad \mathcal{D}_2}{\Gamma \Rightarrow \Delta, w : A \,\square\!\!\rightarrow\, B \qquad w : A \,\square\!\!\rightarrow\, B, \Gamma' \Rightarrow \Delta', z : A} Cut$$
$$\Gamma, \Gamma' \Rightarrow \Delta, \Delta', z : A$$

Further, by application of hp-substitution, another derivation $\mathcal{D}_5$

$$\frac{\dfrac{\mathcal{D}_1(z/y)}{z : A, \Gamma \Rightarrow \Delta, w : A \,\square\!\!\rightarrow\, B} \quad \dfrac{\dfrac{\mathcal{D}_2}{w : A \,\square\!\!\rightarrow\, B, \Gamma' \Rightarrow \Delta', z : A} \quad \dfrac{\mathcal{D}_3}{y : A, y : \square_w(A \supset B), \Gamma' \Rightarrow \Delta'}}{w : A \,\square\!\!\rightarrow\, B, \Gamma' \Rightarrow \Delta'} L\square\!\!\rightarrow}{z : A, \Gamma, \Gamma' \Rightarrow \Delta, \Delta'} Cut$$

The two derivations are then used as premisses of a third cut of reduced weight as follows

$$\frac{\dfrac{\Gamma, \Gamma' \Rightarrow \Delta, \Delta', z : A \qquad z : A, \Gamma, \Gamma' \Rightarrow \Delta, \Delta'}{\Gamma^2, \Gamma'^2 \Rightarrow \Delta^2, \Delta'^2} Cut}{\Gamma, \Gamma' \Rightarrow \Delta, \Delta'} Ctr^*$$

In the second case there is a derivation of the form

$$\cfrac{\cfrac{\mathcal{D}_1}{\Gamma \Rightarrow \Delta, w:A\square\!\!\rightarrow B, x:A} \quad \cfrac{\mathcal{D}_2}{\Gamma \Rightarrow \Delta, w:A\square\!\!\rightarrow B, x:\square_w(A\supset B)}}{\cfrac{\Gamma \Rightarrow \Delta, w:A\square\!\!\rightarrow B}{\Gamma, \Gamma' \Rightarrow \Delta, \Delta'}}R\square\!\!\rightarrow_2 \quad \cfrac{\cfrac{\mathcal{D}_3}{w:A\square\!\!\rightarrow B, \Gamma' \Rightarrow \Delta', z:A} \quad \cfrac{\mathcal{D}_4}{y:A, y:\square_w(A\supset B)}}{\cfrac{w:A\square\!\!\rightarrow B, \Gamma' \Rightarrow \Delta'}{}Cut}}$$

The cut is converted into six cuts of reduced height or weight of cut formula as follows: First, the derivation (call it $\mathcal{D}_5$)

$$\cfrac{\cfrac{\mathcal{D}_1}{\Gamma \Rightarrow \Delta, w:A\square\!\!\rightarrow B, x:A} \quad \cfrac{\cfrac{\mathcal{D}_3}{w:A\square\!\!\rightarrow B, \Gamma' \Rightarrow \Delta', z:A} \quad \cfrac{\mathcal{D}_4}{y:A, y:\square_w(A\supset B), \Gamma' \Rightarrow \Delta'}}{w:A\square\!\!\rightarrow B, \Gamma' \Rightarrow \Delta'}Cut}{\Gamma, \Gamma' \Rightarrow \Delta, \Delta', x:A}$$

with a cut of reduced height. Then the derivation (call it $\mathcal{D}_6$)

$$\cfrac{\cfrac{\mathcal{D}_2}{\Gamma \Rightarrow \Delta, w:A\square\!\!\rightarrow B, x:\square_w(A\supset B)} \quad \cfrac{\cfrac{\mathcal{D}_3}{w:A\square\!\!\rightarrow B, \Gamma' \Rightarrow \Delta', z:A} \quad \cfrac{\mathcal{D}_4}{y:A, y:\square_w(A\supset B), \Gamma' \Rightarrow \Delta'}}{\cfrac{w:A\square\!\!\rightarrow B, \Gamma' \Rightarrow \Delta'}{}Cut}}{\cfrac{\Gamma, \Gamma' \Rightarrow \Delta, \Delta', x:\square_w(A\supset B)}{x:A, \Gamma, \Gamma'^2 \Rightarrow \Delta, \Delta'^2}L\square\!\!\rightarrow \quad \cfrac{\mathcal{D}_4(x/\cdot)}{x:A, x:\square_w(A\supset \cdot)}}$$

with two cuts, the upper of reduced height, and the lower of reduced weight; finally the derivation

$$\cfrac{\cfrac{\mathcal{D}_5 \quad \mathcal{D}_6}{\Gamma^2, \Gamma'^3 \Rightarrow \Delta^2, \Delta'^3}Cut}{\Gamma, \Gamma' \Rightarrow \Delta, \Delta'}Ctr^*$$

with a cut or reduced weight and repeated applications of contraction. ∎

To ensure the consequences of cut elimination we need to establish another crucial property of the system. We say that a labelled system has the *subterm property* if every variable occurring in any derivation is either an eigenvariable or occurs in the conclusion.[5] Clearly, the rules of **G3LC** do not, as they stand, satisfy the subterm property, but we can prove that, without loss of generality, proof search can be restricted to derivations that have the subterm property.

PROPOSITION 10. *Every derivable sequent has a derivation that satisfies the subterm property.*

**Proof.** By induction on the height of the derivation. For the inductive step, the conclusion is clear if the last step is one of the rules in which all the labels in the premisses satisfy the subterm property. For the other rules (in this specific calculus, rules *Ref* and $R\square\!\!\rightarrow_1$), consider the violating cases in which the premisses contain a label which is not in the conclusion. Using hp-substitution, it can be replaced to a label in the conclusion and thus obtain a derivation of the same height that satisfies the subterm property. ∎

By the above result, in the following we shall always restrict attention to derivations with the subterm property.

---

[5]This property is called *analyticity* in Dyckhoff and Negri (2012).

## 4 Lewis' axioms and rules

The axiomatic system for counterfactuals **VC**, regarded by Lewis as the "official logic of counterfactuals" (Lewis, 1973a, p. 132), is captured by **G3LC** since Lewis' axioms are provable in **G3LC** and the inference rules of **VC** are admissible. For brevity, proofs are here omitted. The results stated in this section, together with a proof of soundness of **G3LC** with respect to Lewis' semantics, provide an indirect proof of completeness. In Section 5 a direct completeness proof for **G3LC** with respect to Lewis semantics is however presented.

PROPOSITION 11. *The following rules are admissible in* **G3LC**:

1. *Modus Ponens:* $\dfrac{\vdash A \quad \vdash A \supset B}{\vdash B}$

2. *Deduction within Conditionals: for any* $n \geq 1$

$$\dfrac{\vdash A_1 \& \ldots \& A_n \supset B}{\vdash ((D \mathbin{\Box\!\!\to} A_1) \& \ldots \& (D \mathbin{\Box\!\!\to} A_n)) \supset (D \mathbin{\Box\!\!\to} B)}$$

3. *Interchange of logical equivalents: if* $\vdash A \supset\subset B$ *and* $\vdash \Phi(A)$ *then* $\vdash \Phi(B)$, *where* $\Phi$ *is an arbitrary formula in the language.*

All the axioms of **VC** are derivable in **G3LC**, i.e. for each axiom $A$ the sequent $\Rightarrow x : A$ is derivable in the calculus where $x$ is an arbitrary label.

PROPOSITION 12. *The following axioms are derivable in* **G3LC**:

1. *Propositonal tautologies,*

2. $A \mathbin{\Box\!\!\to} A$,

3. $(\neg A \mathbin{\Box\!\!\to} A) \supset (B \mathbin{\Box\!\!\to} A)$,

4. $(A \mathbin{\Box\!\!\to} \neg B) \lor (((A \& B) \mathbin{\Box\!\!\to} C) \supset\subset (A \mathbin{\Box\!\!\to} (B \supset C)))$,

5. $(A \mathbin{\Box\!\!\to} B) \supset (A \supset B)$,

6. $(A \& B) \supset (A \mathbin{\Box\!\!\to} B)$.

## 5 Completeness

In this section a direct completeness proof for **G3LC** with respect to Lewis semantics is presented. The proof has the overall structure of the completeness proof for labelled systems for modal and non-classical logics given in Negri (2009) and Negri (2014a), but the semantics is here based on comparative similarity systems rather than Kripke models.

DEFINITION 13. Let $\mathcal{W}$ be the set of variables (labels) used in derivations in **G3LC**. A **comparative similarity system** $\mathcal{S}$ is an assignment to every $w \in \mathcal{W}$ of a two-place relation $\preceq_w$ with the aforementioned conditions:

1. *Transitivity*: If $x \preceq_w y$ and $y \preceq_w z$ then $x \preceq_w z$,

2. *Strong connectedness*: Either $x \preceq_w y$ or $y \preceq_w x$,

3. *L-Minimality*: If $x \preceq_w w$ then $x = w$.

An **interpretation** of the labels in $\mathcal{W}$ in $\mathcal{S}$ a map $[\![\cdot]\!] : \mathcal{W} \to \mathcal{S}$. A **valuation** of atomic formulas in $\mathcal{S}$ is a map $\mathcal{V} : AtFrm \to \mathcal{P}(\mathcal{S})$ that assigns to each atom $P$ the set of elements of $\mathcal{W}$ in which $P$ holds. Instead of writing $w \in \mathcal{V}(P)$, we adopt the standard notation $w \Vdash P$.

> Valuations are extended to arbitrary formulas by the following inductive clauses:
> $\mathcal{V}_\bot$: $x \Vdash \bot$ for no $x$.
> $\mathcal{V}_\&$: $x \Vdash A\&B$ iff $x \Vdash A$ and $x \Vdash B$.
> $\mathcal{V}_\vee$: $x \Vdash A \vee B$ iff $x \Vdash A$ or $x \Vdash B$.
> $\mathcal{V}_\supset$: $x \Vdash A \supset B$ iff if $x \Vdash A$ then $x \Vdash B$.
> $\mathcal{V}_{\Box_w}$: $x \Vdash \Box_w A$ iff for all $y$, if $y \preceq_w x$ then $y \Vdash A$.
> $\mathcal{V}_{\Box\to}$: $x \Vdash A \,\Box\!\!\to B$ iff either $z \Vdash A$ for no $z$, or $y \Vdash A$ and $y \Vdash \Box_x(A \supset B)$ for some $y$.

DEFINITION 14. A labelled formula $x : A$ (resp. a relational atom $xR_w y$) is **true** for an interpretation $[\![\cdot]\!]$ and a valuation $\mathcal{V}$ in a system $\mathcal{S}$ iff $[\![x]\!] \Vdash A$ (resp. $[\![y]\!] \preceq_{[\![w]\!]} [\![x]\!]$). A sequent $\Gamma \Rightarrow \Delta$ is true for an interpretation $[\![\cdot]\!]$ and a valuation $\mathcal{V}$ in a system $\mathcal{S}$ if, whenever for all labelled formulas $x : A$ and relational atom $xR_w y$ in $\Gamma$ it is the case that $[\![x]\!] \Vdash A$ and $[\![y]\!] \preceq_{[\![w]\!]} [\![x]\!]$, then for some $w : B$ in $\Delta$, $[\![w]\!] \Vdash B$. A sequent is **valid** in a system $\mathcal{S}$ iff it is true for every interpretation and valuation in $\mathcal{S}$.

THEOREM 15. (**Soundness**) *If a sequent is derivable in* **G3LC** *then it is valid in every comparative similarity system $\mathcal{S}$.*

THEOREM 16. (**Completeness**) *Let $\Gamma \Rightarrow \Delta$ be a sequent in the language of* **G3LC**. *If it is valid in every comparative similarity system, it is derivable in* **G3LC**.

**Proof.** Immediate by Proposition 11, Proposition 12, Theorem 15, and Lewis' own completeness proof (Lewis 1973a, pp. 118-134). ∎

Completeness can be established also as a corollary of the following:

THEOREM 17. *Let $\Gamma \Rightarrow \Delta$ be a sequent in the language of* **G3LC**. *Then either it is derivable in* **G3LC** *or it has a countermodel in $\mathcal{S}$.*

For brevity, the proof is here omitted.

# 6 Decidability

In general cut elimination alone does not ensure terminating proof search in a given calculus. The exhaustive proof search used in the proof of Theorem 17 is not a decision method nor an effective method of finding countermodels when proof search fails, as it may produce infinite branches and therefore infinite countermodels. By way of example, consider the following branch in the search for a proof of the sequent $\Rightarrow w : \Box_x \neg \Box_x A \supset \Box_x B$ (this is analogous to the case for S4 discussed in Negri and von Plato 2011, Section 11.5):

$$
\cfrac{
 \cfrac{
  \cfrac{
   \cfrac{
    \cfrac{
     \cfrac{
      \cfrac{
       \cfrac{
        \cfrac{
         \cfrac{\vdots}{wR_xy, yR_xz, wR_xz, zR_xt, w : \Box_x\neg\Box_x A \Rightarrow t : A, z : A, y : B}
        }{wR_xy, yR_xz, wR_xz, w : \Box_x\neg\Box_x A \Rightarrow z : \Box_x A, z : A, y : B} R\Box_x
       }{wR_xy, yR_xz, wR_xz, w : \Box_x\neg\Box_x A, z : \neg\Box_x A \Rightarrow z : A, y : B} L\supset
      }{wR_xy, yR_xz, wR_xz, w : \Box_x\neg\Box_x A \Rightarrow z : A, y : B} L\Box_x
     }{wR_xy, yR_xz, w : \Box_x\neg\Box_x A \Rightarrow z : A, y : B} \; Trans
    }{wR_xy, w : \Box_x\neg\Box_x A \Rightarrow y : \Box_x A, y : B} R\Box_x
   }{wR_xy, w : \Box_x\neg\Box_x A, y : \neg\Box_x A \Rightarrow y : B} L\supset
  }{wR_xy, w : \Box_x\neg\Box_x A \Rightarrow y : B} L\Box_x
 }{w : \Box_x\neg\Box_x A \Rightarrow w : \Box_x B} R\Box_x
}{\Rightarrow w : \Box_x\neg\Box_x A \supset \Box_x B} R\supset
$$

Clearly the search goes on forever because of the new accessibility relations that are generated by applications of the right rules for the indexed modalities, together with *Trans*. A finite countermodel may nevertheless be exhibited by a suitable truncation of the otherwise infinite countermodel provided by the completeness proof.

Following the method of finitization of countermodels generated by proof search in a labelled calculus, presented for intuitionistic propositional logic in Negri (2014a) and for multi-modal logics in Garg et al. (2012), a *saturation* condition for branches on a reduction tree is defined. Intuitively, a branch is saturated when its leaf is not an initial sequent nor a conclusion of $L\bot$, and when it is closed under all the rules except for $R\Box_x$ in case it generates a loop modulo new labelling. To obtain the finite countermodel, define a partial order through the reflexive and transitive closure of the similarity relation together with a relation that witnesses such loops. Let $\downarrow\Gamma$ ($\downarrow\Delta$) be the union of the antecedents (succedents) in a branch from the endsequent up to $\Gamma \Rightarrow \Delta$.

Let us define the following sets of formulas:

$$\mathcal{F}^1_{\Gamma \Rightarrow \Delta}(w) \equiv \{A \mid w : A \in \downarrow\Gamma\} \cup \{\Box_x A \mid y : \Box_x A, yR_x w \in \Gamma\}$$

$$\mathcal{F}^2_{\Gamma \Rightarrow \Delta}(w) \equiv \{A \mid w : A \in \downarrow\Delta\}$$

and let $w \leq_{\Gamma \Rightarrow \Delta} y$ iff $\mathcal{F}^i_{\Gamma \Rightarrow \Delta}(w) \subseteq \mathcal{F}^i_{\Gamma \Rightarrow \Delta}(y)$ for $i = 1, 2$.

DEFINITION 18. *A branch in a proof search up to a sequent $\Gamma \Rightarrow \Delta$ is* **saturated** *if the following conditions are satisfied:*

1. If $w$ is a label in $\Gamma, \Delta$, then $w = w$ and $wR_x w$ are in $\Gamma$.
2. If $wR_x y$ and $yR_x z$ are in $\Gamma$, then $wR_x z$ is.
3. If $wR_w x$ is in $\Gamma$, then $x = w$ is.
4. If $w, x, y$ are labels in $\Gamma, \Delta$, then either $wR_x y$ or $yR_x w$ is in $\Gamma$
5. There is no $w$ such that $w : \bot$ is in $\Gamma$.

6. If $w : A\&B$ is in $\downarrow\Gamma$, then $w : A$ and $w : B$ are in $\downarrow\Gamma$.
7. If $w : A\&B$ is in $\downarrow\Delta$, then either $w : A$ or $w : B$ is in $\downarrow\Delta$.
8. If $w : A \vee B$ is in $\downarrow\Gamma$, then either $w : A$ or $w : B$ is in $\downarrow\Gamma$.
9. If $w : A \vee B$ is in $\downarrow\Delta$, then $w : A$ and $w : B$ are in $\downarrow\Delta$.
10. If $w : A \supset B$ is in $\downarrow\Gamma$, then either $w : A$ is in $\downarrow\Delta$ or $w : B$ is in $\downarrow\Gamma$.
11. If $w : A \supset B$ is in $\downarrow\Delta$, then $w : A$ is in $\downarrow\Gamma$ and $w : B$ is in $\downarrow\Delta$.
12. If $w : \Box_x A$ and $wR_x y$ are in $\Gamma$, then $y : A$ is in $\downarrow\Gamma$.
13. If $w : \Box_x A$ is in $\downarrow\Delta$, then either

    a. for some $y$, there is $wR_x y$ in $\Gamma$ and $y : A$ is in $\downarrow\Delta$, or
    b. for some $y$ such that $y \neq w$, there is $yR_x w$ in $\Gamma$ and $w \leq_{\Gamma \Rightarrow \Delta} y$.

14. If $w : A \mathbin{\square\!\!\rightarrow} B$ is in $\Gamma$, then either $z : A$ is in $\downarrow\Delta$ for $z$ in $\Gamma, \Delta$, or for some $y$, $y : A, y : \Box_w(A \supset B)$ is in $\Gamma$.
15. If $w : A \mathbin{\square\!\!\rightarrow} B$ is in $\downarrow\Delta$, then $y : A$ is in $\downarrow\Gamma$ and either $z : A$ or $z : \Box_w(A \supset B)$ is in $\downarrow\Delta$ for $z$ in $\Gamma, \Delta$.

Notice that this definition blocks the proof search in the example above when it produces the formula $t : \Box_x A$ because of clause 13.b (since we then have $t \leq_{\Gamma \Rightarrow \Delta} z$). The finite countermodel is defined by the sets $\downarrow\Gamma, \downarrow\Delta$.

PROPOSITION 19. *The finite countermodel defined by the saturation procedure is a comparative similarity system.*

Notice further that by the subterm property the number of distinct formulas in the sequents of an attempted proof search is bounded. Since duplication of the same labelled formulas is not possible by hp-admissibility of contraction, the following holds:

THEOREM 20. *The system* **G3LC** *allows a terminating proof search.*

**Proof.** Let $F$ be the set of (unlabelled) subformulas of the endsequent and consider a string of labels $w_0 R_x w_1, w_1 R_x w_2, w_2 R_x w_3, \ldots$ generated by the saturation procedure. For an arbitrary $x_j$ consider the values of the sets $\mathcal{F}^i(x_k)$ for $k < j$ at the step in which $x_j$ was introduced. Clearly $\mathcal{F}^i(x_j) \not\subseteq \mathcal{F}^i(x_k)$ or else $x_j$ would not have been introduced. So each new label corresponds to a new subset of $F \times F$. Since the number of these subsets is finite, also the length of each chain of labels must be finite. ∎

## 7 Conclusion

This paper presented **G3LC**, a Gentzen-style sequent calculus for David Lewis' logic of counterfactuals **VC**, and proved it sound and complete with respect to Lewis' semantics. In **G3LC**, substitution of labels and left and right weakening and contraction are height-preserving admissible and cut is admissible. Moreover, all the rules are invertible. Finally, a decidability result follows, based on a bounded procedure of root-first proof search that for any given sequent either provides a derivation or a countermodel.

In his book *Counterfactuals*, Lewis presents a class **V** of axiomatic systems for conditional logics, among which is **VC**. A detailed deductive analysis of the entire class, as well as of conditional logics that are based on alternative versions of Lewis' semantics, is left for further work.

The first tableau proof systems for counterfactuals have been presented by de Swart (1983). These systems can be read either as Beth-tableaux systems, with rules for signed formulas, or as sequent systems, and they cover Stalnaker's system **VCS** and Lewis's system **VC**. The primitive connective chosen in de Swart's work is $\leq$, with the formula $A \leq B$ read as "*A is at least as possible as B*". We use instead the counterfactual conditional $A \,\square\!\!\rightarrow B$, read as "*If A were the case, then B would be the case*". These two connectives are interdefinable, as shown by Lewis, but a different choice of the primitive connective clearly gives origin to different proof systems. De Swart gives direct and constructive completeness proofs by using the calculi for defining a systematic proof search procedure that either gives a proof or a finite countermodel. Also in our system the completeness proof is direct and constructive, but the countermodel is constructed directly from the syntactic elements contained in a failed proof search branch, whereas in the Beth-tableaux approach the possible worlds are defined by nodes in the open search tree. There are other important differences which highlight the usefulness of the labelled approach that we have followed.

De Swart's system has, in addition to the standard classical propositional rules, a number $m \cdot n$ of distinct rules $F \leq (m, n)$ for each $m$, $n$, where $m$ and $n$ are positive integers that denote, respectively, the number of signed formulas of the form $F(A \leq B)$ or of the form $T(A \leq B)$ considered as principal formulas of the rule. Each such rule has the effect of discarding all the other formulas, which results in a lack of invertibility. It follows that in the proof search procedure what needs to be explored is not a single tree, but a set of trees.[6] Lastly, in our approach the rules are motivated through a robust meaning explanation that respects the general guidelines of inferentialism, as emphasized in Negri and von Plato (2015). On the contrary, the rules of the unlabelled approach seem to involve a not fully explicable genesis, being found "by the method of trial and error" (cf. de Swart 1983, p. 6). The inherent risk in the lack of a full methodological transparency became evident in a later correction by Gent (1992), who gave an example of a valid formula not derivable in de Swart's system and proposed an alternative sound a complete system for **VC**, while maintaining the main features of de Swart's original system.

Also Lellman and Pattinson (2012) present an unlabelled sequent calculus for Lewis' logic with the binary connective "at least as possible as" as primitive. The calculi are obtained through a procedure of *cut-elimination by saturation* which consists in closing a given set of rules under cut adding new rules. As a result, an optimal PSPACE complexity and Craig interpolation are established.

---

[6] In an example detailed in de Swart (1983, p. 10–11), a proof search for a sequent that contains only two formulas of the form $F(A \leq B)$ and $F(B \leq D)$ results, because of all the combinatorial possibilities, in the construction of 24 different partial trees.

The work by Olivetti et al. (2007) presents a labelled sequent calculus for Lewis conditional logics and is close to the present approach as it follows the methodology of Negri (2005). However, it rests crucially on the limit assumption. In so far as Lewis' preferred interpretation of the counterfactual conditional rejects the limit assumption (see Lewis, 1973a, pp. 20-21), the strategy followed in the present paper appears to be a more faithful proof-theoretic analysis of Lewis' work.

## BIBLIOGRAPHY

[1] Dyckhoff, R. and S. Negri (2012) Proof analysis in intermediate logics. *Archive for Mathematical Logic*, vol. 51, pp. 71–92, 2016.

[2] Dyckhoff, R. and S. Negri (2013) A cut-free sequent system for Grzegorczyk logic with an application to the Gödel-McKinsey-Tarski embedding. *Journal of Logic and Computation*, vol. 26, pp. 169-187.

[3] Dyckhoff, R. and S. Negri (2015) Geometrization of first-order logic, *The Bulletin of Symbolic Logic*, vol. 21, pp 123–163.

[4] Fitting, M. (2007) Modal proof theory. In P. Blackburn, J. Van Benthem, and F. Wolter (eds) *Handbook of Modal Logic*, pp. 85–138, Elsevier.

[5] Garg, D., V. Genovese, and S. Negri (2012) Countermodels from sequent calculi in multi-modal logics. *LICS 2012*, IEEE Computer Society, pp. 315–324.

[6] Gasquet, O. and A. Herzig (1996) From Classical to Normal Modal Logics. In H. Wansing (ed) *Proof Theory of Modal Logic* Applied Logic Series vol. 2, pp. 293-311, Springer.

[7] Gent, I. P. (1992) A sequent- or tableau-style system for Lewis's counterfactual logic VC. *Notre Dame Journal of Formal Logic*, vol. 33, pp. 369–382.

[8] Giordano, L., V. Gliozzi, N. Olivetti, and C. Schwind (2008) Tableau calculi for preference-based conditional logics: PCL and its extensions. *ACM Transactions on Computational Logic*, vol. 10, no. 3/21, pp. 1–45.

[9] Kripke, S. (1963) Semantical analysis of modal logic I. Normal modal propositional calculi. *Zetschrift für mathematische Logik und Grundlagen der Math.*, vol. 9, pp. 67–96.

[10] Lellman, B. and D. Pattinson (2012) Sequent Systems for Lewis' conditional logics. In L. Farinas del Cerro, A. Herzig and J. Mengin (eds.), *Logics in Artificial Intelligence*, pp. 320-332.

[11] Lewis, D. (1973a) *Counterfactuals*. Blackwell.

[12] Lewis, D. (1973b) Counterfactuals and comparative possibility. *Journal of Philosophical Logic*, vol, 2, pp. 418–446, 1973.

[13] Negri, S. (2003) Contraction-free sequent calculi for geometric theories, with an application to Barr's theorem. *Archive for Mathematical Logic*, vol. 42, pp. 389–401.

[14] Negri, S. (2005) Proof analysis in modal logic. *Journal of Philosophical Logic*, vol. 34, pp. 507–544.

[15] Negri, S. (2009) Kripke completeness revisited. In G. Primiero and S. Rahman (eds.), *Acts of Knowledge - History, Philosophy and Logic*, pp. 247–282, College Publications.

[16] Negri, S. (2014) Proof analysis beyond geometric theories: from rule systems to systems of rules. *Journal of Logic and Computation*, doi: 10.1093/logcom/exu037.

[17] Negri, S. (2014a) Proofs and countermodels in non-classical logics. *Logica Universalis*, vol. 8, pp. 25-60.

[18] Negri, S. and J. von Plato (1998) Cut elimination in the presence of axioms. *The Bulletin of Symbolic Logic*, vol. 4, pp. 418–435.

[19] Negri, S. and J. von Plato (2001) *Structural Proof Theory*. Cambridge University Press.

[20] Negri, S. and J. von Plato (2011) *Proof Analysis*. Cambridge University Press.

[21] Negri, S. and J. von Plato (2015) Meaning in use. In H. Wansing (ed) *Dag Prawitz on Proofs and Meaning*, pp. 239–257, Trends in Logic, Springer.

[22] Negri, S. and G. Sbardolini (2015) Proof analysis for Lewis counterfactuals. *The Review of Symbolic Logic*, in press.

[23] Olivetti, N., G. L. Pozzato and C. Schwind (2007) A Sequent Calculus and a Theorem Prover for Standard Conditional Logics. *ACTM Transactions on Computational Logics (TOCL)*, vol. 8/4, 22/pp. 1–51.

[24] Priest, G. (2008) *An Introduction to Non-Classical Logic*, Cambridge University Press.

[25] de Swart, H. C. M. (1983) A Gentzen- or Beth-type system, a practical decision procedure and a constructive completeness proof for the counterfactual logics VC and VCS. *The Journal of Symbolic Logic*, vol. 48, pp. 1–20.

# On the "no deadlock criterion": from Herbrand's theorem to *Geometry of Interaction*

Paolo Pistone

> ABSTRACT. Herbrand's theorem provides a characterization of first-order validity which allows, in a sense, to "eliminate" quantifiers: one has to test for the absence of "deadlocks" in a sequence of unification problems induced by quantifier-free formulae. Similarly, *Geometry of Interaction* provides a characterization of validity for linear logic which, in a sense, allows to "eliminate" logical connectives: one has to test for the absence of "shortcircuits" in the nets representing possible proofs.
> 
> Hence these two interpretations seem to escape the usual circularity affecting definitions of validity, where quantifiers are explained by "meta-quantifiers", implications by "meta-implications" etc. We briefly present these two perspectives and discuss an approach to validity based on a "no deadlock criterion".

## 1 Introduction

In the philosophy of logic it is often argued (see for instance [Pra71, Dum91, Cel06]) that the usual explanations of the logical constants, as relying on model-theoretic or proof-theoretic notions of validity, are circular in the following sense: the rules involved in the explanation are essentially of the same form of the rules to be explained.

For instance, here's how Prawitz comments this "shortcoming" in the model-theoretic explanation of first order quantifiers:

> Whether, e.g., a sentence $\exists x \neg P(x)$ follows logically from a sentence $\neg \forall x P(x)$ depends according to this definition on whether $\exists x \neg P(x)$ is true in any model $(D, S)$ in which $\neg \forall x P(x)$ is true. And this again is the same as to ask whether there is an element $e$ in $D$ that does not belong to $S$ whenever it is not the case that every $e$ in $D$ belongs to $S$, i.e. we are essentially back to the question whether $\exists x \neg A(x)$ follows from $\neg \forall x A(x)$. [Pra71]

The conditions for deriving (quantified) consequences from a quantified statement are stipulated in such a way that one has to derive (quantified) consequences from a quantified statement in order to verify that such conditions hold. In a word, quantifiers are explained by appeal to (meta-) quantifiers.

In this paper we do not enter into the epistemological challenges involved in these remarks; rather, we consider two related interpretations of, respectively, first-order logic and propositional logic, which seem to escape this circularity.

The first example is provided by Herbrand's theorem (1930), which allows to express the validity conditions for a first-order formula as conditions whose verification involves checking the validity of *quantifier-free* formulae, hence eliminating quantifiers.

The second example we consider is *Geometry of Interaction*, launched in 1988 [Gir89b] in order to provide a purely mathematical description of Gentzen's *Hauptsatz*. In *GoI* correct proofs (i.e. those *wirings* which represent actual sequent calculus derivations) can be characterized by a geometrical criterion making no reference to logical rules.

Both perspectives involve a characterization of validity by means of a "no deadlock" or "no shortrip" criterion, i.e. a criterion which demands to check for the absence of circular dependencies in a possible proofs. This intuition seems to indicate a new approach to validity (that we do not develop here in detail): a valid formula is one that can be asserted or defended without running into "deadlocks", i.e. circular expectations.

## 2 Herbrand's theorem

**An equivalent of Gentzen's *Hauptsatz*** Herbrand's theorem roughly asserts that a first-order formula $A$ is valid if and only if a certain quantifier-free formula is a tautology. Herbrand was originally looking for a "finitary" (in the sense of Hilbert's program) version of Löwenheim's theorem, which asserts, again roughly, that a formula $A$ of first-order logic is not valid if and only if its negation is satisfied by a countable model.

Herbrand's result is equivalent to Gentzen's *Hauptsatz*, and constitutes, with it, one of the first *structural* results in proof-theory. Indeed, if $A$ is valid, then a proof of $A$ can be recovered from a quantifier-free tautology by means of three rules corresponding, respectively, to the introduction rules for the quantifiers and the contraction rule.

As remarked by Van Heijenoort in [VH82],

> Le système basé sur les trois règles de Herbrand est, historiquement, le premier exemple de ce qu'on appelle aujourd'hui les systèmes sans coupure; il jouit aussi de ce qu'on appelle la propriété de la sous-formule. [VH82]

For a survey of the applications and developments of Herbrand's theorem in connection with cut-elimination theorems, see [Kre51, Gir82, Koh08].

**Herbrand expansions** Let us take a first-order formula in prenex form

$$A = \exists x_1 \forall y_1 \exists x_2 \forall y_2 B(x_1, x_2, y_1, y_2) \tag{1}$$

The *Herbrandized form* $A_H$ of $A$ is the formula below

$$A_H = \forall f \forall g \exists x_1 \exists x_2 B\big(x_1, x_2, f(x_1), g(x_1, x_2)\big) \tag{2}$$

where the universal variables $y_i$ are replaced by functional terms containing the existential variables "above" $y_i$ (where an existential variable $x_j$ is above a universal variable $y_i$ if the quantifier $\forall y_i$ occurs in $A$ in the scope of the quantifier $\exists x_j$).

This transformation allows to "permute universal quantifiers upwards" as function quantifiers, preserving validity.

Let T be the first-order language generated by a finite stock of constants (indeed just one constant 0), variables $x, y, z, \ldots$ and a finite stock of symbols for $n$-ary functions (in the case above a unary function f and a binary function g).

In order to define sequences of terms of T, one starts with an initial non empty set $\mathcal{CC}_0$ of closed terms of T and, for $k \in \mathbb{N}$, defines the set $\mathcal{CC}_{k+1}$ as the set of the terms formed by applying a function symbol of T to terms in $\mathcal{CC}_k$. A "suite de champ fini" ([Her67]) is a sequence of terms $t_n$ such that $t_0 \in \mathcal{CC}_0$ and, for any $k \in \mathbb{N}$, $t_{k+1} \in \mathcal{CC}_{k+1} - \mathcal{CC}_k$.

Let $A'_H$ be the quantifier-free part of $A_H$, where first-order terms are now taken in T:

$$B(x_1, x_2, \mathtt{f}(x_1), \mathtt{g}(x_1, x_2)) \qquad (3)$$

Let $n \geq 1$, and let $s$ be a map associating, with every existential variable $x$, a "suite de champ fini" $s_n^x$; the $n$-th *Herbrand's expansion* $A_{H,s}^n$ of $A$ is the quantifier-free formula

$$A_{H,s}^n := A'_H[x_1 \mapsto s_0^{x_1}, x_2 \mapsto s_0^{x_1}] \vee \cdots \vee A'_H[x_1 \mapsto s_n^{x_1}, x_2 \mapsto s_n^{x_2}] \qquad (4)$$

In particular, let $id$ be the trivial map associating, with each existential variable $x$, the sequence which is constantly equal to $x$; then we note by $A_H^n$ the formula $A_{H,id}^n$.

Herbrand's theorem can now be formulated as follows:

**THEOREM 1** (Herbrand's theorem, 1930). *$A$ is valid if and only if, for a certain $p \geq 1$ and a certain $s$, $A_{H,s}^p$ is a tautology.*

**A recursive interpretation of formulae** The proof of Herbrand's theorem allows, as remarked in [Kre51], to devise a primitive recursive interpretation of first-order formulae in which quantifiers are not interpreted by means of quantifiers.

A *substitution* $\theta$ is a map from first-order variables to elements of T. Given a first-order formula $A$, the formula $A\theta$ is obtained by applying $\theta$ to all variables occurring free in $A$.

The interpretation of a first order formula $A$ is given by the primitive recursive sequence $A_H^n$ of quantifier-free formulae. Then, an equivalent formulation of theorem (1) is the following:

**THEOREM 2.** *$A$ is valid if and only if, for a certain $p \geq 1$ and a certain substitution $\theta$, $A_H^p \theta$ is a tautology.*

First remark that, letting $cl(A)$ indicate the universal closure of $A$ (obtained by closing universally all free variables and function symbols occurring in $A$), then, for all $n$ and $\theta$, one can easily derive $A$ from $cl(A_H^n \theta)$ by using only right introduction rules for the quantifiers and contraction rules.

Hence, in order to assess the validity of $A$, one has to test whether $A_H^n \theta$ is a tautology for a certain substitution $\theta$, for $n = 0, n = 1, n = 2, \ldots$. This is where Herbrand introduces a fortunate idea: for every $n$, the verification that $A_H^n \theta$ is a tautology can be done by solving a system of equations over the language T.

For instance, in order to check whether $(P(t,u) \vee \neg P(t',u'))\theta$ is a tautology for some $\theta$, where $P$ is an atomic predicate, it suffices to look for a $\theta$ equalizing the two equations

$$t = t' \tag{5}$$
$$u = u' \tag{6}$$

[Her67] contains indeed the first formulation of *unification theory* ([Rob65]), the theory which deals with solving systems of equations over first-order terms. A system of equations $E$ over T is a finite set of equations $t_1 = u_1, \ldots, t_n = u_n$, where $t_1, u_1, \ldots, t_n, u_n \in$ T. A *unifier* for $E$ is a substitution $\theta$ such that, for all $1 \leq i \leq n$, $t_i\theta$ is syntactically equal to $u_i\theta$. The unification problem for a system $E$ is the problem of finding a unifier for $E$.

The first algorithm to decide the unification problem was given in [Rob65], though its main ideas can already be found in [Her67] (p. 96). In particular, given an equation $t = u$, one must consider two main cases:

- if $t = f(t_1, \ldots, t_n)$ and $u = g(u_1, \ldots, u_m)$ one must verify that $f = g$ (and hence $n = m$), and solve the system made of the syntactically simpler equations $t_1 = u_1, \ldots, t_n = u_n$;

- if $t = x$ is a variable, then two subcases arise:

  1. if the variable $x$ does not occur in $u$, then one can take the equation as a definition of $x$ and replace all other occurrences of $x$ in other equations by $u$;

  2. if the variable $x$ occurs in $u$, then the system cannot be solved: for instance, the equation $x = f(x,y)$ cannot be solved. Indeed, it one took this equation as a "circular" definition of $x$, then the algorithm would end into the "deadlock"

  $$x = f(x,y) = f(f(x,y),y) = f(f(f(x,y),y),y) = \ldots \tag{7}$$

  when trying to eliminate this equation by applying clause 1.

EXAMPLE 3. The system made of the equation

$$g(f(x),z) = g(y,g(f(x),y)) \tag{8}$$

is unifiable: take the substitution $\theta = (x \mapsto x, y \mapsto f(x), z \mapsto g(f(x),f(x)))$

EXAMPLE 4. The system made of the equations

$$g(f(x),z) = g(y,g(f(x),y))$$
$$x = y \tag{9}$$

is not unifiable, as it leads to the "deadlock" equation

$$x = f(x) \tag{10}$$

In sum, the assessment of the validity of a formula $A$ is obtained, through the recursive interpretation by means of the sequence $A_H^n$, in two steps:

1. look for a positive integer $p$ such that $A_H^p \theta$ is a propositional tautology, for some $\theta$, by progressively testing unification problems;

2. if such a $p$ is found, derive $A$ from $cl(A_H^p \theta)$.

If $A$ turns out to be valid, then the two steps above correspond to the two parts of a cut-free proof of $A$: first, a propositional (cut-free) derivation of $A_H^N$; second, a sequence of introduction rules for quantifiers and contraction rules.

For a simple example take the formula $D$ below

$$D = \exists x \forall y (P(x) \Rightarrow P(y)) \tag{11}$$

The first expansion of $D$, i.e. the formula $D_H^1 = P(x) \Rightarrow P(\mathtt{f}(x))$ is not a tautology: the system

$$x = \mathtt{f}(x) \tag{12}$$

cannot be solved, as it leads to the deadlock

$$x = \mathtt{f}(x) = \mathtt{f}(\mathtt{f}(x)) = \ldots \tag{13}$$

However, the second expansion $D_H^2$, i.e. the formula

$$(P(x) \Rightarrow P(\mathtt{f}(x))) \vee (P(y) \Rightarrow P(\mathtt{f}(y))) \tag{14}$$

becomes a tautology as soon as one chooses $\theta(x) = 0$ and $\theta(y) = \mathtt{f}(0)$.

Take now the invalid formula $C$ below

$$\forall x \exists y P(x,y) \Rightarrow \exists y \forall x P(x,y) \tag{15}$$

whose Herbrandized (prenex) form is

$$\forall f \forall g \exists x \exists y (P(x, f(x)) \Rightarrow P(g(y), y)) \tag{16}$$

In order to see that $C$ is not valid, we turn to its first expansion $D_H^1$, which produces the system

$$x = \mathtt{g}(y) \tag{17}$$
$$y = \mathtt{f}(x) \tag{18}$$

which cannot be solved, as it leads to the deadlock

$$x = \mathtt{g}(y) = \mathtt{g}(\mathtt{f}(x)) = \mathtt{g}(\mathtt{f}(\mathtt{g}(y))) = \mathtt{g}(\mathtt{f}(\mathtt{g}(\mathtt{f}(x)))) = \ldots \tag{19}$$

One can then easily show, by induction, that for every $n$, the $n$-th expansion $D_H^n$ produces a deadlock similar to (19).

**Quantifiers explained away?** Herbrand's theorem provides an interpretation of first-order formulae which assigns them conditions for validity which do not involve quantifiers in a circular way.

Take the invalid formula $C$. The model-theoretic refutation of $C$ is obtained by looking for a *counter-model* of $C$, i.e. a model $\mathcal{M}$ such that, *for every $a$* in the support $M$ of $\mathcal{M}$, *there exists* a $b \in M$ such that $P[a,b]$ is true in $\mathcal{CM}$ but *for no $b$ it holds that for every $a$* $P[a,b]$ is true in $\mathcal{M}$. Hence the condition for verifying that a model $\mathcal{M}$ is a counter-model of $C$ reproduces, in the meta-language, the quantifiers of $C$ as well as their mutual combination.

The refutation of $C$ that we sketched above, on the contrary, allows to "eliminate quantifiers" by introducing the function symbols $\mathtt{f}, \mathtt{g}$ to express the "dependencies" between variables. The equations (17) express, intuitively, the fact that the witnesses for the variable $x$ might depend on the value assigned to the variable $y$ and that the witnesses for the variable $y$ might depend on the value assigned to the variable $x$. Hence a refutation of $C$ is obtained by remarking that these two constraints are reciprocally incompatible: in order to find a value for $x$ one must keep waiting for a value for $y$ and, vice-versa, in order to find a value for $y$ one must keep waiting for a value for $x$ (a typical "deadlock" situation).

## 3  *Geometry of Interaction*

**Proofs as nets** The program of *Geometry of Interaction* (*GoI*) was launched in 1989 by Jean-Yves Girard in order to devise a geometrical semantics of proofs, based on the fine analysis of cut-elimination provided by linear logic ([Gir87]).

At the heart of Girard's original program there was the consideration of the centrality of Gentzen's *Hauptsatz* for the foundations of logic: whereas Hilbert's finitist program failed as it "aimed at an absolute elimination of infinity" ([Gir89b]), the cut-elimination procedure provides a *finite dynamics* by which infinite notions in proofs are progressively eliminated.

> Hilbert's mistake, when he tried to express the infinite in terms of the finite was of a reductionist nature: he neglected the dynamics. The dynamics coming from the elimination of infinity is so complex that one can hardly see any reduction there. But once reductionism has been dumped, Hilbert's claim becomes reasonable: infinity is an undirect way to speak of the finite; more precisely infinity is about finite dynamical processes. [Gir89b]

However, the concrete manipulation of derivations in sequent calculus constitutes a highly complex and somehow "bureaucratic" task. Whence the idea of developing a purely mathematical interpretation of sequent calculus derivations, in order "to find out the geometrical meaning of the *Hauptsatz*, i.e. what is hidden behind the somewhat boring syntactical manipulations it involves" [Gir89b].

A decisive step towards a geometrical interpretation of proofs came from the development, concomitant with the discovery of linear logic, of the notion of *proof-net* ([Gir87]), a graph-theoretic representation of sequent calculus derivations. An important notion associated with that of proof-net is the one of *path*: a path represents, intuitively, a way to travel through the graph, which can be though as a net or a circuit, the conclusions of the derivation corresponding to its external gates. Hence a path can "enter" or "exit" the net through one of its gates, as well as getting stuck moving in circle inside the net (as in the case of a shortcircuit).

The passage from a sequent calculus derivation $d$ to a proof-net $D$ can be defined inductively along the clauses below (which are limited to the multiplicative case[1]), which also define the associated paths:

i. The identity axiom is translated into a graph with two vertices (labeled $A$ and $\sim A$) and an arrow between them:

$$\frac{}{\vdash \sim A, A} \ (Ax) \qquad \mapsto \qquad A \quad \sim A \qquad (20)$$

The two red arrows indicate the two possible paths along $D$.

ii. An application of the cut rule to two derivations $d_1$ and $d_2$ is translated into a graph consisting of $D_1$, $D_2$ and an extra arrow between the vertices labeled by $A$ and $\sim A$:

$$\frac{\vdots d_1 \qquad \vdots d_2}{\vdash \Gamma, A \quad \vdash \Gamma', \sim A} \ (cut) \qquad \mapsto \qquad D_1 \quad D_2 \quad A \quad \sim A \qquad (21)$$

iii. An application of the left introduction rule for (linear) implication to derivations $d_1, d_2$ is translated into a graph consisting of $D_1, D_2$, a new vertex labeled $\sim (A \multimap B)$ and new arrows linking this vertex to the vertices (respectively of $D_1$ and $D_2$) labeled $A$ and $\sim B$:

$$\frac{\vdots d_1 \qquad \vdots d_2}{\vdash \Gamma, A \quad \vdash \Gamma', \sim B} \ (\multimap L) \qquad \mapsto \qquad \underset{\sim (A \multimap B)}{D_1 \quad D_2} \quad + \quad \underset{\sim (A \multimap B)}{D_1 \quad D_2} \qquad (22)$$

The two graphs above indicate two distinct ways to define a path along the same proof-net.

iv. An application of the right introduction rule for (linear) implication to a derivation $d$ is translated into a graph consisting of $D$, a new vertex labeled $A \multimap B$ and new arrows linking this vertex to the vertices labeled $\sim A$ and $B$:

---

[1] We omit here the treatment of the general case (i.e. full linear logic, allowing to interpret intuitionistic and classical logic) which requires the introduction of the notion of *box* (see [Gir87]).

$$\begin{array}{c} \vdots\, d \\ \vdash \Gamma, \sim A, B \\ \hline \vdash \Gamma, A \multimap B \end{array} (\multimap R) \qquad \mapsto$$

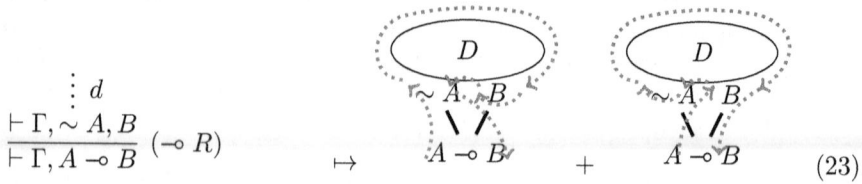

(23)

Again, the two graphs above indicate two distinct ways to define a path along the same proof-net.

We defined a map $d \mapsto D$ which associates, with a sequent calculus derivation, a proof-net $D$ and a set $P_D$ of paths along $D$. We obtain then a first definition of proof-nets:

DEFINITION 5 (proof-net (1)). A proof-net is a graph $D$ which is obtained by translating a sequent calculus derivation $d$ following the clauses above.

Let, for simplicity, $D$ be a proof-net having exactly one gate $A$ (corresponding to a derivation $d$ with exactly one conclusion). In the terminology of [Gir87], a path entering $D$ through gate $A$ and exiting $D$ from $A$ after transiting through every vertex of $D$ exactly twice (in opposite direction) is called a *longtrip*; all other paths (which might enter through gate $A$ and never exit or might never pass through a gate) are called *shortrips* and correspond intuitively to "shortcircuits" in the net.

A fundamental property of proof-nets is then that every path in $P_D$ is a longtrip. Hence, one will never be able to design a graph inducing a "shortcircuit" if one follows, in the construction of the graph, the inductive translation from sequent calculus.

Starting from this remark, one can generalize as follows: one defines an arbitrary graph constructed with the *links* appearing in the definition above as a *proof-structure*. For any such graph $G$, the set of paths $P_G$ is still well-defined. Remark that not all proof-structures are proof-nets: for instance the proof-structure below

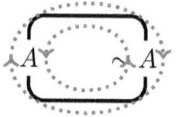

(24)

does not come from any sequent calculus derivation (as it induces two shortrips).

The property above can now be used to obtain a second definition of proof-net:

DEFINITION 6 (proof-net (2)). A *proof-net* is a proof-structure $G$ such that every path in $P_G$ is a longtrip.

Thus one has, on the one hand, an inductive definition of proof-net, as the translation of a sequent calculus derivation, and, on the other hand, a purely geometrical definition of proof-net, with no reference to sequent calculus. The main theorem of the theory asserts then that the two definitions (5) and (6) are equivalent: the graphs characterized by the "no shortrip" criterion are exactly those which come from sequent calculus (i.e. that can be "sequentialized").

Hence this second characterization of proof-nets provides a notion of correctness for proofs which makes no reference to logical rules: it is rather based on a global property of their graph-theoretic representation.

**The unification semiring** Several formulations of *GoI* exist in the literature (starting from the original one in [Gir89a], based on $C^*$-algebras). Here we adopt a "finitist" formulation, based on Herbrand's unification, that can be found in [Gir95, Gir13, ABPS14, Bag14].

Let us take again our first-order language T. A *flow* is an expression of the form

$$t \leftharpoonup u \qquad (25)$$

where $t, u \in$ T are terms having exactly the same variables. Flows are the fundamental bricks to build paths. A set of flows is called a *wiring*. A path is obtained by composing flows following the law below:

$$(t \leftharpoonup u) \cdot (t' \leftharpoonup u') = t\theta \leftharpoonup u'\theta \qquad \theta \in MGU(u, t') \qquad (26)$$

where $MGU(u, t')$ denotes the set of *most general unifiers* of $u, t'$[2]. Remark that, in case $MGU(u, t)$ is empty (i.e. $u, t'$ are not unifiable), then composition fails (this is indeed a partial operation).

With the partial composition law (26), a product on wirings can be defined by

$$V \cdot W = \{t \leftharpoonup u | t \leftharpoonup u = (t_1 \leftharpoonup u_1) \cdot (t_2 \leftharpoonup u_2), t_1 \leftharpoonup u_1 \in V, t_2 \leftharpoonup u_2 \in W\} \qquad (27)$$

with neutral element $\mathtt{id} = \{x \leftharpoonup x\}$, inducing a structure of semiring (see [Bag14])[3].

Given a derivation $d$ of the sequent $\vdash \Gamma$, we first associate with each occurrence of formula $A$ in $\Gamma$ a unary function symbol $\mathtt{p}_A(x) \in$ T. The subformulae of $A$ can then be defined by means of two unary function symbols $\mathtt{l}, \mathtt{r}$, corresponding to "left" and "right". For instance, if $A = B \multimap C$, then

$$\mathtt{p}_B(x) := \mathtt{p}_A(\mathtt{l}(x)) \qquad \mathtt{p}_C(x) := \mathtt{p}_A(\mathtt{r}(x)) \qquad (28)$$

Hence any two distinct occurrences of formulae $A, B$ in the sequents in $d$ correspond to first-order terms $\mathtt{p}_A(x), \mathtt{p}_B(x)$ which are, as the reader can easily verify, not unifiable[4].

---

[2] Indeed, a central result on first-order unification is that, if two terms $t, u$ are unifiable (i.e. if the system $\{t = u\}$ is unifiable), then they have a *most general unifier*, i.e. a unifier $\theta$ such that all other unifier $\theta'$ can be decomposed as $\theta \circ \theta''$, for some substitution $\theta''$. In a sense, a m.g.u. is a "mother of all unifiers". Moreover, all m.g.u. for $t, u$ are equivalent up to a permutation of variables.

[3] If once considers flows of the form $\lambda(t \leftharpoonup u)$, where $\lambda$ is a complex coefficient, then wirings can be written under the form $\sum_i \lambda_i(t_i \leftharpoonup u_i)$ and form a $C^*$-algebra (called *unification algebra*) of operators acting over the Hilbert space $\ell^2(\mathtt{T})$. This is how one can recover the original formulation of *GoI* from this "finitist" version.

[4] To be more precise, one wants rather these terms to be not *matchable*. Matching is a variant of unification where one consider terms "up to permutation of variables". To achieve this, one must introduce a new unary function symbol $\mathtt{g}$ and replace the term $\mathtt{p}_A(x)$ by the term $\mathtt{q}_A(x) := \mathtt{p}_A(\mathtt{g}(x))$ (see [Gir13]).

The arrows described in the inductive definition of proof-nets can be replaced by flows: to an arrow from an occurrence of formula $A$ to an occurrence of formula $B$ one associates the flow

$$p_A(x) \leftarrow p_B(x) \qquad (29)$$

Hence the definition (5) of a proof-net $D$ immediately induces the definition of a wiring $W_d$ associated with a derivation $d$.

**Execution and cut-elimination** The logic programmer must have noticed that the composition law (26) for flows is just a particular instance of Robinson's *resolution rule* ([Rob65]):

$$\frac{\vdash \Gamma, A(t) \quad \vdash \Gamma', A(u)}{\vdash \Gamma\theta, \Gamma'\theta} (Res, \theta) \qquad \theta \in MGU(t, u) \qquad (30)$$

In logic programming the execution of a program is obtained by generating all possible "resolution paths" starting from a finite set of sequents, i.e. by successively applying resolution wherever possible until a sequent is obtained to which resolution can no more be applied.

In *GoI*, given a wiring W, we can generate all paths by successively composing W with itself (i.e. by tentatively composing each other all flows in W). Hence one can define an *execution operator* $Ex(\mathtt{W})$:

$$Ex(\mathtt{W}) = \mathtt{W} + \mathtt{W}^2 + \mathtt{W}^3 + \cdots = \sum_{n}^{\infty} \mathtt{W}^n = (1 - \mathtt{W})^{-1} \qquad (31)$$

which generates all possible paths. Observe that the last equation in (31) holds just in case the infinite series of the iterates of $W$ is convergent; if this is not the case then $Ex(\mathtt{W})$ is not defined (diverging executions correspond then to diverging computations).

Let us consider a cut between two derivations $d_1, d_2$:

$$\frac{\begin{matrix}\vdots\, d_1 & \vdots\, d_2 \\ \vdash \Gamma, A & \vdash \Gamma', \sim A\end{matrix}}{\vdash \Gamma, \Gamma'} (cut) \qquad (32)$$

Following definition (5), the wiring $W$ associated with the derivation above is made of the union of $W_{d_1}, W_{d_2}$ and the wiring $\sigma$ (called a *loop*) made of the two flows $p_A(x) \leftarrow p_{\sim A}(x)$ and $p_{\sim A}(x) \leftarrow p_A(x)$. Similarly to the case above, an execution operator $Ex(\mathtt{W}, \sigma)$ can be defined, which generates all paths entering and exiting the net through the gates in $\Gamma \cup \Gamma'$:

$$Ex(\mathtt{W}, \sigma) = \sigma\mathtt{W} + \sigma\mathtt{W}\sigma\mathtt{W} + \sigma\mathtt{W}\sigma\mathtt{W}\sigma\mathtt{W} + \cdots = \sum_{n}^{\infty} (\sigma\mathtt{W})^n = (1 - \sigma\mathtt{W})^{-1} \qquad (33)$$

Remark that such paths can be arbitrarily long: if a path enters $W_{d_1}$ through a gate in $\Gamma$ and exits it through gate $A$, then by $\sigma$ it enters $W_{d_2}$ through gate $\sim A$; at this point either it exits the net through a gate in $\Gamma'$, either he exits through gate $\sim A$, so that by $\sigma$ he enters again $W_{d_1}\ldots$ Indeed this potentially infinite dynamics encodes all the computational complexity of cut-elimination.

The connection between cut-elimination and execution is established by the following important property: first observe that a wiring $W_d$ coming from a derivation $d$ can always be decomposed into the sum $W_0 + \sigma$, where $\sigma$ contains all flows coming from cuts in $d$; now one can prove that, if $d$ reduces to a cut-free derivation $d'$, then[5] the execution $Ex(W_0, \sigma)$ corresponds exactly to the representation $W_{d'}$ of $d'$. Hence, by computing all paths in the representation of $d$ one obtains the representation of the normal form of $d$.

**Nilpotency** In the language of *GoI* the *Hauptsatz* (in its strong version, a.k.a. strong normalization) corresponds to the fact that the generation of paths breaks down after a finite number of iterations (in other words, that all paths are finite). This property is expressed by the *nilpotency* of the wiring $\sigma W$, i.e. the fact that, for a certain positive integer $N$, $(\sigma W)^N = 0$. If $\sigma W$ is nilpotent, then the execution $Ex(W, \sigma)$ is well-defined, as it is given by the finite iteration

$$\sigma W + \sigma W \sigma W + \sigma W \sigma W \sigma W + \cdots + (\sigma W)^{N-1} \tag{34}$$

In *GoI* one can prove (see [Gir89a, Gir95]) that, for all wirings $W_0 + \sigma$ coming from sequent calculus derivations, $\sigma W_0$ is nilpotent. This theorem provides a geometrical counterpart to the *Hauptsatz*; in particular it implies that execution is well-defined for wirings coming from sequent calculus.

To give an example, let us consider the following derivation $d$:

$$\cfrac{\cfrac{\vdash \sim A, A \quad \vdash B, \sim B}{\vdash \sim A, B, \sim (A \multimap \sim B)}(L \multimap) \quad \vdash \sim C, C}{\cfrac{\vdash \sim A, \sim (\sim B \multimap C), C, \sim (A \multimap \sim B)}{\cfrac{\vdash \sim A, \sim (\sim B \multimap C), \sim C \multimap \sim (A \multimap \sim B)}{\vdash A \multimap \sim (\sim B \multimap C), \sim C \multimap \sim (A \multimap \sim B)}(R \multimap)}(R \multimap)}(L \multimap) \tag{35}$$

The paths in $W_d$ can be visualized from the proof-net $D$:

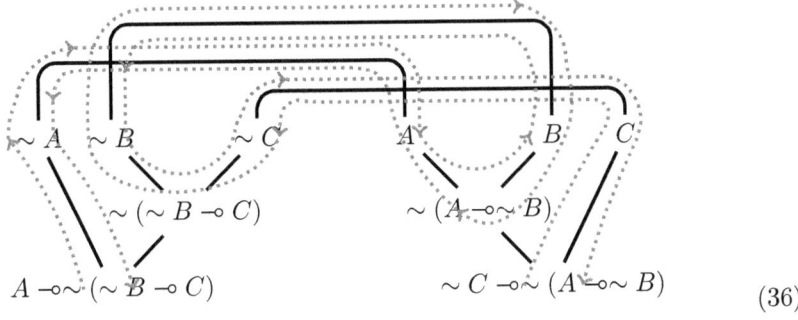

(36)

---

[5]This is shown in [Gir89a] for full linear logic in the case $d$ does not contain among its conclusions formulae of the form $?A$ or $\exists X A$.

Now the nilpotency of $W_d$ (remark that, since there are no cuts, one chooses $\sigma = \mathrm{id}$) expresses the fact that, if a path enters through one of the gates of $D$, it will not be stuck inside the net (since all paths are finite), and will end up exiting $D$ through another gate after a finite amount of time. Hence the "no shortrip" criterion of proof-nets can be expressed by nilpotency[6].

One can exploit a "nilpotency criterion" to show that certain wirings do not come from sequent calculus derivations, and to prove that certain logical principles are not valid. For instance, the wiring $U_0 + \sigma$, where $U_0 = \sigma = \{p_A(x) \leftharpoonup p_{\sim A}(x), p_{\sim A}(x) \leftharpoonup p_A(x)\}$, arising from the proof-structure

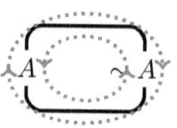

(37)

is not nilpotent, since, for all $n \geq 1$ $(\sigma U_0)^n = \sigma U_0 = \{p_A(x) \leftharpoonup p_A(x), p_{\sim A}(x) \leftharpoonup p_{\sim A}(x)\}$.

A more interesting case is given by the "incorrect" proof-structure below, of conclusion the invalid formula $F = (A \multimap_{\sim} (B \multimap_{\sim} C)) \multimap_{\sim} (B \multimap_{\sim} (A \multimap C))$:

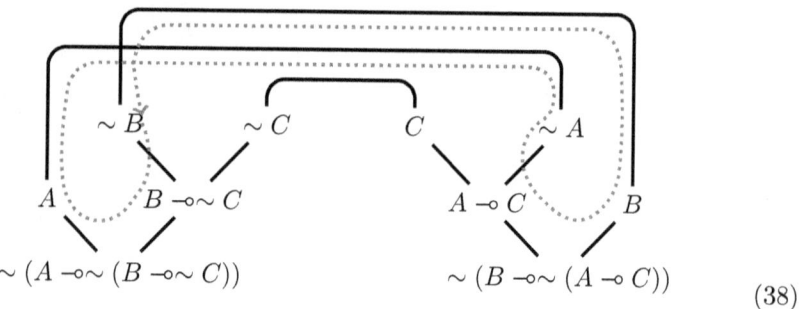

(38)

from the "shortcircuit" in (38) one gets an infinite path in the GoI interpretation, implying that the associated wiring $V_0 + \sigma$ cannot be nilpotent. Now, since the net above is the only one with conclusion $F$, one concludes that $F$ is not valid.

**Connectives explained away?** We briefly recalled an interpretation of proofs which allows to characterize *correct* proofs by means of a geometrical condition.

Take the invalid formula $F$. The model-theoretic refutation of $F$ is obtained by looking for a *valuation*, i.e. an assignment of truth-values to formulae, such that $A \Rightarrow \neg(\neg B \Rightarrow \neg C)$[7] turns out true (meaning that, *if $A$ is true, then if $B$ is not true, then $C$ is not* true) and $\neg(B \Rightarrow \neg(A \Rightarrow C))$ turns out not true (meaning that, *it is not true that, if $B$ is true, then it is not true that, if $A$ is true, then $C$ is true*).

---

[6] Remember that the shortrip criterion was defined for a net having only one gate. In the case of the net (36) it suffices to "close" one gate: by gluing together the two paths one obtains a single longtrip.

[7] Where linear connectives are (faithfully) translated into classical ones.

The refutation of $F$ obtained by means of the "no shortrip criterion" (or the "nilpotency criterion") allows, on the contrary, to get rid of the circular reference to connectives: the combination of connectives in the formula provides instructions to construct paths, though the correctness criterion depends on an abstract property of such paths, making no reference to their construction. The formula $F$ is invalid since a possible proof of it would contain a shortcircuit.

## 4 Conclusion: shortrips and deadlocks

This presentation was thought to highlight some similarities between the idea of validity by a "no deadlock criterion"[8], arising from Herbrand's interpretation of quantification and the idea of validity by a "no shortrip criterion", arising from the *GoI* interpretation of proofs.

**Quantifier nets** A precise connection between the two approaches comes from the extension of the proof-net interpretation to first-order quantifiers (see [Gir91]). We do not enter here into the details; the fundamental idea is that, with the introduction of a universal quantification $\forall y A(y)$, one has to add new paths (called *jumps*) linking the gate $\forall y A(y)$ with all vertices in the net in which the *eigenvariable* $y$ occurs free. The typical case is the one of a vertex $\sim A(y)$ which is premiss of a gate $\exists x \sim A(x)$: then the jump produces a path from the gate $\forall y A(y)$ to the gate $\exists x \sim A(x)$ which closely imitates Herbrand's "jump" of universal variables over existential variables, that would be given by an equation of the form

$$y = \mathtt{f}(x) \tag{39}$$

Remark that the fact of using linear logic (where the contraction rule cannot be used) eliminates the appeal to the expansions $A_H^n$. Hence, in the linear frame, validity can be directly captured by the "no deadlock criterion" provided by unification. The deadlocks in the unification problem induced by a formula $A$ are then directly translated into shortrips in the proof-structures of conclusion $A$. Hence the two characterizations of validity coincide in this case.

**A "no deadlock" approach to validity** The "no deadlock" explanations of validity here sketched do not rely on properties defined (as usual) by induction over formulae. Moreover, validity as "absence of deadlocks" cannot be reduced to the mere "existence" of a proof: both approaches focus rather on the conditions that must obtain for such a proof to be found.

In Herbrand's theorem the task of proving a first-order formula is first reduced to the one of proving one among a sequence of quantifier-free formulae; then, the latter task is reduced in turn to that of showing a certain system of first-order equations to be solvable. Indeed, if this is not the case, i.e. if a deadlock occurs during unification, then a proof can surely not be found, since, in such a proof, the variables should depend on each other in a circular way.

In the *GoI* interpretation, one expands the domain of proofs to "pre-proofs" (like the proof-structures, i.e. syntactic objects that might or might not be proofs): this

---

[8]Actually, the expression "deadlock" can be found in the literature on *GoI*: for instance in [Gir90] wirings inducing no shortrips are called "deadlock-free".

allows to considers also "pre-proofs" in which the dependencies between formulae (or "gates") can be circular. Finally one recovers, in an elegant way, those "pre-proofs" which are actual proofs as those which avoid circular dependencies. Moreover, this characterization has a dynamical content, as correct proofs are exactly those which allow for a terminating cut-elimination procedure (made possible by the absence of "shortrips").

The development of a systematic "no deadlock" approach to validity, accounting for the intuitions suggested in this paper, appears then as an interesting, and relatively new, direction for future research.

## BIBLIOGRAPHY

[ABPS14] C. Aubert, M. Bagnol, P. Pistone, and T. Seiller. Logic programming and logarithmic space. http://arxiv.org/abs/1406.2110, 2014.

[Bag14] Marc Bagnol. *On the unification semiring*. PhD thesis, Aix-Marseille Université, 2014.

[Cel06] Carlo Cellucci. The question Hume didn't ask: why should we accept deductive inferences. In C. Cellucci and Paolo Pecere, editors, *Demonstrative and non-demonstrative reasoning*, pages 207–235. Edizioni dell'Università, 2006.

[Dum91] Michael Dummett. *The logical basis of metaphysics*. Columbia University Press, 1991.

[Gir82] Jean-Yves Girard. Herbrand's theorem and proof-theory. In J. Stern, editor, *Proceedings of the Herbrand Symposium, Logic Colloquium '81*. North-Holland, 1982.

[Gir87] Jean-Yves Girard. Linear logic. *Theoretical Computer Science*, 50(1):1–102, 1987.

[Gir89a] Jean-Yves Girard. Geometry of interaction I: interpretation of system f. In Ferro, Bonotto, Valentini, and Zanardo, editors, *Logic colloquium*, 1989.

[Gir89b] Jean-Yves Girard. Towards a geometry of interaction. *Contemporary Mathematics*, 92, 1989.

[Gir90] Jean-Yves Girard. Geometry of interaction II: deadlock-free algorithms. In *International Conference on Computational Logic, Tallinn*, 1990.

[Gir91] Jean-Yves Girard. Quantifiers in linear logic II. In *Atti del Congresso* Nuovi problemi della logica e della filosofia della scienza, *Viareggio, gennaio 1990*, 1991.

[Gir95] Jean-Yves Girard. Geometry of interaction III: accomodating the additives. In *Advances in Linear Logic, London Mathematical Society, Lecture Note Series*. Cambridge University Press, 1995.

[Gir13] Jean-Yves Girard. Geometry of interaction VI: a blueprint for transcendental syntax. Under consideration for publication in Mathematical Structures in Computer Science, 2013.

[Her67] Jacques Herbrand. Investigations in proof theory. In Jean Van Heijenoort, editor, *From Frege to Gödel: a source book in mathematical logic, 1879-1931*. Harvard University Press, 1967.

[Koh08] Ulrich Kohlenbach. Herbrand's theorem and extractive proof-theory. *Gazette des Mathématiciens*, 1(118):29–40, 2008.

[Kre51] Georg Kreisel. On the interpretation of non-finitist proofs - part I. *The Journal of Symbolic Logic*, 16(4), 1951.

[Pra71] Dag Prawitz. Towards a foundation of a general proof theory. *Logic, Methodology and Philosophy of Science*, VI, 1971.

[Rob65] Alan Robinson. A machine-oriented logic based on the resolution principle. *Journal of the ACM*, 12(1):23–41, 1965.

[VH82] Jean Van Heijenoort. L'oevre logique de Jacques Herbrand et son contexte historique. In J. Stern, editor, *Proceedings of the Herbrand Symposium, Logic Colloquium '81*, 1982.

# Wittgenstein's Struggles with the Quantifiers

Jan von Plato

ABSTRACT. Frege was the first one to see that the proper treatment of quantifiers in logic requires a rule of generalization. It is shown that Wittgenstein never saw this matter, despite its clear presentation in the *Principia*, and even interpreted the role of free-variable expressions wrongly. He was led to avoiding quantifiers, as is shown by his work, only recently understood in detail, in which the principle of induction in primitive recursive arithmetic is replaced by a principle of uniqueness of functions defined by recursion. In this particular case, though, his ban on assumptions with free variables led to a positive result.

1. From *Begriffsschrift* to the *Tractatus*

A careful reader of Wittgenstein's *Tractatus* will notice the categorical absence of any notion of inference or deduction in it. There is instead the semantical method of truth tables by which it can be determined whether a propositional formula is a tautology. How the method is to be extended to the quantifiers is nowhere explained: At 6.1201, the principle of universal instantiation $(x)fx \supset fa$ is simply called a "tautology."

Frege's perhaps central discovery was that the proper treatment of the universal quantifier requires, besides the above instantiation principle, also a *rule* by which generality can be concluded, his "illuminating" observation in the *Begriffsschrift* (p. 21, with Russell's horizontal notation for formulas in place of Frege's vertical one):

"It is even illuminating that one can can derive $A \supset (\mathfrak{a})\Phi(\mathfrak{a})$ from $A \supset \Phi(a)$ if $A$ is an expression in which $a$ does not occur and if $a$ stands in $\Phi(a)$ only in the argument places."

Wittgenstein does not see that this rule is crucial, as is shown by *Tractatus* 6.1271 where he states that all of logic follows from one basic law, the "conjunction of Frege's *Grundgesetze*." A rule of inference can be no part of such a conjunction, and there is no full predicate logic without the rule.

The logic of Frege and Russell was classical and therefore existence could be defined in terms of universality. To universal instantiation corresponds then the tautologous implication $fa \supset (\exists x)fx$. Now, if one wanted to reason about existence, a rule of inference would be needed, one that is dual to Frege's rule of generalization. It is a rather embarrassing fact that the first formal statement of a rule of existential instantiation in logic seems to be as late as in the well-known book *Grundzüge der theoretischen Logik* of 1928, by Hilbert and Ackermann but apparently mostly written by Paul Bernays the discovery of the rule included. It is a direct dual to Frege's: If $F(a) \supset B$ for an arbitrary $a$, then $(\exists x)F(x) \supset B$. In practice, one would apply the rule whenever an existential assumption $(\exists x)F(x)$ needs to be put into use: assume an instance $F(y)$ with an eigenvarable $y$, i.e., one that is generic in the sense that no assumptions that contain $y$ free have been made, beyond $F(y)$. If now $B$ follows from $F(y)$, it follows from $(\exists x)F(x)$. This intuitive procedure becomes, when put in terms of Gerhard Gentzen's natural deduction, the formal rule of existence elimination, given in a *pure form* in the sense that it does not involve other connectives, unlike Frege's and Bernays' rules for $\forall$ and $\exists$. The corresponding pure form of generalization is: If $A(y)$ can be derived from assumptions that do not contain the eigenvariable $y$ free, then $\forall x A(x)$ can be concluded.

Russell tells in the preface to his book *The Principles of Mathematics* of 1903 that he had seen Frege's *Grundgesetze der Arithmetik* but added that he "failed to grasp its importance or to understand its contents," the reason being "the great difficulty of his symbolism" (p. xvi). Upon further study, he wrote a lengthy appendix with the title *The logical and arithmetical doctrines of Frege* (pp. 501–522), though with just a disappointing half a page dedicated to the formalism of logic. He notes the appearance of the universal quantifier in Frege (p. 519):

> He has a special symbol for assertion, and he is able to assert for all values of $x$ a propositional function not stating an implication, which Peano's symbolism will not do. He also distinguishes, by the use of Latin and German letters, respectively, between *any* proposition of a certain propositional function and *all* such propositions.

Frege's Latin and German letters stand for free and bound variables.

The universal quantifier makes its next appearance in Russell's famous 1908 paper on the theory of types. Its section II is titled *All and any*. Mathematical reasoning proceeds through *any*: "In any chain of mathematical reasoning, the objects whose properties are being investigated are the arguments to *any* value of a propositional function." Still, reasoning with just free variables would not do. Next Russell goes on to introduce a formal notation for the universal quantifier, $(x)\phi x$, presumably the first such notation in place of Frege's notch in the assertion sign, if we disregard the $\Pi_x$ notation in Schröder's algebraic logic. The explanation, though, is a disappointment, for it is stated that $(x)\phi x$ denotes the proposition "$\phi x$ is always true," a hopeless mixing of a proposition with an assertion that would never have occurred in Frege. Later, in the more formal section VI of the paper, this is corrected when the Fregean assertion sign $\vdash$ is put to use.

Russell's first example of a quantificational inference is: from $(x)\phi x$ and $(x)(\phi x$ implies $\psi x)$ to infer $(x)\psi x$:

> In order to make our inference, we must go from '$\phi x$ is always true' to $\phi x$, and from '$\phi x$ always implies $\psi x$' to '$\phi x$ implies $\psi x$,' where the $x$, while remaining any possible argument, is to be the same in both.

As can be seen, the rule is applied by which instances can be taken from a universal, after which the propositional rule of implication elimination can be applied. Then, since $x$ is "any possible argument," $\psi x$ is always true, by which $(x)\psi x$ has been inferred. Here we have a clear case of the introduction of a universal quantifier. A further remarkable feature of Russell's example is its purely hypothetical character. He does read the universal propositions in the "is always true" mode, but the argument begins with: "Suppose that we know $(x)\phi x$," thus, we have here a *universal assumption* that is put into use by the rule of universal elimination.

Russell ends his discussion of *all and any* in section II by praising Frege:

> The distinction between *all* and *any* is, therefore, necessary to deductive reasoning and occurs throughout in mathematics, though, so far as I know, its importance remained unnoticed until Frege pointed it out.

Russell's final word on logic is contained in the first volume of *Principia Mathematica* that appeared in 1910 and was co-authored with A. Whitehead. I take Russell to have been the driving force behind the enterprise and refer only to him even if details of *Principia* may have originated with Whitehead. The presentation of logic in *Principia* is somewhat different from Frege and the 1908 formulation that followed Frege, in the sense that both quantifiers appear as primitives. The reason is that Russell wants to have all quantifiers at the head of formulas. To this end, he uses the rules for prenex normal form as definitions, as in (p. 130): $\sim(x)\phi x \equiv (\exists x)\sim\phi x$.

Part I, titled "Mathematical logic," begins with section A on "the theory of deduction" (pp. 90–126), followed by a "theory of apparent variables," i.e., of bound variables (pp. 127–160).

The quantifier axiom is existential introduction: $\vdash \phi x \supset (\exists z)\phi z$

The rule of inference is universal generalization (p. 132): "When $\phi y$ may be asserted, where $y$ may be any possible argument, then $(x)\phi x$ may be asserted." The arbitrariness of $y$ is further explained by: "if we can assert a wholly ambiguous value $\phi y$, that must be because all values are true." We see in the latter again, as in Frege, that the explanation goes from the truth of the universal proposition to any of its instances, not the other way around.

The first example of quantificational inference is the derivation of the principle of universal instantiation: $\vdash (x)\phi x \supset \phi y$

Finally, in this summary of Russell's work, we notice his use of Peano's notion of "formal implication" $\phi x \supset_x \psi x$ that is defined in *Principia* as (p. 139):

$$\phi x \supset_x \psi x \equiv (x)(\phi x \supset \psi x)$$

The *Principia* made it clear that the notion of tautology does not extend to the quantifiers. It is incomprehensible that Wittgenstein didn't realize this crucial limitation in the *Tractatus* by which his early philosophy of logic and mathematics collapses; the only possible explanation is that the impatient philosopher never made it as far as to page 132 of the *Principia*. Russell, in turn, seems to have lost all interest in and understanding of logic after having finished the *Principia*: In the preface to the second edition, in 1927, he makes the embarrassing remark that Sheffer's stroke, the single connective by which one can axiomatize classical propositional logic, is "the most definitive improvement resulting from work in mathematical logic during the past fourteen years" (p. xiv). His promotion of the logically defective *Tractatus*, equally limited to classical propositional logic, belongs to the same category of judgments.

2. Generality and existence in later Wittgenstein

Wittgenstein turned back to philosophy around 1928, greatly interested in the philosophy of mathematics. He went to Cambridge in 1929 and became a lecturer and later professor, and prepared long manuscripts on the basis of his lectures that have been published many years after his death in 1951. He also dictated shorter pieces to his students and friends, such as one known as *The Blue and Brown Books*, with several more of these still to be published today.

Wittgenstein's first works in his "second period" as a philosopher of logic and mathematics include two specific achievements, both of them somewhat cryptic and clarified only decades later. The first is a constructivization of Euler's proof of the infinity of primes, reconstructed in detail in Mancosu and Marion (2003). The second discovery derives from Wittgenstein's careful reading of Skolem's 1923 paper on primitive recursive arithmetic.

The book manuscripts, such as the *Philosophische Grammatik* that was written around 1933, contain lengthy discussions of themes related to logic. Regarding the quantifiers, it emerges from these discussions that Wittgenstein was at great pains at understanding them: As in the *Tractatus*, there is no trace of the rule of universal introduction, but quantifiers are instead simply logical expressions of a certain form. Generality is first taken as a "logical product" and existence as a "logical sum," the latter written, with $f$ a predicate, as (p. 269): $fa \vee fb \vee fc \vee \ldots$

Generality covers all cases, but its explanation as a "product" of instances becomes infinitistic, and that was not acceptable for Wittgenstein (p. 268). In the absence of a rule of generalization, one gets at most that a universality implies any of its instances. Likewise, existence cannot be a summing up of all the disjunctive possibilities for its introduction, because there is an infinity of such. The dual to universal generalization is existential elimination and in its absence, one gets only that an instance implies existence.

Wittgenstein's struggles with the existential quantifier are manifest in the *Grammatik* where he discusses at length an example, in translation the phrase *The circle is in the square*, illustrated by a drawing of a rectangle and a circle inside (p. 260). It is clearly correct to say that there is a circle in the square, but the statement does not fix which: it is not any one specific circle, so what circle is it? Wittgenstein sees that there is a generality behind existence and ponders on the matter page after page; all this because he does not know that there should be a rule of existential elimination, the one Bernays wrote in an axiomatic form and Gentzen as a pure rule of natural deduction. Wittgenstein's "generic circle" is correctly presented through the eigenvariable of an existential instantiation. The difference to generality in the rule of universal generalization is subtle: Given $\exists x F(x)$, to *assume* $F(y)$ for arbitrary $y$ is not the same as to *assume* $F(y)$ *provable* for arbitrary $y$, and only the latter leads to $\forall x F(x)$. One can see that Wittgenstein is at pains at arriving to an understanding of the example, and of existence more generally (as on p. 243: "How do I know that $(\exists x)fx$ follows from $fa$?"). After this, dozens of pages of the *Grammatik* (pp. 257–288) are devoted to the dual notion of generality, but the upshot is: no amount of philosophical reflection in Wittgenstein can replace the lack of explicit principles of quantificational inference.

3. Indirect existence proofs

A direct statement of the infinity of primes could be: For any $n$, there is an $m$ such that $m > n$ and $m$ is prime. The logical form of Euler's argument is: Assume that there is a number $n$ such that for any $m > n$, $m$ is divisible. A contradiction follows. From Euler's argument, we could at most infer that for any $n$, it is impossible that there should not be a number $m$ such that $m > n$ and $m$ is prime; Still, no way of actually producing a prime greater than $n$ need have been given by the proof. Wittgenstein turned the indirect inference into a direct one. The context was a manuscript of Heirich Behmann's in which the latter claimed to be able to convert any classical proof into a constructive one. After criticism by, inter alia, Gödel, Behmann withdrew publication. The full story of the Behmann affair is found in Mancosu (2002)

The nature of indirect existence proofs was debated a lot in the 1920s, because of the intuitionistic criticisms of such classical proofs by Brouwer. Wittgenstein's interpretation was that two notions of existence are in fact involved, and that there is no content in denying the law of excluded middle: One just adapts different rules of proof and the sense of the theorems is different. One of these could be called classical existence, the other constructive existence.

So far, so good. However, considering the absence of quantifier rules in Wittgenstein, it is not surprising that he got some of the properties of universal and existential quantification wrong. He certainly understood the law of excluded middle and the related law of double negation. In the case of indirect existence proofs, the latter can be put in the form of $\neg\neg \exists x A(x) \supset \exists x A(x)$, a law that fails intuitionistically. The properties of intuitionistic logic were not perfectly understood in the early 1930s in general, and here Wittgenstein seems to have committed a specific mistake even though I have so far not found it directly in any text of his: Instead, his pupil Reuben Louis Goodstein followed his lectures in Cambridge in 1931–34 and started work on a topic to which I shall soon turn. In the meanwhile he published an article titled *Mathematical systems*, in the well-known philosophical journal *Mind* in 1939. It was a statement of what he took to be Wittgenstein's philosophy of mathematics. The article contains many exclamations and positions that should perhaps best be described as silly, but there are even indications that Wittgenstein was not displeased with it, contrary to some writings of other pupils of his.

In the paper, Goodstein maintains that the inference from $\neg \exists x \neg A(x)$ to $\forall x A(x)$ is intuitionistically legitimate. The converse implication is intuitionistically provable, so with the claimed inference, the universal quantifier could be defined by the existential one. Instead, this particular argument against intuitionism and for the "strict finitism" of Wittgenstein and Goodstein is just fallacious: In Goodstein (1951, p. 49), written under Wittgenstein's influence around 1940, it is stated that "some constructivist writers maintain that... a 'reduction' proof of universality is acceptable." In Goodstein (1958, p. 300), we find again that Brouwer rejects indirect existence proofs, here $\neg(\forall x)\neg P(x) \Rightarrow (\exists x)P(x)$, "whilst retaining the converse implication $\neg(\exists x)\neg P(x) \Rightarrow (\forall x)P(x)$." In other words, if $(\exists x)\neg P(x)$ turns out impossible, a reduction gives $(\forall x)P(x)$; certainly not anything Brouwer or any other constructivist thinker would have ever proposed.

The reason for the above misunderstanding is somewhat subtle. The intuitionistically invalid implication $\neg \exists x \neg A(x) \supset \forall x A(x)$ is perhaps at a first sight rather close to $\neg \exists x A(x) \supset \forall x \neg A(x)$. The latter is intuitionistically provable, in fact one of the first examples of intuitionistically correct inference that Gentzen gave when he presented the calculus of natural deduction in his thesis (1934–35). One could think that it makes no difference to have $\neg A(x)$ under the negated existence, and $A(x)$ under the universal, instead of the other way around, but this is not in the least so: With $\neg A(x)$ in place of $A(x)$, we do get $\neg \exists x \neg A(x) \supset \forall x \neg\neg A(x)$, but the double negation cannot be deleted.

Wittgenstein was not alone with his problems: The correspondence between Arend Heyting and Oskar Becker gives ample illustration of how difficult it was to get intuitionistic logic right, even for people who tried hard (see Van Atten 2005).

A tentative conclusion can be drawn from this little story: A part of the motivation of Wittgenstein's refusal of the quantifiers, even the intuitionistic ones, in favour of a strict finitism, was based on misunderstanding the nature of the intuitionistic quantifiers.

4. From induction to recursion

In 1945, there appeared in the *Proceedings of the London Mathematical Society* a long article titled "Function theory in an axiom-free equation calculus." The bearing idea of the work was to recast primitive recursive arithmetic in an even stricter mould than the quantifier-free calculus of Skolem (1923): Even the venerated principle of arithmetic induction had to go, replaced by a principle by which two recursive functions defined by the same equations are the same (p. 407): "If two functions signs '$a$', '$b$' satisfy the same introductory equations, then '$a = b$' is a proved equation." A footnote added to this principle tells the following: "This connection of induction and recursion has been previously observed by both Wittgenstein and Bernays." The author of the paper, this time not in the least silly, was Wittgenstein's student Goodstein. The full story of his paper can be recovered through the correspondence he had with Paul Bernays. In the opening letter of 29 July 1940, he writes:

> The manuscript which accompanies this letter gives some account of a new formal calculus for the Foundations of Mathematics on which I have been working for the past six years.

Unfortunately, the original version of the paper is not to be found. The most we know are some comments by Bernays such as the following from his first letter to Goodstein, of 28 November 1940:

> Generally my meaning is that your attempt could be quite as well, and perhaps even better appreciated, if you could deliver it from the polemics against the usual mathematical logics which seem to me somewhat attackable, in particular as regards your arguments on the avoidability of

quantifiers. Of course in your calculus, like in the recursive number theory, quantifiers are not needed. But with respect to the "current works on mathematical philosophy" the thesis that "the apparent need for the sign '$(x)$' arose from a confusion of the two different uses ... of variable signs" can hardly be maintained.

Bernays mentions also that he had presented in 1928 at the Göttingen Mathematical Society "the possibility of taking instead of the complete induction the rule of equalizing recursive terms satisfying the same recursive equations," a discovery he left unpublished. Bernays' first letter to Goodstein is ten pages long, typewritten single-spaced, and it displays his full command of Goodstein's calculus. Goodstein was enormously impressed as can be seen from his letters and thankfully revised his paper and cleared it of polemics, adding all the references to a literature that had been unknown to him; quite embarrassingly, even the extensive treatment of primitive recursive arithmetic in the first volume of the *Grundlagen der Mathematik*, Section 7, pp. 287–343 belonged there.

When Wittgenstein's book manuscript *Philosophische Grammatik* came out in 1969, one could find his discovery of the way from proof by induction to proof by recursion equations clearly stated, and developed to some extent mainly through a few examples (*Grammatik*, pp. 397–450). The text was written between 1932–34, the years during which Goodstein attended Wittgenstein's lectures. The crucial discovery comes out on the very first page devoted to the topic (*Grammatik*, p. 397), where Wittgenstein considers the associative law for sum in elementary arithmetic, denoted by $A$:

$$(a+b)+c = a+(b+c) \qquad A$$

Skolem's 1923 paper on primitive recursive arithmetic, Wittgenstein's source for the topic of elementary arithmetic, gives the standard inductive proof for $A$, based on the recursive definition of sum by the recursion equations:

$$a+0 = a$$
$$a+(b+1) = (a+b)+1$$

If one counts the natural numbers from 1 on, the second equation gives the base case of the inductive proof. For the step case, one assumes $A$ for $c$ and proves it for $c+1$, i.e., $(a+b)+(c+1) = a+(b+(c+1))$. The left side is by the recursion equation equal to $((a+b)+c)+1$, then applying the inductive hypothesis to $(a+b)+c$ one gets $((a+b)+c)+1 = ((a+(b+c))+1$, and finally by two applications of the recursion equation in the opposite direction $((a+(b+c))+1 = a+((b+c)+1) = a+(b+(c+1))$.

In *Grammatik*, p. 397, Wittgenstein gives the proof as follows:

What Skolem calls the recursive proof of $A$ can be written as follows:

$$\left. \begin{array}{rcl} a+(b+1) &=& (a+b)+1 \\ a+(b+(c+1)) &=& a+((b+c)+1) = (a+(b+c))+1 \\ (a+b)+(c+1) &=& ((a+b)+c)+1 \end{array} \right\} B$$

We have to put emphasis on Wittgenstein's words "can be written," for this is not Skolem's proof by induction, but another proof that Wittgenstein goes on to explain in the following words:

In the proof [B], the proposition proved clearly does not occur at all.–
One should find a general stipulation that licenses the passage to it.
This stipulation could be expressed as follows:

$$\left.\begin{array}{ll} \alpha & \varphi(1) = \psi(1) \\ \beta & \varphi(c+1) = F(\varphi(c)) \\ \gamma & \psi(c+1) = F(\psi(c)) \end{array}\right\} \quad \overset{\Delta}{\varphi(c) = \psi(c)}$$

When three equations of the forms $\alpha, \beta, \gamma$ have been proved, we shall say: "the equation $\Delta$ has been proved for all cardinal numbers."

Here we see the essence of the argument: Two functions $\varphi$ and $\psi$ that obey the same recursion equations, are the same function. Wittgenstein himself writes (*Grammatik*, p. 398):

I can now state: The question whether $A$ holds for all cardinal numbers shall mean: Do equations $\alpha, \beta,$ and $\gamma$ hold for the functions

$$\varphi(\xi) = a + (b + \xi), \ \psi(\xi) = (a + b) + \xi$$

Wittgenstein's principle can be considered, as in the letter of Bernays quoted above, a "rule of equalizing recursive terms." Taken as a rule, it is a derivable rule in primitive recursive arithmetic. In the other direction, given the premises of the induction rule, here $\varphi(1) = \psi(1)$ and $\varphi(y) = \psi(y) \supset \varphi(y+1) = \psi(y+1)$ for an arbitrary $y$, the conclusion by which $\varphi(x) = \psi(x)$ holds for arbitrary $x$ can be recovered from Wittgenstein's uniqueness principle for recursion equations (as shown in von Plato 2014a,b).

Wittgenstein's book does not reveal the motive for preferring proofs by recursion equations to proofs by induction, but in 1972, Goodstein published a paper "Wittgenstein's philosophy of mathematics" in which the matter is explained. In reference to the *Grammatik* that had come out three years earlier, Goodstein recalls Skolem's inductive proof and then adds (p. 280):

In his lectures Wittgenstein analysed the proof in the following way. He started by criticizing the argument as it stands by asking what it means to *suppose* that (1) [associativity] holds for some value $C$ of $c$. If we are going to deal in suppositions, why not simply suppose that (1) holds for any $c$.

Goodstein now gives a very clear, intuitive explanation of why Wittgenstein's method works: With $c = 0$, $(a + b) + 0 = a + b = a + (b + 0)$. Thus, the ground values of Wittgenstein's $\varphi$- and $\psi$-functions are the same, here $\varphi(0) = \psi(0)$ with the natural numbers starting from 0 instead of 1 as in the 1930s. For the rest, when $c$ grows by one, $\varphi(c)$ and $\psi(c)$ obtain their values in the same way, here, both growing by 1, by which $(a + b) + c$ and $a + (b + c)$ are always equal. Wittgenstein's cryptic remarks in the *Cambridge Lectures* of 1939 (ed. Diamond, p. 287) get now an explanation: He indicates in brief words why his method works, namely by equal ground value and equal growth of $\varphi$ and $\psi$.

As the above-quoted clear recollection on the part of Goodstein shows, Wittgenstein was led to propose a finitism that was even stricter than that of Skolem, in that *assumptions with free variables* were to be banned. These assumptions are a crucial component in inductive inference, where one assumes a property $A(n)$ for an arbitrary natural number $n$ then shows that the successor of $n$ has the property, expressed as $A(n+1)$. However, the assumption $A(n)$ is a far cry from assuming, say in the case of associativity, that the inductive predicate "holds for any $c$" as Goodstein suggests at the end of the quote. It is the simplest error in inference with the quantifiers to assume $A(x)$, then to conclude $\forall x A(x)$: The eigenvariable condition in universal generalization is that $x$ must not occur free in any assumption on which its premiss $A(x)$ depends, but here one must keep in mind that if $A(x)$ itself is an assumption, it depends on itself so to say, thus, $x$ is free in an assumption. More generally: To assume $A(x)$ is not the same as to assume $A(x)$ provable and only the latter gives $\forall x A(x)$. No amount of philosophical reflection in Wittgenstein can replace the command over quantificational inferences that results from Gentzen's pure formulation of the quantifier rules in terms of natural deduction.

5. Turing's scruples

Wittgenstein lectures on the foundations of mathematics in Cambridge during the first half of 1939 have been reconstructed by Cora Diamond in 1975, on the basis of four sets of notes by participants. These lectures were graced by the presence of Alan Turing who, as a reader of the lectures soon notices, had something to comment on almost every lecture. Turing was to be absent from one lecture for which reason Wittgenstein announced that "it is no good my getting the rest to agree to something that Turing would not agree to" (pp. 67–68). The lectures show no progress on the part of Wittgenstein as regards the understanding of the principles of quantificational logic. The remarks about generality, existence, and the circle-in-the-square example are in substance the same as in 1933 (as on pp. 268–269). Moreover, Wittgenstein's pretense – witnessed by Bernays' comments on Goodstein's lost manuscript – has not changed (p. 270): "If Russell gives an interpretation of arithmetic in terms of logic, this removes some misinterpretations and creates others. There are gross misunderstandings about the uses of 'all', 'any', etc." Sad to say, these misunderstandings were all Wittgenstein's, caused by his apparent inability to learn from what others had accomplished.

Turing kept, remarkably, silence in front of the *multorum ignorantia* at the points of the lectures in which the quantifiers were discussed. His reaction is instead seen in a manuscript he was working on in the early 1940s. It bears the title *The reform of mathematical notation and phraseology* and can be seen in manuscript form on the pages of the Turing archive. Two of the central points were: 1. "Free and bound variables should be understood by all and properly respected." 2. "The deduction theorem should be taken account of."

He then gives examples of constants and variables and adds: "The difference between the constants and the free variables is somewhat subtle. The constants appear in the formula as if they were free variables, but we cannot substitute for them. In these cases there has always been some assumption made about the variable (or constant) previously."

The deduction theorem is the main way of handling free variables: "This process whereby we pass from $P$ proved under an assumption $H$ to 'If $H$ then $P$' may be called 'absorption of hypotheses'. The process converts constants or 'restricted variables' into free variables."

His example, slightly rephrased, is: Let the radius $a$ and volume $v$ of a sphere be given. Then $v = \frac{4}{3}\pi a^3$.

> The 'deduction theorem' states that in such a case, where we have obtained a result by means of some assumptions, we can state the result in a form in which the assumptions are included in the result, e.g., 'If $a$ is the radius and $v$ is the volume of the sphere then $v = \frac{4}{3}\pi a^3$. In this statement $a$ and $v$ are no longer constants.

There are passages in the manuscript version of Turing (1948), available at the Turing archives, which suggest that Turing had at least some knowledge of Gentzen's system of natural deduction. It is a pity he did not use it in the explanation of free and bound variables: In the example, there is a typical "Let" phrase about given $a$ and $v$, an instance of the form $S(x,y)$ that states that $x$ is the radius and $y$ the volume of a sphere. Eigenvariables $a$ and $v$ are put in place of $x$ and $y$ to get the assumption $S(a,v)$ and then the result $v = \frac{4}{3}\pi a^3$ derived. The deduction theorem introduces the implication $S(a,v) \supset v = \frac{4}{3}\pi a^3$ with no assumptions about $a$ or $v$ left, so that generalization gives $\forall x \forall y (S(x,y) \supset y = \frac{4}{3}\pi x^3)$. The situation is the same with induction: Once an assumption $A(n)$ has been made and $A(n+1)$ proved, implication introduction, or "the deduction theorem" in Turing's axiomatic terminology, is used to conclude $A(n) \supset A(n+1)$, no more dependent on the assumption $A(n)$, so that the second premiss of induction $\forall x(A(x) \supset A(x+1))$ can be inferred. – Here we have it, had Turing just cared to explain the correct use of free-variable assumptions to Wittgenstein, but there are no comments by anyone in the last lecture that discusses briefly Wittgenstein's form of primitive recursive arithmetic.

## 6. Conclusion

The quantifiers are as old as logic itself, through Aristotle's theory of the four quantifiers *every, no, some*, and *not some*, what they mean when prefixed to the indefinite form of predication *A is a B*, and what the correct forms of inference are. Even if Frege was proud to present a formalization of the syllogistic inferences in terms of predicate logic, as the final example of his new notation in the *Begriffsschrift*, no formal quantifiers in the modern sense are needed for their theory, ones that would bind variables. Quite amazingly, all of Wittgenstein's logical discourses remained on a similar pre-Fregean level, unaffected by Frege's most central discovery, namely the way inference to generality is made possible.

References:

van Atten, M. (2005) The correspondence between Oskar Becker and Arend Heyting. In V. Peckhaus, ed, *Oskar Becker und die Philosophie der Mathematik*, pp. 119–142, Fink Verlag, Munich.

Frege, G. (1879) *Begriffsschrift, eine nach der arithmetischen nachgebildete Formelsprache des reinen Denkens*. Nebert, Halle.

Frege, G. (1893) *Grundgesetze der Arithmetik, begriffschriftlich abgeleitet*, vol. 1, Pohle, Jena.

Gentzen, G. (1934-35) Untersuchungen über das logische Schliessen. *Mathematische Zeitschrift*, vol. 39, pp. 176-210 and 405-431.

Goodstein, R. (1939) Mathematical systems. *Mind*, vol. 48, pp. 58–73.

Goodstein, R. (1945) Function theory in an axiom-free equation calculus. *Proceedings of the London Mathematical Society*, vol. 48, pp. 58–73.

Goodstein, R. (1951) *Constructive Formalism*. Leicester U.P.

Goodstein, R. (1958) On the nature of mathematical systems. *Dialectica*, vol. 12, pp. 296-316.

Goodstein, R. (1972) Wittgenstein's philosophy of mathematics. In A. Ambrose and M. Lazerowitz, eds, *Ludwig Wittgenstein: Philosophy and Language*, pp. 271–286, Allen and Unwin, London.

Hilbert, D. and W. Ackermann (1928) *Grundzüge der theoretischen Logik*. Springer.

Hilbert, D. and P. Bernays (1934) *Grundlagen der Mathematik I*. Springer.

Mancosu, P. (2002) On the constructivity of proofs: a debate among Behmann, Bernays, Gödel, and Kaufmann. In *Reflections of the Foundations of Mathematics*, eds. W. Sieg et al., pp. 349–371. ASL Lecture Notes in Logic.

Mancosu, P. and M. Marion (2003) Wittgenstein's constructivization of Euler's proof of the infinity of primes. In *The Vienna Circle and Logical Empiricism*, ed F. Stadler, pp. 171-188, Kluwer.

von Plato, J. (2014a) Generality and existence: quantificational logic in historical perspective. *The Bulletin of Symbolic Logic*, vol. 20 (2014), pp. 417–448.

von Plato, J. (2014b) Gödel, Gentzen, Goodstein: the magic sound of a G-string. *The Mathematical Intelligencer*, vol. 36 (2014), pp. 22–27. Republished in M. Pitici, ed, *The Best of Mathematics Writing 2015*, pp. 215–227, Princeton University Press, in press.

Russell, B. (1903) *The Principles of Mathematics*. Cambridge.

Russell, B. (1908) Mathematical logic as based on the theory of types. *American Journal of Mathematics*, vol. 30, pp. 222–262.

Skolem, T. (1923) Begründung der elementaren Arithmetik durch die rekurrierende Denkweise ohne Anwendung scheinbarer Veränderliche mit unendlichem Ausdehnungsbereich. As reprinted in Skolem's *Selected Works in Logic*, 1970, pp. 153–188.

Turing, A. (1944) The reform of mathematical notation and phraseology. Manuscript in the Turing Archives.

Turing, A. (1948) Practical forms of type theory. *The Journal of Symbolic Logic*, vol. 13, pp. 80–94.

Whitehead, A. and B. Russell (1910) *Principia Mathematica*, vol. I. Cambridge. Second edition 1927.

Wittgenstein, L. (1922) *Tractatus Logico-Philosophicus*. Routledge, London.

Wittgenstein, L. (1969) *Philosophische Grammatik*. Blackwell, London.

Wittgenstein, L. (1975) *Wittgenstein's Lectures on the Foundations of Mathematics Cambridge, 1939*. Ed. C. Diamond, Cornell.

# PART III

# PHILOSOPHY OF NATURAL SCIENCES

# Typicality in Statistical Mechanics: An Epistemological Approach

Massimiliano Badino

ABSTRACT. The use of typicality has recently enjoyed an increasing popularity among physicists interested in the foundations of statistical mechanics. However, it has been the target of a mounting philosophical critique mainly challenging its explanatory value. After an initial stage of intense dialogue, the debate seems now to have reached a deadlock of mutual incommunicability. Instead of treating typicality as a probabilistic ingredient of an argument, in this paper I unfold the techniques and mathematical practices related with this notion and show that typicality works as a way to combine these techniques in a consistent epistemic story of equilibrium.

Keywords: typicality, statistical mechanics, Boltzmann, celestial mechanics, explanation.

## 1 A Troubled Notion

In recent years, the debate on the foundations of equilibrium statistical mechanics has increasingly focused upon the notion of typicality (see for example [1], [2], [3], [4], [5]). Briefly said, typicality is a way to explain the central problem of statistical mechanics, that is why systems such as gases tend to evolve toward a state of equilibrium and stay there for indefinitely long periods of time. Intuitively, one says that a property is typical when it holds in the vast majority of cases or, alternatively, the cases in which it does not hold are negligible in number. Let $\Gamma$ be the set of accessible states of a thermodynamic system and let $\mu$ be a measure function. If it is possible to divide $\Gamma$ into two disjoint subsets, $T_1$ and $T_2$, such as (1) only the states in $T_1$ have the property $\tau$, and (2) $\mu(T_1) \approx 1$, while $\mu(T_2) \approx 0$, then $\tau$ is a typical property of the system. The basic argumentative line used by the upholders of the typicality approach can be summarized as follows:

1. Let $\Gamma$ the accessible region of a thermodynamic system and let $M_{eq}, M_{neq}$ the subsets of the equilibrium and nonequilibrium macrostates, respectively. These subsets form a partition of $\Gamma$.

2. Let $x$ be a microstate and $x(t)$ its trajectory under the dynamics of the system. In other words, $x(t) = x_1, x_2, x_3, \ldots$ where $x_i \in \Gamma$.

3. A certain measure function $m_L$ exists, called the Lebesgue measure, such that $m_L(M_{eq}) \approx 1$; the microstates in $M_{eq}$ have the property of "being in equilibrium", hence this property is typical in the thermodynamic system.

4. Also the microstates $x(t) = x_1, x_2, x_3, \ldots$ are typically in equilibrium, hence, the trajectory of an arbitrary state is mainly contained in $M_{eq}$.

5. Ergo, the system will tend to equilibrium and remain there, because equilibrium is typical.

This straightforward argument has enjoyed a large approval among physicists and aroused an equally large discontent among philosophers. The former like especially its simplicity and its generality. In fact, it has also been extended to interpret Bohmian quantum mechanics ([6], [7], [8]). By contrast, the latter consider the argument above seriously flawed. There are three kinds of criticisms against typicality.

First, the definition of a typical property depends essentially on the size of the macrostate, which in turn depends on the definition of a suitable measure function (step (3) in the argument). In statistical mechanics, the convention is to use the so-called Lebesgue measure. Philosophers object that there is no argument, either philosophical or physical, to claim that Lebesgue measure must enjoy any preference and be considered as the "natural" one. Second, until step (4), the argument only deals with statements concerning measure of macrostates, but the conclusion is a statement about the physical behavior of observable systems. It seems, that (5) concerns the probability that a system will behave in a certain way, so that the argument would require a leap from statements about measures to statements about physical probabilities ([9], [10, 182-191]). Third, no purely measure-theoretical consideration on the macrostates would ever suffice without some dynamical assumption ([1]). In the argument presented above, this assumption is expressed in step (4), where it is supposed that the trajectory contains the same ratio of equilibrium/nonequilibrium states as in the total accessible region.

The effect of these critiques has been to virtually interrupt the dialogue between philosophers and physicists. The eminently logical character of the philosophical analysis has appeared to physicists too detached from their actual foundational problems in statistical mechanics. Thus, many working scientists tend to consider this analysis as hairsplitting and uninformative. On the other side, philosophers have quickly dismissed typicality. From the point of view of traditional philosophical analysis, typicality appears as mere hand-waving at best, or as circular at worst.

In this paper I argue that the problem is partly due to philosophers' conception of explanation. Generally, philosophers working in foundations of statistical mechanics have deployed a Hempelian model according to which an explanation is an argument whose conclusion is equilibrium. Most of the philosophical criticisms against typicality concentrate upon the flaws of arguments containing such notion. I argue, however, that the Hempelian model does not capture what the physicists mean by the explanatory value of typicality. Hence, we have to enlarge our conception of explanation. I submit that typicality provides a *satisfactory causal explanation of the qualitative aspects of equilibrium*. Let me spell out this claim by starting with the final part. By that I mean that typicality only accounts for the general fact that systems exhibit a tendency toward equilibrium, but does not yield any quantitative analysis. Second, by causal explanation is mean that typicality gives us:

1. A set of causal factors for the qualitative aspects of the equilibrium;

2. A formal description of how these factors act.

Here, I adopt Woodward's theory of causal explanation, [11]: the causal factors of an event are those factors that, if properly manipulated, would change the event. Further, condition (2) tells us in which sense we should manipulate the causal factors to obtain a different result. Finally, the satisfactoriness of an explanation does not depend on relations between its parts, but on the resources it uses. I claim that a satisfactory explanation must fulfill the following:

3. Historic-pragmatic value: a sensible use (possibly a reconfiguration) of the traditional practices and techniques deployed in the field.

This element has been totally neglected in philosophical literature on explanation.[1] It is motivated by the almost trivial consideration that explanations do not happen in a vacuum, but are historically situated. Scientists try to construct (and value) explanations that make use of traditional techniques and practices, perhaps providing them with a new meaning and new potentials. Hence, a good explanation must be evaluated relatively to the history of the practices and relatively to the subculture in which it is accepted. In the following sections, I argue that this model illuminates the explanatory value of typicality. I quickly summarize the genealogical lines of the mathematical practices related to the use of typicality in physics (section 2) and I show how these lines converge to the modern approach (section 3).

## 2 Typicality in Physics: A Genealogy

Current use of typicality is not as clear as many of its supporters would wish. To understand the roots of this notion, it may be useful to begin with examining three definitions of typicality adopted in the literature. The first definition comes from a philosophical paper:

> Intuitively, something is typical if it happens in the 'vast majority' of cases: typical lottery tickets are blanks, typical Olympic athletes are well trained, and in a typical series of 1,000 coin tosses the ratio of the number of heads and the number of tails is approximately one. [2, 997-998]

The second definition comes from a historical paper:

> Generally speaking, a set is typical if it contains an "overwhelming majority" of points in some specified sense. In classical statistical mechanics there is a "natural" sense: namely sets of full phase-space volume. [12, 803]

---

[1] A note of clarification: the point of requisite (3) is not to provide an explanatory value to dead and buried theories, but to stress that the explanatory value of any theory depends crucially on what a certain community can do with them.

Finally, the third definition comes from one of the most distinguished upholders of the typicality approach, Joel Lebowitz:

> [A] certain behavior is typical if the set of microscopic states [...] for which it occurs comprises a region whose volume fraction goes to one as [the number of molecules] $N$ grows. [13, 7]

Apart from the different levels of technicality and their specific aims, these definitions point out two traits of typicality. First, it relies on *the separation of two families of events*, those which are "almost certain" and those which are "negligible". This evaluation depends on the relative sizes of the corresponding families. Second, Lebowitz's definition stresses the asymptotic character of typical behaviors: they tend to a certain maximal size as the number of degrees of freedom of the problem approaches infinity. The first element is related to the tradition of celestial mechanics that goes back to the notorious three-body problem. The second element is linked to the combinatorial techniques used in statistical mechanics. There are, as we will see, intersections between these traditions, which explain how they can both feature in the definitions of typicality.

## 2.1 Celestial Mechanics and Topology

Since mid-18th century, mathematicians struggled to show that three bodies interacting according to the gravitational law would never undergo catastrophic collisions or expulsions. The usual strategy to deal with this problem was to solve the equations of motion by means of trigonometric series and to show that these series do not contain diverging (secular) terms. After many failed attempts to provide an explicit solution of the equations of motion, mathematicians grew skeptical that these solutions would ever be discovered. In the second half of the 19th century, it became increasingly clear that there was no way to solve the three-body problem in closed form and other paths were tried.

Instrumental in this change of tack was the adoption of new topological techniques. The undisputed champion of this line of attack was Henri Poincaré [14]. Instead of establishing stability analytically, Poincaré sough for the conditions under which *most of the trajectories* are stable. This method does not require an explicit solution of the equations of motion and do not call for any assumption of randomness. Rather, it aims at classifying trajectories in stable and unstable and then to show under which circumstances the former outnumber the latter [15].

As an example of this procedure, one can consider the famous recurrence theorem [16, III, 847-876]. By a very general topological argument, Poincaré showed that almost all possible mechanical trajectories of a conservative system return, after a very long time, infinitesimally close to their initial state (or, as Poincaré had it, they are Poisson-stable). The set of trajectories that do not behave like that is negligible.

When Poincaré developed his approach, he did not have a precise mathematical notion of "almost-all" or "negligible". This notion became available only in the early 20th century with the development of Henri Lebesgue's theory of measure. The combination of topological and measure-theoretical techniques was successfully put to work on other important problems of celestial mechanics such as the study of singularities and perturbations (for a discussion see [17]). It thus comes as no surprise that contemporary theory of dynamical systems are customarily defined as the study of *typical or generic* properties of systems, that is properties that hold of the vast majority of the possible trajectories. It is important to recognize, though, that these properties are defined asymptotically. Consider, for example, the introduction to one of the most complete and authoritative books on the topic:

> The most characteristic feature of dynamical theories, which distinguishes them from other areas of mathematics dealing with groups of automorphisms of various mathematical structures, is the emphasis on asymptotic behavior [...] that is properties related to the behavior as time goes to infinity. [18, 2]

Typical properties are therefore those properties that come to be recognized as such only in the long run.

## 2.2 Statistical Mechanics

Although much younger and very different in subject matter, kinetic theory—the predecessor of modern statistical mechanics—faced a similar problem as celestial mechanics. The behavior of a gas composed of many molecules colliding mechanically cannot be predicted by solving the equations of motion. In fact, even the knowledge of the initial conditions is out of reach. Thus, from the beginning, statistical mechanics introduced a set of approximation techniques and assumptions in order to make the problem tractable. For example, the collisions between the molecules and the walls bring in a disturbing effect in the sequence of elastic collisions between molecules. This is the so-called "wall-effect". To take into account this effect in the equations of the problem leads to innumerable formal complications, therefore it is usually assumed that the container is big enough that the wall effect remains confined to a negligibly small portion of the whole space. Analogously, basically all arguments in kinetic theory are cast by supposing ideal conditions such as the number of molecules grows to infinity, or the duration of a collision tends to zero and so on.

One of Ludwig Boltzmann's great insights was that the nature of the problem of irreversibility is not affected by the use of these approximation techniques based on asymptotic tendencies. These techniques only cancel out the probabilistic fluctuations and make the results strictly valid. They produce "statistical determinism". For this reason, Boltzmann made ample use of probabilistic arguments and tools constantly framed within asymptotic assumptions [19].

It was clear to Boltzmann that there are two different, albeit related questions: (1) what is the essence of irreversibility and (2) how to formulate this essence in terms of the specific microscopic arrangements and dynamical laws of the molecules. As for the first question, Boltzmann concluded that irreversibility is due to the extremely large number of molecules in complicate collisions. It is this large number that justifies an assumption of equiprobability for the microstates and thus a probabilistic procedure that leads to the equilibrium distribution as the largest one:

> The great irregularity of the thermal motion and the multiplicity of the forces acting on the body from the outside make it probable that its atoms [...] pass through all the possible positions and velocities consistent with the equation of energy. [20], [21, I, 284]

He illustrates this point most effectively in his famous 1877 combinatorial theory [22], [21, II, 164-223]. Boltzmann assumes very many molecules and calculates the numbers of ways in which energy can be distributed over them. It turns out that the overwhelming majority of these ways are represented by a bell-shaped distribution. This is Maxwell's distribution, which represents the state of equilibrium. It's by far the largest in terms of the number of microscopic molecular allocations of energy compatible with it. The remarkable point is that the dominance of the equilibrium state depends crucially on the huge number of degrees of freedom of the problem: the relative size of the equilibrium state respect to the other increases enormously with the number of degrees of freedom. This behavior is characteristic of asymptotic probability laws such as the law of large numbers or the central limit theorem. For this reason, Boltzmann understood the essence of irreversibility as a probabilistic law valid under suitable asymptotic conditions [19].

The second question was harder. If we assume that molecules obey the laws of mechanics, we run into the reversibility problem. Mechanical motion can be inverted and still remain perfectly mechanical, so how are we to understand irreversibility as a form of mechanical motion? Why a sequence of mechanical collisions leading the system from an arbitrary state to equilibrium should occur more often than its reverse, which is matter-of-factly as mechanical? The most important debate on this question took place on the pages of *Nature* in 1894-95 and involved, besides Boltzmann, as distinguished British physicists as Bryan, Burbury, Watson, and Culverwell. Four possible solutions to this question emerged from the debate.

1. The mechanical reversal of a state violates the formal condition on which Boltzmann's theorem of irreversibility was based ($H$-theorem). This solution appeared unacceptable to Boltzmann because it emptied the theorem of any physical meaning and downgraded it to a purely mathematical statement.

2. Mechanical reversal is unstable. The situation is analogous to riding a bicycle backwards: it is mechanically possible, but any small perturbation will destroy the equilibrium. Boltzmann liked this option: a reversal presupposes a perfect coordination between the molecules, which is easy destroyed.

3. In its path from the nonequilibrium state to equilibrium, the trajectory branches off in many possible states. It is true that for each whole path the reverse exists, but at each stage there are more ways to go toward equilibrium than in the opposite direction. This is the idea of the $H$-curve.

4. Microscopic molecular arrangements are molecularly disordered. This is the so-called molecular chaos that Boltzmann introduced in the first volume of his *Gastheorie*.

I will dwell especially upon this last point. Boltzmann's notion of molecular chaos is profound, but not very clear. His basic point is that molecules must be arranged and must behave in a way that leaves all theoretical possibilities open. In other words, any regularity that forces the system out of its typical state of equilibrium must derive from some specially engineered arrangement that made probability considerations invalid:

> If we choose the initial configuration on the basis of a previous calculation of the path of each molecule, so as to violate intentionally the laws of probability, then of course we can construct a persistent regularity. [23, I, 22]

Thus, in making the reversal, we request the molecules to retrace exactly the same sequence of collisions as before. This kind of interventions (or "conspiracy") on the dynamics of the system leads to atypical results. It is important to note that all these solutions of the reversibility objection contain traits characteristic of what is today known as chaos theory. We will see these traits displayed in Lebowitz's paper in the next section. Before concluding this section, however, I want to stress that Boltzmann had clearly in mind also the importance of the notion of negligibility. Poincaré's recurrence theorem is based on the concept of integral invariant, a mathematical technique that Boltzmann had himself introduced and used, albeit imperfectly, since the end of the 1860s [24], [21, I, 49-96]. In the *Gastheorie* he discusses the possibility that a gas, being a conservative and confined mechanical system, passes through its state again and again as prescribed by the recurrence theorem. He finds that this can happen only after an enormous interval of time. He concludes:

> One may recognize that this is practically equivalent to never, if one recalls that in this length of time, according to the laws of probability, there will have been many years in which every inhabitant of a large country committed suicide, purely by accident, one the same day, or every building burned down at the same time—yet the insurance companies get along quite well by ignoring the possibility of such events. If a much smaller probability than this is not practically equivalent to impossibility, then no one can be sure that today will be followed by a night and then a day. [23, II, 254]

Boltzmann was therefore well aware of the topological argument, which aims at distinguishing between typical and negligible events.

## 3 The Explanatory Value of Typicality

In the 20h century, the theory of dynamical systems and statistical mechanics took up and developed the trends outlined above. Measure theory provided a set of concepts and tools to express typical and negligible events. Furthermore, these tools were used to prove asymptotic statements like in the case of Emil Borel's proof of the law of large numbers (1909). George D. Birkhoff's 1931 ergodic theorem can also be considered a sort of law of large numbers applied to statistical mechanics. Birkhoff showed that dynamical systems have the propriety of ergodicity (from which many statistico-mechanical consequences follow) if and only if the set of trajectories that do not remain in an invariant portion of the phase space is negligible (i.e., it has measure-0). Properties that holds typically or generically are said to hold "almost-everywhere" [25].

Another important development of statistical mechanics in the 20th century is Alexander Khinchin's asymptotic approach [26], [25]. Khinchin claimed that the fundamental proposition of statistical mechanics, the irreversible approach to equilibrium, was just the physical formulation of the central limit theorem. Accordingly, the entire theory could be recast in purely probabilistic terms, leaving aside any physical assumption. Khinchin proved a theorem that systems for which the macroscopic parameters can be expressed by particular functions (sum-functions) reach equilibrium in the long run.

Finally, one of the most successful approach to statistical mechanics focuses on "large systems". The basic tenet is that when we examine the behavior of systems under particular asymptotic circumstances (for example the so-called thermodynamic limit where the number of molecules, the energy, and the volume tend to infinity, but the density and the energy density stay finite), we are able to prove kinetic theorems rigorously [27]. The most impressive result obtained by this approach is Lanford's theorem according to which for a particular gas model and in a particular limit, it is practically certain that the system will reach equilibrium [28], [29].

The upholders of typicality belong to this tradition. Most of them have worked within the framework of the large systems approach. Therefore, it is essential to keep in mind this long-term development to evaluate the meaning of the concept of typicality. The supporters of the typicality approach inscribe themselves in the Boltzmannian line of rigorous mathematical arguments framed within an asymptotic conceptual space where fluctuations become negligible. To illustrate this aspect I briefly discuss a paper by Joel Lebowitz. There are three points that I want to emphasize.

First, the notion of typicality serves the general purpose of understanding the transition from the microscopic to the macroscopic level. Remember the quote given above: typicality is a feature that emerges when the number of molecules approaches infinity. Put in other words, typicality discriminates between behaviors associated with a large number of degrees of freedom and behaviors associated with less complex systems. The former exhibit time-asymmetry, the latter do not:

> The central role in time asymmetric behavior is played by the very large number of degrees of freedom involved in the evolution of the macroscopic systems. It is only this which permits statistical predictions to become "certain" ones for typical individual realizations, where, after all, we actually observe irreversible behavior. This typicality is very robust—the essential features of macroscopic behavior are not dependent on any precise assumptions such as ergodicity, mixing or "equal a

priori probabilities", being strictly satisfied by the statistical distributions. [13, 3]

This is a point often neglected by philosophers. Typicality is not just shorthand for "very high probability", i.e., another probabilistic notion subject to probabilistic conditions. Typicality is a feature of systems with many degrees of freedom, systems that are handled by certain techniques. More importantly, the high number of degree of freedom plays a real causal role in Woodward's sense. Like in Boltzmann's combinatorics and in Khinchin's probabilistic approach, the equilibrium state dominates over the others *because* there are many particles. Were there just a few of them, the equilibrium would be not so overwhelmingly more probable. Hence, it is by manipulating the number of degrees of freedom that we can make an effect on equilibrium.

The second point is related to the first: Lebowitz introduces a distinction between the qualitative and the quantitative aspects of irreversibility. As said above, the qualitative aspect depends only on the large number of degrees of freedom. From this, the typicality explanation of irreversibility follows. However, this aspect does not yield the hydrodynamical-like equations to predict the concrete behavior of a macroscopic system. For this we need more specific microscopic models, which, however, depend very little on the details of the microscopic dynamics. It is at this level that we find ergodicity, mixing and chaotic dynamics:

> I believe that these models capture the essential features of the transition from microscopic to macroscopic evolution in real physical systems. In all cases, the resulting equations describe the typical behavior of a single macroscopic system chosen from a suitable initial ensemble i.e. there is a vanishing dispersion of the values of the macroscopic variables in the limit of micro/macroscale ratio going to zero. [13, 17]

Again, it is crucial to notice that these models lead to time-asymmetric behavior only because they are applied to a large number of degrees of freedom. As such, chaotic dynamics or ergodicity are time-symmetric:

> This is an important distinction (unfortunately frequently overlooked or misunderstood) between irreversible and chaotic behavior of Hamiltonian systems. The latter, which can be observed in systems consisting of only a few particles, will not have a uni-directional time behavior in any particular realization. [13, 25]

The third point concerns Lebowitz's way of dealing with the reversibility objection. He argues that a reversal of the microscopic motion is conceivable but "effectively impossible to do [...] in practice." To support this claim he uses three arguments, all related to chaos dynamics. The first is that such reversal would be unstable under external perturbations. The second is that mechanical reversal requires a "perfect aiming" and

> [i]t can therefore be expected to be derailed by even smaller imprecisions in the reversal and/or tiny random outside influences. This is somewhat analogous to those pinball machine type puzzle where one is supposed to get a small metal ball into a particular small region. You have to do things just right to get it in but almost anything you do gets it out into larger region. [13, 9]

Lebowitz deploys the example of the pinball, but he might as well mention the example of riding a bicycle backwards: it is the same kind of mechanical situation. Finally, he points out a hidden assumption in the dynamics for typical behavior:

> For the macroscopic systems we are considering the disparity between relative sizes of the comparable regions in the phase space is unimaginably larger. The behavior of such systems will therefore be as observed, in the absence of any "grand conspiracy". [13, 9]

The idea that there must be some artificial intervention for such a system to exhibit an atypical behavior reminds immediately Boltzmann's remark about intentional violations of the laws of probability.

These quotes prove the kinship between the typicality approach and the tradition encompassing celestial mechanics, Boltzmann's statistical mechanics, and the large systems approach. But they also allow us to draw a more general philosophical conclusion. Typicality provides for a plausible *epistemic story* of the qualitative aspects of equilibrium by ascribing it to causal factors i.e., the high number of degrees of freedom, whose action is described by combinatorics and measure-theoretical concepts. It is not a probabilistic ingredient to be added to an argument, although it makes use of a probabilistic argumentative pattern ("given a suitable definition of probability, if the probability of one event is overwhelmingly larger than all alternatives, one can neglect the latter"). More importantly, typicality is a reflective way to classify, organize, and reconfigure a set of theoretical practices as diverse as topological methods, approximations procedures and statistical techniques. It derives from the mathematical practices outlined above and allows to combine them in an explanation of equilibrium. Part of its popularity is due to its historical-pragmatical value. Thus, typicality works as an *epistemic trope*: it is an assemblage of concepts, methods, and argumentative patterns that organize well-established mathematical practices into a specific epistemic story of equilibrium.

## Acknowledgements

This paper has been written with the support of the GIF Research Grant No. I-1054-112.4/2009.

# BIBLIOGRAPHY

[1] R. Frigg. Why Typicality Does Not Explain the Approach to Equilibrium. In M. Suarez (ed.), *Probabilities, Causes, and Propensities in Physics*, Springer, Berlin 2007, pages 77-93.
[2] R. Frigg. Typicality and the Approach to Equilibrium in Boltzmannian Statistical Mechanics. *Philosophy of Science* 76: 997–1008, 2009.
[3] R. Frigg and C. Werndl. Demystifying Typicality. *Philosophy of Science* 79: 917–929, 2012.
[4] S. Goldstein. Typicality and Notions of Probability in Physics. In Y. Ben-Menahem and M. Hemmo (eds.), *Probability in Physics*, Springer, Berlin 2012, pages 59-71.
[5] I. Pitowsky. Typicality and the Role of the Lebesgue Measure in Statistical Mechanics. In Y. Ben-Menahem and M. Hemmo (eds.), *Probability in Physics*, Springer, Berlin 2012, pages 41-58.
[6] D. Dürr, S. Goldstein, N. Zanghí. Quantum Equilibrium and the Origin of Absolute Uncertainty. *Journal of Statistical Physics* 67: 843–907, 1992.
[7] D. Dürr. Bohmian Mechanics. In J. Bricmont, D. Dürr, F. Petruccione, M. C. Galavotti, G. Ghirardi and N. Zanghí (eds.), *Chance in Physics. Foundations and Perspectives*, Springer, Berlin 2001, pages 115-132.
[8] C. Callender. The Emergence and Interpretation of Probability in Bohmian Mechanics. *Studies in History and Philosophy of Modern Physics* 38: 351–370, 2007.
[9] M. Hemmo and O. H. Shenker. Measures over Initial Conditions. In Y. Ben-Menahem and M. Hemmo (eds.), *Probability in Physics*, Springer, Berlin 2012, pages 87-98.
[10] M. Hemmo and O. H. Shenker. *The Road to Maxwell's Demon*. Cambridge University Press, Cambridge 2013.
[11] J. Woodward. *Making Things Happen: A Theory of Causal Explanation*. Oxford University Press, New York 2003.
[12] S. B. Volchan. Probability as Typicality. *Studies in History and Philosophy of Modern Physics* 38: 801–814, 2007.
[13] J. L. Lebowitz. Macroscopic Laws, Microscopic Dynamics, Time's Arrow and Boltzmann's Entropy. *Physica A* 193: 1–27, 1993.
[14] J. Barrow-Green. *Poincaré and the Three Body Problem*. American Mathematical Society, Providence 1997.
[15] J. Laskar. The Stability of the Solar System from Laplace to the Present. In R. Taton and C. A. Wilson (eds.), *Planetary Astronomy from the Renaissance to the Rise of Astrophysics*, Cambridge University Press, Cambridge 1995, pages 240-248.
[16] H. Poincaré. *New Methods of Celestial Mechanics*. 3 vols, American Institute of Physics, College Park, MD 1892-1899.
[17] F. Diacu and P. Holmes. *Celestial Encounters. The Origins of Chaos and Stability*. Princeton University Press, Princeton 1996.
[18] A. Katok and B. Hasselblatt. *Introduction to the Modern Theory of Dynamical Systems*. Cambridge University Press, Cambridge 1995.
[19] M. Badino. Mechanistic Slumber vs. Statistical Insomnia: The Early Phase of Boltzmann's H-Theorem (1868-1877). *European Physical Journal – H* 36: 353–378, 2011.
[20] L. Boltzmann. Einige allgemeine Sätze über Wärmegleichgewicht. *Wiener Berichte* 63: 679-711, 1871.
[21] L. Boltzmann. *Wissenschaftliche Abhandlungen*. 3 vols. Barth, Leipzig 1909.
[22] L. Boltzmann. Über die Beziehung zwischen dem zweiten Hauptsatze der mechanischen Wärmetheorie und der Wahrscheinlichkeitsrechnung respective den Sätzen über das Wärmegleichgewicht. *Wiener Berichte* 76: 373-435, 1877.
[23] L. Boltzmann. *Vorlesungen über Gastheorie*. 2 vols. Barth, Leipzig 1896-1898.
[24] L. Boltzmann. Studien über das Gleichgewicht der lebendigen Kraft zwischen bewegten Materiellen Punkten. *Wiener Berichte* 58: 517-560, 1868.
[25] M. Badino. The Foundational Role of Ergodic Theory. *Foundations of Science* 11: 323-347, 2006.
[26] A. Khinchin. *Mathematical Foundations of Statistical Mechanics*. Dover, New York 1949.
[27] O. Penrose. Foundations of Statistical Mechanics. *Reports on Progress in Physics* 42: 1937-2006, 1979.
[28] O. E. Lanford. Time Evolution of Large Classical Systems. In J. Moser (ed.), *Dynamical Systems, Theory and Applications*, Springer, Berlin 1975, pages 1-111.
[29] O. E. Lanford. The Hard Sphere Gas in the Boltzmann-Grad Limit. *Physica A* 106: 70–76, 1981.

# Disentangling Context Dependencies in Biological Sciences

Marta Bertolaso

ABSTRACT.
　The aim of this paper is to disentangle two different kinds of context dependency in biological explanations by looking at explanations of cancer. One kind of context dependency is employed as an *explanans* in the Tissue Field Organization Theory (TOFT), where cell behavior depend on the field (the context). The other kind of context dependency—I argue—underlies both systemic and molecular accounts of cancer, and pertains the identification of the *relata* of the explanation more than the explanation itself. This double nature of context dependency creates an interesting unified picture of explanation, where mechanistic explanations are always possible even though a mechanistic account of the biological world is not. It also sets the conditions for a particular kind of compatibility between TOFT and molecular accounts, in which TOFT is more general and SMT's molecular accounts—when they work—are particular cases of TOFT.

**Keywords:** context dependencies, biological explanations, systems, cancer

## 1 Introduction

Context dependency is a critical feature of scientific explanation in the biological sciences. For example, it has been used as an argument against reductionism [7] and included in an expanded account of mechanisms [9]. Given its deep ontological implications, the notion of context dependency has been caught up in discussions like those on emergent properties or the criticisms to reductionist-mechanistic accounts in biological sciences. A way out of these tensions has been to consider the context's relevance as a methodological or pragmatic recommendation [18, 5]. However, when the context is seen as a mere methodological feature of the biological explanation, mechanistic accounts shift towards multilevel and more complicated accounts in a never-ending inclusive process of new elements [8]. Systemic accounts notably adopt a more holistic stance, where the context plays a relevant explanatory role that is not clarified adequately. My working hypothesis is that different kinds of context dependencies are at stake in these debates. In particular, we should disentangle *conceptual* context dependency from *explanatory* context dependency in biological explanation in order to understand how some kinds of biological explanations work and how the *relata* that structure such explanation are mutually dependent.[1]

## 2 Molecular accounts vs. systemic accounts of cancer

There are, in scientific literature, different interpretations of cancer that are often seen as conflicting. To a first approximation, there is a theory that defends the genetic origin of cancer (Somatic Mutation Theory or SMT [25, 26]) and another one that claims that cancer is a problem of tissue organization (Tissue Organization Field Theory or TOFT [21]). In this paper I will preferably refer to *molecular accounts* of cancer, a class of accounts that contains the Somatic Mutation Theory. Its authors, in fact, appeal to genes and molecules to give an account of the neoplastic proper. Such molecules and their functions, instantiate the most relevant causal elements. The TOFT maintains, instead, an organicist view in which the environmental context is more relevant than genes in the origin and establishment of the phenotype of tumour cells. TOFT can be seen as a prototypical *systemic account* of cancer, and is often defended as an 'antireductionist' position. SMT, on the other hand, is classified as 'reductionist' (and further loaded with the genetic determinism assumption). I will return on these labels in what follows.

To clarify how these accounts work in scientific practice let's focus on the question they aim to answer: Why does a tumour cell behaves like this (and not like that)? The *explanandum* is that a neoplastic cell no further proliferates in an integrated way in its organic environment.

A good example of a molecular account of cancer is the Hierarchical Model of Cancer, proposed as an explanatory model for some types of cancer. In this model, a cell that retains the neoplastic property of proliferating in an aberrant way is framed categorized as a Cancer Stem Cell (CSC).[2] The CSC gives origin to an offspring of cells that differentiate aberrantly, while only sometimes retaining the tumorigenic property of their ancestor. In this case the molecular element is the cell itself whose biological identity is determined by genetic and epigenetic changes.

For the Tissue Organization Field Theory, instead, carcinogenesis is attributable to a process similar to an organogenesis that does not reach completion. The TOFT points out that proliferation should be considered the default state of metazoan cells and the tissue organization the result of a developmental program. The interactions between cells, mediated by membrane proteins that recognize paracrine, mechanical splices or endocrine signals that act at a distance, are responsible for the transmission of signals significant for the cells in terms of proliferation and differ-

---

[1] [6] suggests to think of 'mechanism description' and 'mechanism explanation' as two distinct epistemic acts. This approach is related but still different from the aim of this paper that is much more focused on how mechanistic and systemic accounts are related in accounting for a biological multi-level phenomenology. Cf. also [3]. In this paper I am linking up and relating the analysis and thesis I have presented in Chapters 4 and 5 of that monograph.

[2] Main proponents of the Hierarchical Model would not probably deny that other cells of the tumour (non-CSC) are also neoplastic, they just believe that they are clinically irrelevant due to their short replication potential. This does not change, the relevance for this paper's argument of how CSC are actually identified and play an explanatory role. What I am discussing here, in fact, is an aspect of the stemness concept when applied to tumour cells, not the replication potential of the (tumour) cells as such as highlighted in the section 3. For further discussion on this point see also [4].

entiation. Cell proliferation is thus chronologically removed from the control of the cell cycle that takes place at the subcellular level in the hierarchical organization of metazoans. In this view, the malignant tumour phenotype doesn't have to be understood as an effect that necessarily follows a causal event (for example, a genetic mutation). Rather, it should be seen as uncoupling of intrinsic potentialities of the cells that are, in tumours, executed without adequate contextual control. Cancer-related mutations can of course have a causal role, but this causal role amounts to their effects on tissue organization.[3] The epilogue of the neoplastic process depends, most generally, upon the *persistence of the same conditions that caused the original breakdown* of the functional organization at the tissue level. In sum, the natural history of cancer is told from a very different perspective than the molecular one. The molecular and the systemic accounts share the *explanandum*—i.e., the aberrant behaviour of tumour cells—and they both acknowledge that cancer is not an event but a process. The explanans seems, instead, to differ: genetic alterations and tissue organization respectively.

## 3 Explanatory structures: context in the front vs. context in the back

Now let us take the two accounts of cancer summarized in the previous section and focus on *the role of context and context dependency therein*.

The TOFT has context 'in the front'. In this systemic approach, the context plays an important explanatory role–indeed, it is the *explanans*: cancer is the result of a disruption in the tissue's architecture. TOFT acknowledges higher-level effects as causally relevant to the maintenance of the functional properties of cells. In particular, the systemic explanation uses *time dependencies* and *context dependencies* specified in terms of compromised relationships among coupled biological rhythms and long-range spatial interactions, adding an interesting level of systemic analysis to the overall explanatory account and specifying the relevance of cell interactions at an organismic level. Accordingly, the functional properties of tumour cells are addressed in terms of function loss rather than in terms of function acquisition [2]. Molecular accounts strive to neglect context in their explanations, yet context is somehow there, 'in the back'. The biological context appears, for example, in the definition of Cancer Stem Cells: stemness is, in fact, a context-dependent property. Defining a molecular part entails the contextual dimension. Indeed, in a molecular account, '[B]y simplifying the nature of cancer – portraying it as a cell-autonomous process intrinsic to the cancer cell – [...] cancer development depends upon changes in the heterotypic interactions between incipient tumor cells and their normal neighbours" [11, 67]. As evident from this quotation, we may say that the attempt to find an explanation at the 'lowest possible level'—either in terms of genes or in more general molecular-mechanistic terms—fails, and that molecular accounts appear as moving towards a tissue level.

The atomistic commitment of a strong molecular account strives to focus on *essential* properties (of the part, of the cell). But cancer (as a complex biological

---

[3] This important issue of the field-mediated effects of mutations will be reminded in the closing part of the paper.

process) does not primarily affect essential properties of the cells, but relational ones. it is not enough to tell a stem cell *what it has to do*, but it is fundamental to clarify what it has to do *in a precise point* and not in another one. In fact, there are very interesting convergences between molecular and systemic accounts. One is the concept of *fields of cancerization*. These fields are groups of cells from which specific morphological structures develop through the mediation of biophysical and biochemical cues, mainly through epigenetic changes that in cancer are aberrant [24]. The shared idea is that in cancer the functional stability of such fields is compromised.

In contrast, the systemic account replaces essential properties with an 'essentiality-by-location' principle. It conceptualizes *dynamic properties* of cancer cells more in terms of capacities than in functional mechanistic terms. Such capacities are jointly determined by intrinsic features of the cell and by features of its environmental context. We are dealing with *relational properties* of the cells that do not follow the rules of what is necessary but of what is possible (the kind of possibility entailed in the concept of pre-disposition).

The molecular and the systemic account differ by the structure of the explanations. As mentioned already, TOFT has context in the front, molecular accounts have it in the back. We could say that, like in the negative and positive of a photo, in the molecular account the context remains in the back (in the conceptualization of the *relata*), in the systemic account it is in the front, playing the explanatory role too. But what does having context in the back mean? I shall argue that there is a different kind of context dependency underlying both TOFT and molecular accounts.

## 4  The conceptual context dependency

In the previous section I have argued that TOFT and molecular accounts differ by the role of context in the structures of their explanations. The difference concerns, we may say, a kind of context dependency: *explanatory* context dependency. There is another kind of context dependency, which I will call *conceptual* context dependency, which has to do with the definition and identity of the *relata* of the explanation, more than with the structure of the explanation itself. This kind of context dependency is the one that is shared between systemic and molecular accounts of cancer.

In an interesting discussion of reductionism with Evelyn Fox Keller [12], John Dupré [10] distinguishes the problem of the relata identification and definition from the problem of in what way a reduction can be realized.

The problem of *relata* identification and definition is the problem of the "dependence of the *identity* of parts, and the interactions among them, on higher-order effects". Dupré insists that "the fact that biology–a science–works with concepts that depend on the larger systems of which they are part, as well as on their constituents, is a fatal objection to any attempt to defend a reductionist position about biological explanation" [10, 38]. Whereas Keller thinks that *context* and *interactions* are artificial distinctions and that the "context is simply all those other factors/molecules whose interactions with the object or system in question have not been made explicit and, hence, have not been included in the description" [12, 30], for Dupré, instead, context refers 'to features of an object's environment that are necessary to confer on the object a particular capacity [...]. Interactions are simply the exercise of such capacities with relation to some other entity that will presumably constitute all or part of that context" [10, 45].

The problem of *in what way* a reduction can be realized concerns the appropriate level where explanation should be sought. TOFT's authors [22] stressed arguments that to understand a specific biological phenomenon each hierarchical level must be studied without expecting that the lower levels will necessarily contribute to our understanding a phenomenon cannot be studied independently from the level at which it is observed. They also argued that top-down causality is the most adequate assumption to explain complex mechanisms but top-down and bottom-up accounts respond to two different epistemic concerns.

A way to conceive the difference between explanatory strategies is to say that explanatory strategies rely upon different assumptions about *what is fundamental in explanatory terms*. 'In principle' claims about what is fundamental in scientific explanations sound like this: the "analysis of the specific physical and chemical phenomenology involved in biological processes should, in principle, suffice for an understanding of what endows biological systems with the properties of life" [12, 21]. Dupré says that if we are interested in ecological systems (i.e. biological systems from the point of view of their functional organization) what is under analysis are ultimately systems whose behaviour "is fully determined by the behavior of, and interactions between, the parts. And hence, the elements of behavior that are not so determined are what we don't know when we know everything about the parts and the way they are assembled" [10, 38]. These are, indeed, 'in principle' claims. In fact, Dupré (cit.) and Keller (cit.) themselves pointed out that there is no obvious definition of what is fundamental in explanatory terms in biological sciences. The meaning of fundamental in biology, for example, cannot be clearly equated with simple, nor is it at all obvious that it should be common to all biological entities' explanation.

What should be considered more fundamental in explanatory terms is *neither logically derivable* from the structure of the explanation nor can it be assumed on the basis of *an ontological commitment*. It is related with the intrinsic relationships that structure the scientific explanations, their relata, and the practical character of the scientific enterprise.[4] Whether or not we can explain biological phenomena in molecular terms (what Malaterre would call physical monism [15]), we very often do as a methodological choice. Accordingly, explanatory strategies do not differ by what they consider fundamental: they differ by their commitment about *the right level at which an explanation should be offered*.

Dupré's distinction of two problems about reductionism (how do parts depend on the whole and how a reduction should be performed) parallels the two kinds of context dependency that I have pointed out. Conceptual context dependency is

---

[4]See [2, 4] on the identification of the mesoscopic level in scientific explanation, as a strategy that takes into account all the mentioned constraints and works to identify the explanatory level that maximizes determinsm for the given explanandum.

about the weight of the context in the identity and definition of parts and their interactions. This is a more fundamental ontological problem, and mirrors Dupré's issue of the dependence of parts' identity and interactions among on higher-order effects. Explanatory context dependency concerns the weight that should be given to the context in an explanation. This is a problem of explanation structure, just like Dupré's problem of how a reduction should be carried out.

So there are at least two dimensions of context dependency that can underlie explanatory endeavours. The first dimension concerns the 'certain way' parts are assembled in the studied systems, and the conceptual dependence of parts on the larger system. This is the dimension of conceptual context dependency that determines (a) the level of generalization of the *relata*, which (b) admits the pragmatic focus on different contrast classes. These dimensions are partially independent though related. Moreover, (b) is secondary, from a procedural point of view, to the former aspect (a). Once this process of identification of the explanatory elements is acknowledged, the explanatory picture eventually acquires an interesting unity.

## 5 Mechanism and emergence in biology

I have discussed the structure of explanatory accounts in terms of the explanatory relevance of context dependency. I will now discuss the issue in the wider context of explanations in biology.

In philosophy, the "new mechanistic program" is the most updated way to analyze those scientific approaches that focus on molecular interactions. The new mechanistic program correctly emphasizes that scientists usually explain phenomena by describing the underlying mechanism. Machamer et al. [14] proposed this standard philosophical account that offered a general characterization of mechanism that attempts to capture the way scientists use this word and to show the ways in which mechanisms are involved in the explanation of phenomena. This 'mechanistic philosophy', as defined by Skipper and Millstein [20], has developed into a robust alternative to theory reduction. Unlike the more general idea of a mechanistic worldview, the "mechanismic program" [17] is not primarily concerned with biological ontology, but with the nature of biological explanations. So that, whereas mechanicism is closely aligned with the spirit of reductionism and the unity of science, the mechanismic program focuses on multilevel explanations given in terms of causal mechanisms and seemingly with a non-reductive view of science. In the attempt to reframe functional explanations in mechanistic terms, examples taken from different domains in the life sciences have been used to support mechanistic accounts.

The flexibility and pragmatic fertility of such mechanistic explanations count against any attempt to reject their scientific acceptability. However, mechanistic reductions in life sciences are clearly challenged by context-dependence of molecular features and multiple realization of higher-level features, as well as by temporality and dynamic stability of biological systems. Some authors [16, 15] highlight that any attempt of describing biological explanations in mechanistic terms has to face with the revenant idea of *emergence*, and that it has been scientific practice, more than philosophy, to impose such cumbersome re-emergence of emergence. In these discussions, emergence comes back, not in vitalistic terms, but in the language of non-linear dynamical systems, that is in a language of self-causation, bringing to the fore the systems self-organizing properties. As Silberstein notes [19], such properties 'go beyond' the 'physical' interactions among single elements, without being completely independent from them. The emphasis on those properties enhanced the arguments of the non-reductive physicalism originally rejected by Kim. As Mitchell [16] discusses, the complexity of the temporal dimension challenges Kim's account of emergence and highlights its limits. In fact, the progressive stratification of functional levels represents the history of the system and such evidence substantiates the mentioned entanglement of organizational and evolving dimensions.[5]

Biology has been, since the beginning, the science with most difficulties to define its object of inquiry. Nevertheless, in a more general sense, biology focuses on the specific dynamism of living systems, which is minimally definable in terms of *self-organizing and adaptive dynamic processes characterized by a multi-level regulatory phenomenology*. Such conceptualization, because of the very notion of system, refers to a set of elements in standing relationship (on this point see also [1]). Properties of biological systems are therefore often seen as emergencies held by an organized whole and its parts. It is important to remark that both mechanistic and systemic perspectives have a common root in the challenge of explaining how a *biological system maintains the integrity of its parts in its dynamic evolution*. This means providing an explanation of robust phenomena and understanding their multilevel regulatory phenomenology. It is therefore the persistence of living systems in space and time that poses the most relevant philosophical questions regarding the adequateness and the epistemological status of both mechanistic accounts and systemic explanatory models. This is why in biology we find a particularly strong degree of conceptual context dependency, which subordinates the parts' definition on the wider system they belong to. The wider system holds a normative dimension that allows a judgment of the physiology/pathology of parts' behaviour.

In sum, although there are features of the world (and epistemological constraints) that allow mechanisms to be always found, the world is unlikely to be mechanistically definable. With this in mind, let us finally tackle the issue of compatibility between molecular accounts and systemic accounts of cancer.

## 6 SMT as a particular case of TOFT

Typically, molecular and systemic accounts of cancer are stigmatized, respectively, as reductionist and antireductionist, and the debate ends up by arguing for a radical replacement of SMT with TOFT. This is not the only way of framing the polarity between SMT and TOFT. First, let us clear up some distinctions. My argument on the two kinds of context dependence and Dupré's argument on the two different reductionist issues uncouple the reductionist-antireductionist dichotomy from the SMT-TOFT dichotomy. SMT and TOFT are two explanatory strategies and, as such, propose two different kinds of (methodological) reduction, characterized by different degrees of explanatory context dependency. With their radically differ-

---

[5]The concept of "evolving", in this context, is synonymous to change, i.e. change of the functional structure of the system in persistence of its functioning identity.

ent approaches, these two reductionist strategies overcome the traditional 'nothing but' issue about the definition of biological systems. Fields, cells, genes, are all implicated in cancer and can be privileged levels for explanation (and hence for reduction). On the other hand, TOFT also goes along with a conceptual context dependency claim: an antireductionist reflection on the 'certain way' parts are assembled that has epistemological consequences in terms of conceptualization of the *relata* and their interactions. This opens a deep rethinking of the nature of biological interactions, i.e. on the characterization of the parts-whole relationships.

A possible way of conciliating the two approaches is simply to state that they shed light on different aspects of the same issue. I argue for a much more precise and stringent relationship than such mild compatibilism. When focusing on different levels of the biological organization in explaining carcinogenesis, the SMT and the TOFT show *explanatory independence* while being *epistemologically interdependent*. I take these terms as introduced by Angela Potochnick [18]. "The coexistence of distinct explanations for a single event I call *explanatory independence*. The explanations are independent in the sense that each individually explains the event in question; indeed, each is the best explanation of the event in the context of certain research interests" [18, 12, my emphasis]. "By [epistemic interdependence] I mean that the success of these models depends on diverse sources of information about causes not explicitly represented—information gathered with the help of other tools and other fields of science—and that this dependence is mutual" [18, 17].

Another argument to conciliate molecular and systemic approaches is to consider that the causal relevance (and thus explanatory value) in the process of carcinogenesis is found *sometimes* in genes, and, *more often*, in cell interactions at the tissue level. Sporadic cancers are in fact more appropriately explained by TOFT, while in heritable cancers a genetic account seems inescapable. This is certainly the case, but I don't think that the matter should be exhausted in frequentist terms (i.e., by postulating two mutually exclusive, comparable explanations that are alternatively true case by case).

My position is, instead, that the TOFT and molecular accounts are two explanatory strategies and, as such, propose two different kinds of reduction. In addition, TOFT provides the conditions that discriminate when a molecular account will work and when it will not. I said earlier that some inherited mutations certainly do play a causal role in cancer, but I also mentioned that they do so by the effect they have upon the maintenance (or disruption) of tissue organization. This is what I mean when I argue that when SMT works (as in the inherited cases of cancer), it is just as a particular case of TOFT. Thanks to these epistemological considerations, experimental evidence for a systemic account[6] is not trivially *against* a molecular

---

[6]The organismic perspective that characterizes the systemic account, makes sense of some experimental data overlooked or not satisfactorily explained by a molecular account. Here are some non-exhaustive examples. The most straghtforward piece of evidence is that inheritance of cancer-related mutations generally never exceeds 5% of cases (some evidences from studies on APC and colonrectal cancer seems to fit this interpretation: cf. [13, 25, 23]). Then, spontaneous regression of tumours has been found in the experiments performed using teratocarcinomas and embryonic environments highlighting that the regression of the neoplastic phenotype, (i.e. the return to normality of tumour cells). Regression is contradictory with the assumption (typical of SMT and

account. The privileged status of the tissue level is linked to the dynamic properties that characterize this level of the biological organization. On their part, molecular accounts do not actually omit higher-level causal dependencies. Those causes are subsumed within the conceptual assumptions of the model.

While any reductionism rightfully focuses on causal interactions among parts of the system, a deeper message of the systemic approach focuses on the *relationships* that *make those interactions causally relevant (or less relevant, or irrelevant)*. In this sense, relationships have a top-down causal role. They link up in what I call *biological determinations*. Biological determinations–usually captured in terms of higher-order effects–are instantiations of organizational principles that account for the onset of these higher-level properties, i.e. their robustness.

What TOFT is actually supporting is a theory of fields in biological sciences more than a claim in favour of a privileged explanatory level in cancer research. TOFT shows that any explanatory account of biological behaviour conceptually implies a non-reductionist dimension in the process of identification of the relata of the explanatory account. Even a notion like cancerization field, shared (as we saw above) by molecular accounts and TOFT, can be proposed with very different ontological imports with respect to *how the notion of field affects functional definition of parts and parts' stability*. Looking at the system as a whole and focusing on the functions that emerge as relational dynamic networks, one sees elements that acquire their specific explanatory relevance depending on the level of discussion and on the scientific question posed. For this reason, the mutual dependence of SMT and TOFT has an asymmetry that justifies why TOFT is epistemologically more powerful and comprehensive than SMT and can be generative of other explanatory accounts different from the tissue one.

# 7 Conclusions

In scientific practice what is 'more fundamental' is defined by the 'essentiality-by-location' principle, i.e. through the process of conceptualization of dynamic properties of a biological system's elements. Once this is acknowledged, different approaches are possible in cancer research. Different contrast classes can be identified, but the explanatory models that focus on, for example, TC's contrast classes or X's contrast classes (i.e. in this context the actual event functionally related to TC and X) are not only not incompatible because they may eventually have different *explananda*, but imply each other through the conceptual context dependency (i.e. what a thing is) and the explanatory context dependency (i.e. what parts we select in an explanatory account). The *explanans* are often molecular parts of

---

falling squarely in a molecular account) of cancer's dependency from DNA mutation, since the dominant feature of such mutations brings with it the necessary and sufficient condition that a mechanistic explanation requires. Regression from a neoplastic phenotype is, in fact, observed with a much higher frequency than would be expected were it due to back mutation or secondary suppressor mutations [23]. Another interesting area of evidence concerns *differentiation therapies* (i.e., treatments of malignant cells that lead them resume the process of maturation and differentiation into mature cells). Such therapies are potentially more effective in leukaemia, characterized by an extraordinary simplification of the tissue (i.e., the blood). Such treatments easily lead to think of the privileged explanatory status of the tissue level.

the organism, but are identified by virtue of their relationship with the higher-level macro property or an end state that specifies the *explananda*. The non-reductionist dimension that intrinsically characterizes biological explanations is thus related to the definition of parts and how we understand the structure of the world that is not, instead, mechanistically definable although, as shown, there are intrinsic features of the world, and of the way we know it, that allow mechanisms to be always identified.

# BIBLIOGRAPHY

[1] E. Agazzi. *Scientific Objectivity and its Contexts*. Springer, 2014.

[2] M. Bertolaso. The neoplastic process and the problems with the attribution of function. *Rivista di Biologia/Biology Forum*, 102:273–296, 2009.

[3] M. Bertolaso. *How Science works. Choosing Levels of Explanation in Biological Sciences*. Aracne, Rome, 2013.

[4] M. Bertolaso, A. Giuliani, and S. Filippi. The mesoscopic level and its epistemological relevance in systems biology. In *Recent Advances in Systems Biology*. Nova Science Publishers, Inc., 2013.

[5] G. Boniolo. A contextualized approach to biological explanation. *Philosophy*, 80:219–247, 2005.

[6] G. Boniolo. On molecular mechanisms and contexts of physical explanation. *Biological Theory*, 7:256–265, 2013.

[7] I. Bringdant and A. Love. Reductionism in biology. In *Stanford Encyclopedia of Philosophy*. 2012.

[8] L. Darden and C. Craver. Reductionism in biology. In *Encyclopedia of Life Sciences*. 2009.

[9] M. Delehanty. Emergent properties and the context objection to reduction. *Biology and Philosophy*, 20:715–734, 2005.

[10] J. Dupré. It is not possible to reduce biological explanations to explanations in chemistry and / or physics. In F. J. Ayala and R. Arp, editors, *Contemporary Debates in Philosophy of Biology*. Wiley-Blackwell, Oxford, 2010.

[11] D. Hanahan and R.A. Weinberg. The hallmarks of cancer. *Cell*, 100:57-70, 2000.

[12] E.F. Keller. It is possible to reduce biological explanations to explanations in chemistry and/or physics. In J. Ayala and R. Arp, editors, *Contemporary Debates in Philosophy of Biology*. Wiley-Blackwell, Oxford, 2010.

[13] R. Kemler. From cadherins to catenins: cytoplasmic protein interactions and regulation of cell adhesion. *Trends Genet*, 9:317–21, 1993.

[14] P. Machamer, Lindley Darden, and C.F. Craver. Thinking about mechanisms. *Philosophy of Science*, 67(1):1–25, 2000.

[15] C. Malaterre. Making sense of downward causation in manipulationism: illustrations from cancer research. *History and Philosophy of Life Sciences*, 33:537–561, 2011.

[16] S.D. Mitchell. Emergence: logical, functional and dynamical. *Synthese*, 185:171–186, 2012.

[17] D.J. Nicholson. The concept of mechanism in biology. *Studies in History and Philosophy of Biological and Biomedical Sciences*, 43:152–163, 2012.

[18] A. Potochnik. Explanatory independence and epistemic interdependence: a case study of the optimality approach. *The British Journal For the Philosophy of Science*, 61:213–233, 2010.

[19] S. Silberstein. Reduction, emergence and explanation. In P. Machamer and S. Silberstein, editors, *The Blackwell Guide to the Philosophy of Science*. Blackwell, Maiden MA-Oxford, 2002.

[20] Robert A. Skipper and Roberta L. Millstein. Thinking about evolutionary mechanisms: natural selection. *Studies in History and Philosophy of Science Part C: Studies in History and Philosophy of Biological and Biomedical Sciences*, 36(2):327–347, jun 2005.

[21] C. Sonnenschein and A.M. Soto. *The Society of Cells: Cancer and Control of Cell Proliferation*. Springer-Verlag Inc, New York, 1999.

[22] C. Sonnenschein and A.M. Soto. Response to "In defense of the somatic mutation theory of cancer". *Bioessays*, 33:657–659, 2011.

[23] A.M. Soto and C. Sonnenschein. Emergentism as a default: cancer as a problem of tissue organization. *Journal of Biosciences*, 30:103–118, 2005.

[24] T. Ushijima. Epigenetic field for cancerization. *Journal of Biochemistry and Molecular Biology*, 40:142–150, 2007.

[25] H.S. Wasan, H.S. Park, K.C. Liu, N.K. Mandir, A. Winnett, P. Sasieni, W.F. Bodmer, R.A. Goodlad, and R.A. Weinberg. The genetic origins of human cancer. *Cancer*, 61:1963–1968, 1988.

[26] R.A. Weinberg. *The Biology of Cancer*. Garland Science, London, 2006.

# Mechanistic Causality and the bottoming-out problem

Laura Felline

ABSTRACT. The so-called bottoming-out problem is considered one of the most serious problems in Stuart Glennan's mechanistic theory of causality. It is usually argued that such a problem cannot be overcome with the acknowledgement of the non-causal character of fundamental phenomena. According to such a widespread view, in the mechanistic account causation must go all the way down to the bottom level; a solution to the bottoming-out problem, therefore, requires an appeal to an ancillary account of causation that covers fundamental phenomena. In this paper I reconsider the arguments that led to this conclusion and criticize them. I argue that the no-causality-at-the-fundamental-level solution is in harmony with the causal anti-fundamentalism that characterizes the mechanistic theory. Moreover, contrarily to the dualistic solution put forward by Glennan, the no-causality-at-the-fundamental-level is not an ad-hoc solution. Finally, I provide the sketch for an account of regularities and counterfactuals at the fundamental level that is consistent with the singularist and ontologically parsimonious spirit of the mechanistic account.

## 1 Introduction

The New Mechanistic philosophy promises a fresh perspective on old issues in the philosophy of science. Among such applications, one of the most interesting has been within the issue of causation. There are different ways to understand the role of mechanisms in a theory of causation [21]; here I want to focus on one of the most straightforward proposals, put forward by Stuart Glennan, that "two events are causally connected when and only when there is a mechanism connecting them" ([13], 64).

Glennan has originally put forward his account as an answer to Hume's challenge that we cannot observe the 'secret connection' which binds events together, making it impossible to distinguish genuine causal connections from pure conjunctions. According to Glennan, his mechanistic account falls neither within the Humean approach, which finds such distinction in epistemic criteria, nor within the anti-Humean, which finds it in the notion of physical necessity.

The problem that is usually illustrated as the most urgent for Glennan's proposal is the so-called bottoming-out problem [30], i.e. the problem of accounting for fundamental phenomena that are not underpinned by a mechanism. Glennan's solution to the bottoming-out problem is to bite the bullet and accept a dualistic theory of causation, where mechanistic causation covers all higher-level phenomena and a different kind of causation takes place only at the fundamental level. There is, though, another possible solution, which is to acknowledge that mechanistically fundamental phenomena are not causal phenomena, period. So far such a solution has never been taken in serious consideration because of the belief that, in order to be consistent, the mechanistic view requires causation to go all the way down to the most elementary relations. In this paper I want to reconsider such objection and the rationale behind it.

A premise might be in order before we start. This paper does not aim at advocating in general for the adequacy of the mechanistic account of causality – neither per se, nor against the other competing accounts of causation.[1] Nor it is an aim of this paper to advocate in general for the view that denies causality in fundamental physics. The limited aim of this paper is to reconsider the reasons that have led to the widespread conclusion that Glennan's mechanistic account of causation is not compatible with a no-causality-at-the-fundamental-level solution and that, therefore, it requires an ancillary theory of causation at the fundamental level, where the mechanistic account is not applicable. If I am right in claiming so, we could envisage to re-consider a full-fledged solution to the bottoming-out problem that has been so far too quickly dismissed.

This is the structure of the paper. Section 1 introduces the basic ideas of Glennan's mechanistic account of causality and the bottoming-out problem. In S 2 I consider the main arguments for the claim that no-causality-at-the-fundamental-level is not a viable option within Glennan's mechanistic account and counter to them. If there is no causality at the fundamental level, we need an alternative account of counterfactuals and regularities at the fundamental level, that goes along with the singularism and ontological parsimony that characterise causality at higher-levels. In S 3 I put forward an outline of how such accounts might work.

## 2 Causality and the Bottoming-out Problem: the basics

To begin with, let us first illustrate some salient features of Glennan's view of causation. The mechanistic account of causation is proposed in opposition to a time-honoured view according to which causality is grounded on Laws of Nature. Accordingly:

> Covering Principle: If an event e1 causes an event e2 then there are properties F, G such that (a) e1 instantiates F, (b) e2 instantiates G and (c) "F instantiations are sufficient for G instantiations" is a causal law. ([12] p. 64, quoted in [16])

According to Glennan, causal relations are always grounded in an underlying mechanism: "a mechanical theory of causation suggests that two events are causally connected when and only when there is a mechanism connecting them" ([13], p. 64), where:

> A mechanism for a behavior is a complex system that produces that behavior by the interaction of a number of parts, where the interactions between parts can be characterized by direct, invariant, change-relating generalizations ([15], p. S344)

---

[1] To cite but some of the most recent, dispositional accounts (i.e. [19] and [20]) or process accounts (e.g. [2])

In contrast with the above seen quotation by Fodor, which describes a generalist account of causal relations, Glennan's account of causal relations is a singularist one: what grounds causal relations are individual mechanisms, while causal laws are typically descriptions of the regular behaviour of a mechanism. According to Glennan, the chief virtue of the mechanistic account is that it makes the issue of distinguishing between causal connections and accidental conjunctions a scientific one: in order to show whether two events are genuinely causally connected, it is sufficient to show that there is a mechanism connecting them.

Mechanisms are hierarchical systems, in the sense that each part of a mechanism is a mechanism itself and its behavior is therefore explainable with the description of its components and the interaction between them [16]. This process of regress, though, is not infinite. At one point one reaches a level where mechanistic reasoning and mechanistic explanation have no place. This follows, under an atomist stance, straightforwardly by assumption: if the layers of physical composition have a bottom level, i.e., the level of the most elementary components of the world – then, by assumption, the behaviour of such elements cannot be explained in terms of the interaction between its component parts. But regardless of whether or not one adopts an atomist stance, mechanistic reasoning, and mechanisticexplanations with it, drastically loses its ubiquitous role at the level of current fundamental physics [25]. But, since there is a level of fundamental physics, where phenomena are not underpinned by mechanisms, "how do we explain the causal connection between events at the level of fundamental physics?" ([13], p. 64)

This is the so-called bottoming-out problem. Glennan's reaction to it is that "there should be a dichotomy in our understanding of causation between the case of fundamental physics and that of other sciences (including much of physics itself)" ([13], p. 3). However, the acknowledgement of an ancillary account of causation specific for that domain where the mechanistic account does not work, sounds suspiciously ad hoc. One might as well ask why someone who thinks that the mechanistic account successfully captures the essence of causal relations, should not rather acknowledge that the phenomena which do not fall in this definition are non-causal. In this case, causation would only characterize higher level phenomena, concerning higher-level complex systems, whose behaviour depends on the interactions between their components. So far, the hypothesis of the non-causality of mechanistically fundamental phenomena has never been taken in serious consideration because of the shared belief that, in order to be consistent, the mechanistic view of causation requires causation to go all the way down to the most elementary relations. In the next section I will reconsider such arguments and counter to them.

## 3 Mechanistic Causality without fundamental causality

At this point I should point out that, as a pluralist towards the semantics of causation, I am not at ease with a universal metaphysical account of causal relations.[2] That said, I also think that the dualistic solution is not good for the equilibrium of the mechanistic account and that the the bottoming-out problem should not be a reason to reject the mechanistic account of causation. In the rest of this paper

---

[2]For an example of a pluralist stance in causation, see for instance [18] and [21].

I want to outline the basics of a way to tackle the issue of causality in fundamental physics, by keeping consistency with the mechanistic approach to causality at higher levels.

Glennan considers the possibility that there is no causation at the fundamental level, but discards it because he thinks that higher-level causation should be grounded on causation at the most fundamental level:

> This explanation of the role of causal mechanisms is available so long as the generalizations are mechanically explicable, but here we come to what may seem the key metaphysical issue. If mechanisms are truly going to explain how one event produces another, all of the interactions between parts, at all levels in the hierarchy of mechanisms, will need to be genuinely causally productive. If it were to turn out that these interactions at the fundamental level were not truly interactions, then none of the putative causal relations mediated by mechanisms would be genuine. ([17], p. 811)

A first answer to this worry is that the solution no-causality-at-the-fundamental-level is in harmony with the 'anti-fundamentalist' spirit at the foundations of the New Mechanistic philosophy. Fundamentalism is the view, suggested by the traditional philosophical accounts of causation and causal explanation, that causation must be grounded in the most fundamental physical processes,[3] or that "good explanations can be formulated only at the most fundamental level" ([5], p. 11, n. 13). But if causation does not need to be grounded in the most fundamental physical processes, it should be therefore possible to have causation at higher-levels without causation at the most fundamental level.

In his discussion of Glennan's proposal, Carl Craver ([5], p. 90-91) anticipates such a natural appeal to anti-fundamentalism and objects to it. Craver focuses on the fact that Glennan's mechanistic account aspires to meet Hume's challenge of the non observability of causation, as a secret necessary connection which binds cause and effect. He therefore takes for granted that any account meeting Hume's challenge must describe causation as a necessary connection. As such – the argument goes – causation must go all the way down to the bottom level.

> Suppose that one is trying to understand the necessary connection between X and Y (that is, X →Y) at one level above the fundamental level. Glennan [13] says that the necessity in the connection between X and Y should be understood in terms of the connections between items at the fundamental level, say, X →a →b →Y. Glennan grants that a and b have no necessary connection between them[4] and that talk of causal relevance and manipulation such

---

[3]"the mechanical theory of causation rejects a wide-spread assumption about the nature of causation [...] that whatever causal connections are, they ultimately have something to do with the most fundamental physical processes. The closer we are to fundamental physics, the more our statements are about the true causes of things; the further we stray into the higher level sciences, the more we move away from causal statements and toward mere empirical generalizations." ([13], p. 22)

[4]That there is no necessary connection between a and b follows from the assumption that there is no causation at the fundamental level" in the following way: since in this context necessity is a feature of causation, then if there is no causation, there is no necessity. The fact that there might be necessary connections which are non-causal is here irrelevant – as what Craver is trying to show is that causal relations (as necessary connections) must be built on causal relations (as necessary connections). Craver does not need to claim (and would not be justified to claim) that non-causal connections are always non-necessary because it is irrelevant for the argument he is making. I am grateful to an anonimous referee for pointing out this possible misunderstanding.

a connection may be unintelligible. But how can a necessary causal connection between X and Y be built out of relations in which there is no necessary connection and for which such talk is unintelligible?

Craver's argument assumes that causation is a necessary connection. Once such a characterization is granted, it is hardly deniable that causation at higher level cannot be built out of more fundamental non-causal connections.
However, it is debatable whether necessity naturally fits in the mechanistic answer to Hume's challenge.
First of all, a straightforward consequence of replacing Laws of Nature with mechanisms at the foundations of causation, is exactly that one should give up the concept of necessity. This is evident in general for those accounts of mechanisms in which interactions between parts are regulated by 'regular, change-relating generalizations' (e.g. [17], but apply also to Craver's account of mechanisms [5]). Contrarily to Laws of Nature, regular change-relating generalizations (or, with [5] expression: 'more or less invariant change-relating generalizations') are not exceptionless and cannot therefore constitute a necessary connection.
More generally, however, necessity is not part of the natural categorial framework of the New Mechanistic philosophy. The well-known issues related to the concept of necessary, exceptionless regularities are in fact among the primary motivations for replacing the concept of Laws of Nature with the concept of mechanism at the heart of many discussions in the philosophy of science (e.g. [26], S 3.2). The partial conclusion of this section is that the arguments typically appealed to in order to prove that a mechanistic account of causation necessitates causation at the fundamental level are ineffective.
At this point it is important to notice that the conjecture that fundamental phenomena are non-causal is far from being an ad hoc hypothesis or a terminological stratagem, uniquely motivated by the bottoming-out problem. On the contrary, there are various independent arguments to support such conclusion. First of all, fundamental phenomena are not only problematic for the mechanistic account, but also for Woodward's difference-making account, which is the other most relevant scientifically informed account of causality.
Moreover, independently of a specific account of scientific causation, fundamental laws seem to resist causal interpretations, even in the lightest possible characterization of causality. Of course there are the already well-known arguments to the conclusion that causal notions play no role in this domain (e.g. most notably [11] and [29]). But this resistance is particularly striking when one considers the histories of failed attempts to provide a causal account of fundamental physical phenomena. Consider for instance, length contraction – which has for decades resisted attempts to be explained by means of a mechanist explanation based on a more fundamental theory of matter – the Uncertainty Relations, but also non-local quantum correlations [6] [8] [9] [10].

## 4 Regularities and counterfactuals

Up to this point I hope I have succeeded in showing that Glennan's mechanistic account of causation might be consistent with a no-causality-at-the-fundamental-level view. In this last section I tackle the problem of accounting for regular behaviours and counterfactuals at the fundamental level.

As we have seen, following the mechanistic account, interactions between the component parts of a mechanism are characterized by invariant and change-relating generalizations ([14], p. S344). The regular behaviour related to causal processes is therefore explained through the robustness of the mechanism's parts, in the sense that their behaviour are stable, and it is in principle possible to take out a part of the mechanism and consider its properties in another context. ([13], p.53) Moreover, mechanisms provide a straightforward understanding of counterfactual claims. Rather than being characterized by appealing to an abstract notion of similarity between possible worlds, or unanalysed notions of cause or propensity, counterfactual claims are justified by our knowledge of the model of a mechanism:

> Given a model of a mechanism that exhibits the functional dependence of variables that represent the mechanism's parts and their properties, one evaluates a counterfactual claim by using the model to calculate what would happen if one were to intervene and fix the value of a variable to the antecedent of the counterfactual. ([17], p. 806).

This section approaches therefore the bottoming-out problem as the problem of accounting for regularities and counterfactuals at the fundamental level, without appealing to mechanisms and causality and in a way that is consistent with the mechanistic account of higher-level causality, regularities and counterfactuals. Once again, I am not going to provide a full fledged account of regularities and counterfactuals in fundamental physics – the limited aim of the following discussion is to argue that it is possible to account for regularities and counterfactuals in fundamental physics in a no-causation-at-the-fundamental-level solution to the bottoming-out problem.

We have seen in S 1 that in contrast with the criterion put forward in Fodor's quoted passage (which appeals to the existence of Laws of Nature as universals) Glennan's version of the mechanistic account of causation privileges a singularist view. Mechanisms are, in fact, "particular systems of interacting parts, where these interactions occur at a particular place and time". ([17], p. 809) Causality, therefore, is a relation between individual events and general causal claims are only generalizations of such individual relations. In order for our solution to be consistent with the mechanistic account, the former must therefore be coherent with a singularist stance.

It goes without saying that we cannot appeal to Laws of Nature. This would betray the spirit of singularism and imply giving up the ontological parsimony featuring Glennan's account and that represents two of its strengths. Moreover, in the previous section, I have rejected the characterization of causality as a necessary connection, in any connotation that goes beyond the robustness and justification of counterfactuals that are justified by our knowledge of the mechanisms – so, again, necessity is out of question also at the fundamental level.

Let's first tackle the problem of regularities in fundamental physics. A first way to see the problem is to say that if there is no robust mechanism, nor law, then the fact that different systems in the same initial state behave in the same way calls for an explanation. According to this view, if, in the spirit of singularism, the evolution of a system would only depend on facts that are immanent to such particular instance, it would be a mystery that systems in the same state behave in the same way. Thus, it would seem that we need an explanation of this common behaviour, in the form of some common metaphysical underpinning.

But is it true that we need an explanation in the first place? When can we say that a phenomenon P requires an explanation? Notice that the question here does not concern the necessity of a scientific explanation – the explanation of regularities is not the aim of scientific, but of philosophical investigation – but neither are we concerned by a generic metaphysical explanation. The question is instead whether one could expect an explanation of regularities, of the sort one could expect to come from the domain of the metaphysics of science – a metaphysics which is informed by scientific issues and knowledge.

On the basis of this, my proposal is that in order to claim that P requires an explanation, without which P would remain an 'unexplainable mystery', P must create a tension in our (scientific) representation of the world. This might be either because P is incoherent with some other element of such representation, or even just because we have some reasons to expect non-P to be true, rather than P. Is any of this the case for regularities and the alleged problem of accounting for them philosophically? Let us therefore say, in the spirit of singularism, that each system's behaviour depends only on factors that are immanent to the specific situation. Under such an assumption, I take it that there is nothing logically incoherent in the fact that systems that are similar in the relevant respect, behave in the same way. Here, 'similar in the relevant respect' must be understood in the minimal sense 'the relevant variables take the same values'. For instance, let's take a simplified case and say that I know that a behaviour of system S exclusively depends on the value of variables A and B of S. Let's say that the behaviour of S depends on A and B taking the values a and b in S. There is nothing logically incoherent in the fact that every other system which is not S, but in which A and B take the values a and b, also exhibit that behaviour. (Incidentally, I am obviously not arguing here that one can legitimately infer which variables are relevant for a behaviour, from the analysis of one single case. The epistemic problems concerning regularities are ignored in this argument.)

Maybe, however, it might be said that, as a matter of pure 'metaphysical intuition', it is more 'natural' to expect non-regular behaviours in the world, i.e. that systems that are similar in the same relevant respect, behave (deterministically or stochastically, this is irrelevant) in different ways.

My pure metaphysical intuition, however, does not say so: it is just not clear to me why one should expect that systems that are similar in the relevant aspects should behave differently. Indeed, I would find it a mystery, one in need of an explanation, if such similar systems behaved in different ways! Of course, one might counter-argue that my metaphysical intuition has been instilled by a life-long experience of regularities and that this is why I expect similar behaviour in similar systems. This might be true, however intuition has always the defect of being formed in one way or another by experience, or, anyway, we have no mean to say how an intuition has been formed, nor mean to guarantee a purely intuitive metaphysical judgment. The same doubt, therefore, applies to the opposite intuition as much as to mine.

Once both the charge of inconsistency and that of counter-intuitiveness are ruled out, insisting on the requirement for an explanation of regularities in terms of a metaphysical underpinning seems to me a sort of a 'metaphysical obsession' – a question as pertinent to the scope of philosophy of science as questions like: "why is there change?" or "why is there something rather than nothing?".

Besides considerations on metaphysical possibilities or intuition, there is a more epistemically driven consideration that suggests that regularities do not require an explanation within a metaphysics of science. In real scientific practice, the assumption that similar systems behave in the same way plays the role of an a priori assumption for theoretical research. Here, one typically starts an inquiry from the consideration of a regular common behaviour in a set of systems. Such a common behaviour is then explained by the assumption that the concerned systems are similar in a relevant respect that determines such behaviour. On the other hand, anomalies in the behaviour of some systems within an ensemble, are explained by the assumption that some unknown factor (relevant to the anomalous behaviour) makes such systems different from the others.

To expect an explanation for the fact that similar systems behave in the same way, therefore, means turning the logical explanatory order used in science upside down: you should not try to explain an a priori of your knowledge!

The second issue is to account for counterfactuals. The question is how to account for counterfactuals at the fundamental level if there is no causation, no mechanism and no Laws of Nature. Remember that within Glennan's theory "one evaluates a counterfactual claim by using the model to calculate what would happen if one were to intervene and fix the value of a variable to the antecedent of the counterfactual." At the fundamental level, the justification of counterfactuals here works exactly as it does at higher levels, although the functional dependencies are here not provided by a mechanistic model, but by a mathematical model. The justification of our counterfactual inferences comes therefore from the justification of the model (more on the justification of the model, by top-down or bottom-up strategy can be found in [3]).

## 5 Conclusions

In this paper I have faced the so-called bottoming-out problem in Stuart Glennan's account of causation. In particular, I have argued that a possible solution to the problem – the no-causality-at-the-fundamental-level solution – has been unjustly dismissed as a viable option. I have pointed out three questions that have been (or might be) considered problematic in such an option: how to ground mechanistically understood causal relations on non-causal relations, how to account for regular behaviour, and how to account for counterfactuals without an underlying mechanism at the fundamental level. With respect to the first of these issues, I have answered at some standard objections found in the literature. The two other issues are not explicitly treated in the literature, probably because the first problem has been usually considered deadly for the no-causality-at-the-fundamental-level hypothesis. The analysis here proposed of these problems, therefore, is necessarily more at a programmatic stage than the first one. However, I hope I have shown that the first, most straightforward doubts that a no-causality-at-the-fundamental-level solution might raise with respect to an account of regularities and counterfactuals at a fundamental level might be approachable. In particular, on the one hand the approach to counterfactuals at the fundamental level would be based on the same logic that applies to mechanisms; on the other hand, the approach to regularities consists in rejecting the very same request for an explanation.

# BIBLIOGRAPHY

[1] Bechtel, W., and Abrahamsen, A. (2005). Explanation: A mechanist alternative. Studies in History and Philosophy of Science Part C: Studies in History and Philosophy of Biological and Biomedical Sciences, 36(2), 421-441.
[2] Blondeau, J. and Ghins, M. (2012). Is There an Intrinsic Criterion for Causal Lawlike Statements? International Studies in the Philosophy of Science, 26:4, 381-401
[3] Bokulich, A. (2009). 'How scientific models can explain', Synthese, 180(1), pp. 33-45.Brown, H. R., and
[4] Pooley, O. (2006). Minkowski space-time: a glorious non-entity. Philosophy and Foundations of Physics, 1, 67-89.
[5] Craver, C. F. (2007). Explaining the brain. Oxford University Press.
[6] Dorato, M. and Felline, L. (2010). Scientific explanation and scientific structuralism. In Scientific structuralism (pp. 161-176). Springer Netherlands.
[7] Egg, M., and Esfeld, M. (2014). Non-local common cause explanations for EPR. European Journal for Philosophy of Science, 4(2), 181-196.
[8] Felline, L. (2010). Remarks on a structural account of scientific explanation. In EPSA philosophical issues in the sciences (pp. 43-53). Springer Netherlands.
[9] Felline, L. (2011). Scientific explanation between principle and constructive theories. Philosophy of Science, 78(5), 989-1000.
[10] Felline, L. (2015). Mechanisms meet structural explanation. Synthese, 1-16.
[11] Field, H. (2003). Causation in a physical world. Oxford handbook of metaphysics, 435-460.
[12] Fodor, Jerry (1989). Making Mind Matter More. Philosophical Topics, 17(1): 59–74.
[13] Glennan, S. S. (1996). Mechanisms and the nature of causation. Erkenntnis, 44(1), 49-71.
[14] Glennan, S. (2000). Rethinking mechanistic explanation. Philosophy of Science, 69(S3), S342–S353.
[15] Glennan, S. (2002). 'Rethinking mechanistic explanation', Philosophy of Science, 69(S3), pp. S342-S353.
[16] Glennan, S. (2010). 'Mechanisms, causes, and the layered model of the world', Philosophy and Phenomenological Research, 81(2), pp. 362-381.
[17] Glennan, S. (2011). 'Singular and general causal relations: A mechanist perspective', in P. M. Illari, F. Russo and J. Williamson (eds.), 2011, Causality in the Sciences, Oxford: Oxford University Press, pp. 789-817.
[18] Hitchcock, Christopher. (2007). 'How to be a causal pluralist'. In Thinking about Causes: From Greek Philosophy to Modern Physics., Woters, G. and Machamer, P. (eds), pp. 200–221. Pittsburgh: University of Pittsburgh Press.
[19] Hüttemann, A. (2007). Causation, Laws and Dispositions. In Kistler, M. and Gnassounou, B. (eds.), Dispositions and Causal Powers. Ashgate.
[20] Huttemann, (2013). 'A Disposition-Based Process Theory of Causation'. In Mumford, S. and Tugby, M. (eds.), Metaphysics and Science. Oxford. 101.
[21] Illari, P., and Russo, F. (2015). Causality: philosophical theory meets scientific practice. Clarendon Press. Oxford.
[22] Illari, P. M., and Williamson, J. (2012). What is a mechanism? Thinking about mechanisms across the sciences. European Journal for Philosophy of Science, 2(1), 119-135.
[23] Illari, P. M., and Williamson, J. (2010). Function and organization: Comparing the mechanisms of protein synthesis and natural selection. Studies in History and Philosophy of Science Part C: Studies in History and Philosophy of Biological and Biomedical Sciences, 41(3), 279-291.

[24] Janssen, M. (2009). Drawing the line between kinematics and dynamics in special relativity. Studies In History and Philosophy of Science Part B: Studies In History and Philosophy of Modern Physics, 40(1), 26-52.
[25] Kuhlman, M. and Glennan, S. (2015) On the Relation between Quantum Mechanical and Neo-Mechanistic Ontologies and Explanatory Strategies.
[26] Machamer, P., Darden, L., and Craver, C. (2000). Thinking about mechanisms. Philosophy of Science, 67, 1–25.
[27] Machamer, P. (2004). Activities and causation: The metaphysics and epistemology of mechanisms. International Studies in the Philosophy of Science, 18(1), 27-39.
[28] McKay Illari, P. and Williamson, J. (2010). Function and organization: Comparing the mechanisms of protein synthesis and natural selection. Studies in History and Philosophy of Biological and Biomedical Sciences, 41, 279–291.
[29] Norton, J. D. (2008). Why constructive relativity fails. The British Journal for the Philosophy of Science, 59(4), 821-834.
[30] Williamson, J. (2013). How can causal explanations explain?. Erkenntnis, 78(2), 257-275.
[31] Woodward, J. (2007). Causation with a human face. Causation, physics, and the constitution of reality, 66-105.

# Quantity of Matter or Intrinsic Property: Why Mass Cannot Be Both

Mario Hubert

ABSTRACT. I analyze the meaning of mass in Newtonian mechanics. First, I explain the notion of primitive ontology, which was originally introduced in the philosophy of quantum mechanics. Then I examine the two common interpretations of mass: mass as a measure of the quantity of matter and mass as a dynamical property. I claim that the former is ill-defined, and the latter is only plausible with respect to a metaphysical interpretation of laws of nature. I explore the following options for the status of laws: Humeanism, primitivism about laws, dispositionalism, and ontic structural realism.

## 1 Primitive Ontology

Any scientific theory must explicitly state what it is about. In particular, every fundamental physical theory must explain the aspect of the world to which its mathematical formalism refers. Albert Einstein reminds us of this truism:

> Any serious consideration of a physical theory must take into account the distinction between the objective reality, which is independent of any theory, and the physical concepts with which the theory operates. These concepts are intended to correspond with the objective reality, and by means of these concepts we picture this reality ourselves. (Einstein, Podolsky and Rosen 1935 [3], p. 777)

This seemingly innocent quote contains a strong metaphysical claim and a non-trivial epistemological assertion. On the one hand, Einstein presupposes a world existing in- dependently of any human being. There is an objective reality irrespective of the way we perceive or make judgments about it. On the other hand, we can form physical theories in order to account for the behavior of objects in the world. Physics, in particular, uses mathematics as its central language. And here lies the challenge physics has to meet, since the mathematical entities, like numbers or functions, do not refer to anything in the world unless they are interpreted as doing so. An even greater problem arises when the mathematical entities refer to objects that cannot be directly perceived by our sense organs, for there is always a grain of doubt about their existence.

But physical theories, by means of mathematics, are all we have to explain and predict the behavior of objects, such as electrons, tables, stars, and galaxies. And not all the mathematical entities of a physical theory stand on an equal footing.

First, the theory has to postulate basic material entities that are supposed to be the constituents of all the objects around us. Without this requirement, a physical theory is empty.

This requirement for any fundamental physical theory started to get lost in the formation of quantum mechanics. As such, Einstein continued to call attention to it in his philosophical writings. And more than half a century later, the mathematical physicists Dürr, Goldstein, and Zanghì formed the notion of a *primitive ontology* (original paper from 1992 reprinted in Chap. 2 of Dürr, Goldstein, and Zanghì 2013 [2]) to remind us that quantum mechanics has to postulate certain basic objects in order to be a meaningful theory.

By definition, a primitive ontology consists of the fundamental building blocks of matter in three-dimensional space. It cannot simply be inferred from the mathematical formalism of the theory. Instead, it must be postulated as its referent. So all objects, like tables and chairs, are constituted by the elements of the primitive ontology, and the behavior of these elements determines the behavior of the objects. Maudlin (2015) [7] emphasizes that with the help of a primitive ontology, a physical theory establishes a connection between theory and data. In particular, every measurement-outcome can eventually be explained in terms of a primitive ontology, and the measurement apparatus has no special status with respect to the measured system.

So what do the elements of a primitive ontology look like? This depends on the physical theory we use. In quantum mechanics, for instance, there are three famous options, which actually lead to three different *theories* and not only to three different *interpretations* of the same theory: Bohmian mechanics presupposes a particle ontology; GRWm, a continuous distribution of matter; and GRWf, flashes, that is, a discrete distribution of events in space-time. As in the Bohmian case, the primitive ontology of Newtonian mechanics consists only of particles.[1] Particles are point-size objects sitting on points of Newton's absolute space. A point in space can either be occupied by a particle, or it can stay empty. And therefore two or more particles cannot share the same point in space at the same time.

In order to account for the behavior of the primitive ontology, a physical theory has to introduce *dynamical* entities. The predominant dynamical elements of Newtonian mechanics are mass and forces. The standard story is that particles have mass, and in virtue of having mass they exert certain forces between one another. Mass and forces play a different role to particles. While particles constitute all physical objects, mass and forces constrain the motion of the particles.

In this paper, I focus on the ontological role of mass—forces will be treated only in so far as they elucidate the role of mass. There are two standard ways to interpret the ontological status of mass: it can be the measure of the quantity of matter or an intrinsic property of particles. I argue in the next section that the quantity of matter has to be defined in a different way. In Section 3, I explain that the

---

[1] I consider Newtonian mechanics an action-at-a-distance theory. Interpreting this theory as postulating a gravitational field in addition to the particles would open Pandora's box about the ontological status of fields in general. A detailed treatment of classical fields is beyond the scope of this paper.

status of properties depends on the metaphysics of laws of nature. There are three predominant positions: Humeanism, primitivism about laws, and dispositionalism. In each theory mass plays a different ontological role.

## 2   Mass and Quantity of Matter

Newton starts his *Mathematical Principles of Natural Philosophy* with a definition of the quantity of matter:

> **Definition 1**
>
> *Quantity of matter is a measure of matter that arises from its density and volume jointly.*
>
> If the density of air is doubled in a space that is also doubled, there is four times as much air, and there is six times as much if the space is tripled. The case is the same for snow and powders condensed by compression or liquefaction, and also for all bodies that are condensed in various ways by any causes whatsoever. [...] Furthermore, I mean this quantity whenever I use the term "body" or "mass" in the following pages. It can always be known from a body's weight, for—by making very accurate experiments with pendulum—I have found it to be proportional to the weight, as will be shown below. (Newton 1999 [17], pp. 403-404)

If mass is defined as density times volume, then the notion of mass has no physical content or explanatory value, since the density itself is defined as mass per volume. Ernst Mach harshly criticizes Newton's definition on this point:

> Definition 1 is, as has already been set forth a pseudo-definition. The concept of mass is not made clearer by describing mass as the product of the volume into density as density itself denotes simply the mass of unit volume. The true definition of mass can be deduced only from the dynamical relations of bodies. (Mach 1919 [14], p. 241)

Newton does not give a definition of density; nor is density examined in the scholium following the definitions. It seems that Newton assumes that the reader has a pre-knowledge or an intuition about density such that Definition 1 is more of a rule showing how mass, volume, and density are related rather than a logical definition. As Mach correctly states, a definition of mass in the above sense does not work, which leads to the following two questions:

1. What does "quantity of matter" mean?
2. Is mass connected to the quantity of matter?

Concerning the first question, a primitive ontology of particles allows us to count the particles in a certain volume, and it is natural to take this as the *definition* of quantity of matter without getting into any redundancy. For instance, the quantity of matter of a table consists then of the number of particles, which form the table.

Since the particles themselves have no internal structure, it does not make sense to assign a quantity of matter to each. At least, it is not meaningful to assign different quantities of matter to particles so that a particle $A$ carries a quantity of matter $a$, and particle $B$ carries a quantity of matter $b$, with $a \neq b$.

Besides, physics in general and Newtonian mechanics in particular do not need a separate or independent notion of quantity of matter in order to be applied to the world. Statistical mechanics, which relies on counting the number of particles, does not run into conceptual or empirical problems despite lacking an additional notion of the quantity of matter. So it is more precise and parsimonious to define quantity of matter by the number of particles.

Concerning the second question, Newton himself confesses in the last sentence of the quote above that we only have epistemic access to mass when weighing an object, and from the weight we can deduce the quantity of matter. Mach goes a step further, stating that "[t]he true definition of mass can be deduced only from the dynamical relations of bodies." In this regard, mass is not related to the quantity of matter of an object. Instead, its true and only meaning is dynamical.

## 3 Mass as a Dynamical Property

The dynamical role of mass is captured in Newton's first and second law of motion.

### Law 1

*Every body perseveres in its state of being at rest or of moving uniformly straight forward, except insofar as it is compelled to change its state by forces impressed.*

### Law 2

*A change in motion is proportional to the motive force impressed and takes place along the straight line in which that force is impressed.* (Newton 1999 [17], p. 416)

The first law states that the natural motion of a particle is inertial motion, that is, either staying at rest or moving with constant velocity in a straight line. The only thing that can change this motion is the influence of external forces. The second law then shows exactly how the forces act on the particle: first, the stronger the force the greater the acceleration, and, second, the acceleration is parallel to the external force.

Newton's second law is nowadays mathematically formulated as a differential equation. Consider $N$ particles $P_1, \ldots, P_N$ at positions $\vec{q}_1 \ldots, \vec{q}_N$; their trajectories $\vec{q}_1(t), \ldots, \vec{q}_N(t)$ fulfill the differential equation

$$\vec{F}_i\left(\vec{q}_1(t), \ldots, \vec{q}_N(t), \dot{\vec{q}}_1(t), \ldots, \dot{\vec{q}}_N(t), t\right) = m_i \ddot{\vec{q}}_i(t), \qquad (1)$$

where $\vec{F}_i$ is the force on the $i$-th particle, $\dot{\vec{q}}_i$ its velocity, $\ddot{\vec{q}}_i$ its acceleration, and $m_i$ its inertial mass. Clearly, the above differential equation makes precise what Newton put into words. And it includes the content of his first law, too: the absence of forces results in inertial motion. So there is actually only one law of motion that generates all classical trajectories of particles, namely, the above differential equation (1).

Still, the law of motion is not complete. We need a precise formulation of the forces involved. On the fundamental level, one important force is gravitation:

$$\vec{F}_i(\vec{q}_1, \ldots, \vec{q}_N) = \sum_{j \neq i} G\, m_i m_j \frac{\vec{q}_j - \vec{q}_i}{\|\vec{q}_j - \vec{q}_i\|^3} \qquad (2)$$

with the gravitational constant G and the gravitational masses $m_i$ and $m_j$ of the particles $P_i$ and $P_j$ respectively. The inertial and gravitational masses are a priori physically distinct quantities: the former is a feature of all particles and must be considered in all kinds of interactions; the latter is a specific quantity as part of the law of gravitation (2). It is an empirical fact that inertial mass equals gravitational mass, and, therefore, we can treat them as one quantity. Note also that in Newton's theory there are no massless particles, because equation (1) breaks down if we insert $m = 0$. So mass is an essential feature of particles in Newtonian mechanics.

Construed as a dynamical property, there are three ways in which physics describes mass:

1. mass is an intrinsic property of particles;

2. mass is just a parameter of the laws of motion;

3. mass is a coupling constant.

I claim that these three interpretations can only be made precise with respect to some metaphysical framework. In what follows, we discuss mass in the framework of Humeanism, primitivism about laws, and dispositionalism. Finally I interpret mass within the theory of ontic structural realism, which I regard as an instance of dispositionalism.

## Humean Supervenience

Humean supervenience, the modern form of Hume's metaphysics, was first posited by David Lewis:

> It is the doctrine that all there is to the world is a vast mosaic of local matters of fact, just one little thing and then another. [...] We have geometry: a system of external relations of spatiotemporal distance between points. Maybe points of spacetime itself, maybe point-sized bits of matter or aether fields, maybe both. And at those points we have local qualities: perfectly natural intrinsic properties which need nothing bigger than a point at which to be instantiated. For short: we have an arrangement of qualities. And that is all. All else supervenes on that. (Lewis 1986 [11], pp. ix-x)

The ontology of Humean supervenience is characterized by the contingent distribution of local matters of particular facts: the Humean mosaic. There is a net of spatiotemporal points that are connected only by external metrical relations, and, at those points, certain qualities can be instantiated by at least one of three entities that Lewis regards as fundamental: space-time itself, particles, or values of fields. Given some initial distribution of those qualities there is nothing in the ontology that constrains its further development.

Obviously, our world contains regularities, but on the Humean view this is just a contingent fact. In order to avoid giving an enormously long list of particular facts describing these regularities, (Lewis 1994 [12], p. 478) introduces his best system account of laws of nature. According to his proposal, the laws of nature are theorems of the best deductive system, which combines or balances simplicity and strength in describing the temporal development of local matters of particular facts throughout space and time. A long list of these facts would be highly informative but very complex, whereas a single law of nature would be very simple but probably not contain enough information. So the best system comprises a certain finite number of laws of nature as theorems, which offer the perfect compromise.

In Lewis's Humeanism, mass can be part of the ontology of the mosaic: in which case it is a "natural intrinsic property" instantiated at points of space-time. This move, however, poses a serious metaphysical problem: mass becomes a categorical property that is defined as independent of the causal role it plays in the world. Hence, mass has a primitive identity or quiddity, which allows it to play a different causal role in another possible world. For example, in another possible world mass could play the role of charge. This would be the very same property that we call mass, but it would act like charge does in our world.

This seems absurd and leads to the problem—called humility—of our not having epistemic access to the true identity of mass, because all we can know are the causal consequences of properties. So, given two worlds that coincide in the temporal development of all their particles, it would, first, be metaphysically possible for these worlds to be different with respect to the quiddity of their categorical properties, and, second, it would be impossible for us to know which world we inhabited. Lewis bites the bullet and accepts this metaphysical burden in favor of a sparse ontology with no modal connections.

Ned Hall (see Sec. 5.2 of Hall 2009 [9]) proposes a different strategy for conceptualizing the status of mass. He interprets the Humean mosaic as consisting solely of point-sized particles standing in certain spatiotemporal relations (see also Loewer 1996 [13] and Esfeld 2014 [5]). The particles do not have intrinsic properties, let alone categorical ones, and all non-modal facts about the world are just the positions of these particles. Mass enters the scene as part of the best system describing the temporal development of the particles, as part of the fundamental laws of nature in the Humean sense, and as part of the differential equations that describe the trajectories of particles. In a description of the history of the world that balances simplicity and informativeness, mass functions as a parameter in this best system.

For Newtonian mechanics, restricted to gravitational interaction, the best system may be interpreted as consisting of the equations (1) and (2). As a way of speaking or as a convenient metaphor, we can ascribe these parameters to the particles themselves such that every particle $P_i$ is characterized by a magnitude $m_i$. But this interpretation of mass does not change or add anything to the ontology. The distribution of propertyless particles is the entire ontology; everything else, including mass, supervenes on this mosaic. As such, there are no categorical properties in the ontology, and the problem of quiddity or humility does not arise in this version of Humean supervenience.

One general critique attacks Humean supervenience on a point that Humeans regard as one of its greatest virtues: the sparse ontology that lacks modal connections. It is unsatisfactory that there are no facts about *why* we see regularities in our world. Even the laws of nature as part of the best system cannot explain *why* particles follow a Newtonian trajectory. For particles just move as they do, in a contingent way. All a Humean can do is give a good description or summary of the regularities, and if the regularities change she has to change her description too. Consequently, we have to include modal connections in the ontology.

## Primitivism about Laws

Primitivism about laws regards the existence of the laws of nature as a primitive fact, where the laws themselves govern the behavior of the primitive ontology. One famous adherent of this position is Tim Maudlin:

> To the ontological question of what makes a regularity into a law of nature I answer that lawhood is a primitive status. Nothing further, neither relations among universals nor role in a theory, promotes a regularity into a law. [...] My analysis of laws is no analysis at all. Rather I suggest we accept laws as fundamental entities in our ontology. Or, speaking at the conceptual level, the notion of a law cannot be reduced to other more primitive notions. The only hope of justifying this approach is to show that having accepted laws as building blocks we can explain how our beliefs about laws determine our beliefs in other domains. Such results come in profusion. (Maudlin 2007 [7], pp. 17-18)

As stated by Maudlin, the entire ontology is made up of the primitive ontology plus the laws of nature. It is a primitive fact that there are laws of nature, and that particles move according to these laws. The task of physics, then, is to discover these laws. For instance, it is a primitive fact that equations (1) and (2) hold in a Newtonian universe, and here we come to an answer regarding *why* a particle follows a Newtonian trajectory: because there are such laws.

What is the role of mass in this framework, then? It is just a parameter of the Newtonian laws of motion referring to nothing at all in the primitive ontology. Mass is not a parameter that results from the best description, as in the Humean case; rather it is an essential parameter of the laws of nature leading to correct trajectories.

The notion of a parameter is slightly inappropriate here, because it invites us to think about mass as being adjusted or altered under certain circumstances. But the only circumstance available to us is the universe as a whole. There is a primitive ontology consisting of N particles and the laws of nature. And it happens to be the case that the law is formulated such that there are $N$ constants $m_1, \ldots, m_N$.

So it seems more appropriate to interpret $m_1, \ldots, m_N$ as *constants of nature* on a par with the gravitational constant G or Planck's constant $\hbar$. Recognizing masses as constants of nature is clearer and more in the spirit of primitivism than dubbing them parameters. It is likely that the idea of mass as a parameter came from the application of Newton's laws to real life cases, where it had to be adjusted to describe the physical bodies of a given subsystem.

Similarly to Hall, one can pretend that the parameter mass is "located" at the particle's position and speak as if it were an intrinsic property of particles. In this sense mass still has a purely nomological role, but this way of speaking may aid our intuition.

Primitivism about laws retrieves modal connections as part of ontology in the form of laws. Maudlin does not state how laws are connected to the primitive ontology. There seems to be an intuition that laws "govern" or "direct" the behavior of particles, but these phrases are purely metaphorical (Loewer 1996 [13], p. 119). In the above quote, there is no attempt to explain these metaphors: "lawhood is a primitive status." Nevertheless, one can ask, "How can a law as an abstract entity govern anything in the world? How can particles or any material body 'obey' these laws?" Primitivism about laws just answers, "It is a primitive fact." Nonetheless, one position that tries to answer these questions by introducing an underlying mechanism is dispositionalism.

## Dispositionalism

This strategy tries to recover modal connections by introducing dynamical properties into physical systems, which are called *dispositions* or *powers* (for instance, Bird 2007 [1]). Accordingly, a physical system behaves the way it does because it has a certain property or disposition to do so. This idea can be applied to the primitive ontology of Newtonian mechanics. Mass is then an intrinsic property of particles. It is intrinsic in the sense that the mass of one particle does not depend on the masses of other particles.

Mass, interpreted as a disposition, does not give an intrinsic identity to particles. The identity of particles stems from their location in space. Since Newtonian mechanics relies on an absolute background space, where every point in space is by definition distinguished from any other point in space, it is sufficient to ground the identity of particles on their position in absolute space.[2] The role of mass is solely a dynamical one; that is to say, it constrains the motion of particles.

Moreover, it is essential for mass to have the same causal-nomological role in all possible worlds; it is not a categorical property, and consequently it does not bear the problems of either quiddity or humility. This causal-nomological role is expressed by Newton's laws (1) and (2). In other words, Newton's laws are grounded in the ontology by the intrinsic masses of particles. Our epistemic access to mass as a disposition is possible through observation of what it does in the world, that is, its causal-nomological role; and the laws of nature (1) and (2) are a concise expression of its effects.

---

[2] As argued in Esfeld, Lazarovici, Lam and Hubert 2015 [7], this can be already done in a relational space.

A crucial feature of dispositions is their need for certain triggering conditions in order to be manifested in the world. Zooming into Newton's second law (1) we can see the following: the manifestation of the mass $m_j$ of particle $P_j$ at time $t$ is its acceleration $\vec{a}_j(t)$ given the positions and velocities of all the particles (including $P_j$) at time $t$. So the positions and velocities of *all* particles are triggering conditions for the manifestation of mass. In the case of gravitation, the mass $m_j$ cancels out on both sides of (1), and we deduce that the acceleration $\vec{a}_j(t)$ does not mathematically depend on $m_j$. Yet, $\vec{a}_j(t)$ is the manifestation of the mass of the $j$-th particle, though it is independent of the precise value of $m_j$.

## Ontic Structural Realism

The interpretation of mass as a coupling constant does not seem to fit either of the metaphysical schemes discussed above. What is the ontological status of mass distinct from its being an intrinsic property of particles or a constant in the laws of motion? Mass, interpreted as a *coupling* constant, emphasizes the dynamical relations between particles. Particles move as they do because they stand in certain relations described by Newton's laws (1) and (2), and the role of mass is then to quantify these relations.

A metaphysical approach that supports this view is ontic structural realism (OSR). According to the original idea of OSR, the world consists purely of structures, all the way down to the fundamental level (Ladyman and Ross 2007 [10], French 2014 [8]). If there happen to be physical objects in the ontology, they are interpreted as nodes of structures. And this is the weak point of OSR, because the existence of structures without objects to instantiate them is implausible. Esfeld (2009) [4] therefore suggests that OSR requires objects as the relata of structures, and he interprets the structures as being modal. That is to say, they constrain the temporal development of the objects instantiating them.

Esfeld's proposal qualifies OSR as an instance of dispositionalism. The only difference lies in the nature of the dynamical entities. Intrinsic properties are no longer responsible for the dynamical constraints; this task is fulfilled solely by relations between the elements of the primitive ontology.

It is now straightforward to apply this idea to Newtonian mechanics. Particles are the objects that stand in certain spatiotemporal relations resulting from their positions in absolute space, and in addition to these spatiotemporal relations they stand in certain dynamical relations. The latter relations are the modal structure. In the case of gravitation, this structure functions according to (2), and the manifestation of this structure is the acceleration of particles according to (1). Note that the spatial relations between the particles are not modal, because these relations alone have no causal-nomological role in the dynamical behavior of particles.

So mass cannot be interpreted as an intrinsic property of particles in this framework; rather, it is a parameter that specifies the gravitational structure regarded as an additive bipartite particle–particle relation according to (2), and, in this sense, the particles are coupled. In other words, the motion of one particle changes the motion of other particles in the universe because taken together particles instantiate a dynamical structure. A crucial feature of this dynamical structure is that it is reducible to or separable into direct relations between two particles; this reduction fails in the quantum case, which requires a non-separable holistic structure as proposed by Esfeld, Lazarovici, Hubert and Dürr 2014 [6]. In sum, the notion of a coupling constant points to two aspects of mass: on the one hand, mass is a constant in the laws of motion, and, on the other hand, this notion anticipates dynamical relations between particles.

## 4 Conclusion

The aims of this paper were twofold. First, I showed that the notion of a primitive ontology can be fruitfully used in classical mechanics. Second, I argued that the status of mass depends on the metaphysics of the laws of nature. It subsequently became clear that mass has to be interpreted as a dynamical entity introduced by Newton's laws of motion. The metaphysical theories that I discussed allow mass to be construed in three different ways: it may be regarded as a parameter, as an intrinsic property, or as a coupling constant. I tried to remain neutral with respect to the "best" interpretation. A thorough evaluation of the different positions remains to be undertaken.

## Acknowledgments

I wish to thank Michael Esfeld, Dustin Lazarovici, and an anonymous referee for many helpful comments on previous drafts of this paper. This work was supported by the Swiss National Science Foundation, grants no. PDFMP1_132389.

## BIBLIOGRAPHY

[1] A. Bird. *Nature's Metaphysics: Laws and Properties*. New York: Oxford University Press, 2007.
[2] D. Dürr, S. Goldstein, and N. Zanghì. *Quantum Physics without Quantum Philosophy*. Heidelberg: Springer, 2013.
[3] A. Einstein, B. Podolsky, and N. Rosen. Can quantum-mechanical description of physical reality be considered complete? *Physical Review*, 47(10):777–80, 1935.
[4] M. Esfeld. The modal nature of structures in ontic structural realism. *International Studies in the Philosophy of Science*, 23(2):179–94, 2009.
[5] M. Esfeld. Quantum Humeanism, or: physicalism without properties. *The Philosophical Quarterly*, 64(256):453–70, 2014.
[6] M. Esfeld, D. Lazarovici, M. Hubert, and D. Dürr. The ontology of Bohmian mechanics. *The British Journal for the Philosophy of Science*, 65(4):773–96, 2014.
[7] M. Esfeld, D. Lazarovici, V. Lam, and M. Hubert. The physics and metaphysics of primitive stuff. *The British Journal for the Philosophy of Science*, advance access, 2015. doi: 10.1093/bjps/axv026.
[8] S. French. *The Structure of the World: Metaphysics and Representation*. Oxford: Oxford University Press, 2014.
[9] N. Hall. Humean reductionism about laws of nature. Manuscript, 2009. URL http://philpapers.org/rec/halhra.
[10] J. Ladyman and D. Ross. *Every Thing Must Go: Metaphysics Naturalized*. New York: Oxford University Press, 2007.
[11] D. Lewis. *Philosophical Papers*, volume 2. New York: Oxford University Press, 1986.
[12] D. Lewis. Humean supervenience debugged. *Mind*, 103(412):473–90, 1994.
[13] B. Loewer. Humean supervenience. *Philosophical Topics*, 24(1):101–27, 1996.
[14] E. Mach. *The Science of Mechanics: A Critical and Historical Account of Its Development*. Chicago: The Open Court Publishing Co., 4th edition, 1919.
[15] T. Maudlin. *The Metaphysics Within Physics*. New York: Oxford University Press, 2007.
[16] T. Maudlin. The universal and the local in quantum theory. *Topoi*, 34(2):349–58, 2015.

[17] I. Newton. *The Principia: Mathematical Principles of Natural Philosophy.* Berkeley: University of California Press, 1999. Translated into English by I. Bernard Cohen and Anne Whitman.

# 'Geometry as a Branch of Physics': Philosophy at Work in Howard P. Robertson's Contributions to Relativity Theories

Roberto Lalli

ABSTRACT. A historical analysis of the epistemological views held by the mathematical physicist Howard P. Robertson is attempted. The specific features of Robertson's methodological prescriptions to define sound relationships between geometry and experience will be brought out by comparing Robertson's terminologies with those employed by other thinkers who addressed similar issues. It will be shown that Robertson's explicit epistemological claims can be better understood as reflections on his daily work in theoretical physics. The analysis will lead to suggest that Robertson's ontological commitments cannot be described as a form of explanatory realism as has been claimed.

## 1 Introduction

"Is space really curved?" With this question Howard Percy Robertson (1903-1961) opened his contribution to the anthology *Albert Einstein Philosopher-Scientist* entitled 'Geometry as a Branch of Physics,' in which the well-known American mathematical physicist and cosmologist summarized his thoughts on the epistemology of geometry in connection with the theory of general relativity.[1]

The issue concerning the ontological existence of curved space was of course not new. The relationship between the theoretical framework of general relativity, its empirical confirmations and the ontology of space and time had been at the heart of debates on the philosophical interpretations of general relativity since the theory was first formulated in November 1915. One might legitimately maintain that Robertson's eighteen-page paper did not, and could not, add much to the thirty-year discussion between authoritative philosophers belonging to different traditions including logical empiricism, neo-Kantianism and realism.[2] Yet, Robertson's essay is of interest because it was an effort to translate in philosophical language a series

---

[1] Howard P. Robertson, "Geometry as a Branch of Physics," in *Albert Einstein: Philosopher-Scientist*, ed. Paul A. Schilpp (Evanston: Library of Living Philosophers, 1949), pp. 315-332.

[2] For the early philosophical interpretations of relativity theories with specific reference to their implications for the concepts of space and time, see Michael Friedman, *Foundations of Space-Time Theories: Relativistic Physics and Philosophy of Science* (Princeton: Princeton University Press, 1987); Klaus Hentschel, *Interpretationen und Fehlinterpretationen der speziellen und der allgemeinen Relativitätstheorie durch Zeitgenossen Albert Einsteins* (Boston: Birkhäuser, 1990); and Thomas A. Ryckman, *The Reign of Relativity: Philosophy in Physics 1915–1925* (New York: Oxford University Press, 2005).

of methodological prescriptions that shaped Robertson's practice as an expert of general relativity and group theory. The analysis of Robertson's essay and a comparison between its conclusions and the arguments he actually employed in both published papers and private letters allows for a historical scrutiny of the interconnections between Robertson's explicit epistemological positions and his daily work, intended both as the employment of specific theoretical tools and the choices he made among different approaches to theory construction, especially in the field of cosmology.

In order to investigate these interconnections I structure the paper as follows. In the first section, Robertson's explicit epistemological positions exposed in his 1949 essay are outlined. To bring out the specific features of Robertson's methodological claims, in the second section I compare them to the views held by those authoritative philosophers of the period who addressed the same topic in seemingly similar manners. In the third section, I will put Robertson's epistemological stances in connection with what he actually did in his major contributions to the development of relativistic cosmology. In the concluding remarks, I maintain that Robertson's ontological views cannot be interpreted as a simple form of explanatory realism.

## 2  Robertson's methodology of physical geometry

Before going into the details of Robertson's essay, I recall some of the most important scientific achievements Robertson accomplished in the course of his rather short career, which stretched between the late 1920s and the late 1950s. After having completed his postdoctoral studies in Göttingen and Munich acquiring a strong expertise in differential geometry and group theory, in the late 1920s Robertson became one of the most influential experts of general relativity theory and relativistic cosmology in the United States. His most well-known result is the rigorous derivation of the so-called Friedmann-LemaÓtre-Robertson–Walker (FLRW) metric (also called Robertson–Walker metric) between 1929 and 1935, which contains all the geometries associated with the assumptions of homogeneity and isotropy of three-dimensional space.[3] In 1933, Robertson wrote the long review article 'Relativistic Cosmology,' which promoted the expanding universe as the most reliable model for a theoretical description of the universe. In doing that, Robertson introduced many elements that are still part of the standard cosmological model.[4] He also served as

---

[3] Robertson derived the line element that bears his name in various papers. The most quoted is Howard P. Robertson, "Kinematics and world structure," *Astrophysical Journal,* 82 (1935): 284–301. In this paper, Robertson did not assume the validity of general relativity, but derived the metric only from very general hypotheses concerning the isotropy and homogeneity of space. On the centrality of the FLRW metric in the current standard model of relativistic cosmology see, e.g., Steven Weinberg, *Gravitation and Cosmology: Principles and Applications of the General Theory of Relativity* (New York: Wiley and Sons, 1972), pp. 407-418; George F. R. Ellis, "Standard Cosmology," in *Cosmology and Gravitation: Proceedings of the $5^{th}$ Brazilian School of Cosmology and Gravitation,* ed. M. Novello (Singapore: World Scientific, 1987), pp. 83-151.

[4] Howard P. Robertson, "Relativistic Cosmology," *Reviews of Modern Physics,* 5 (1933): 62-90. For comments and analyses on its relevance in the history of cosmology, see, George F. R. Ellis, "The Expanding Universe: A History of Cosmology from 1917 to 1960," in *Einstein and the History of General Relativity, Einstein Studies, Vol. 1,* ed. Don Howard and John Stachel (Boston: Birkhäuser, 1987), pp. 367-431; and George F. R. Ellis, "Editorial Note: H. P. Robertson,

the main referee of various American journals for papers concerning general relativity, unified field theories and cosmology.[5] Robertson, eventually, left the world of pure research in physics, for he became involved in military activities as a scientific advisor during and after World War II. He continued to teach mathematical physics at Caltech until 1961, when he died due to some complications following a car accident.[6] Historians of physics agree that Robertson held a leadership position in the field of general relativity in the period christened by Jean Eisenstaedt the "low water mark of general relativity," which roughly went from mid-1920s to mid-1950s.[7]

In his essay 'Geometry as a Branch of Physics,' Robertson aimed at analyzing the relationship between deduction and observation in the problem of physical space from what he defined a "neutral mathematico-physical viewpoint in a form suitable for incorporation into any otherwise reliable philosophical position."[8] Starting from the assumption that geometry was a purely deductive science built on a set of axioms and logical processes, Robertson's explicit target was to define the methodology that allowed one to choose the most appropriate geometry for the description of the physical space in accord with the available empirical data.

At first, Robertson discussed congruence geometries, defined as those geometries in which the intrinsic relations between elements of a configuration are unaffected by the position and orientation of the configuration. Referring to the reflections on the relationship between geometry and experience exposed by Hermann von Helmholtz, Robertson recognized that congruence geometries had often been considered the only acceptable choices for the description of the physical space because they entail the free mobility of rigid bodies without deformation, then giving meaning to the definition and comparison of distances.[9]

The nineteenth-century development of non-Euclidean geometries had shown that Euclidean geometry was only one special case of congruence geometries, each characterized by different values of the constant curvature $K$. As it is well known, the intuitive idea of curvature can be grasped by thinking to a two-dimensional surface embedded in a three-dimensional Euclidean space. The curvature, then, corresponds to the curvature of the surface in the third dimension. Robertson reminded the reader that starting from the axioms of the congruence geometry it is possible to derive general formulas that establish exact relationships between mathematical concepts such as distance, angle, and area, and then to determine the value of the

---

Relativistic Cosmology," *General Relativity and Gravitation*, 44 (2012): 2099-2114.

[5] The correspondence about the refereeing activities of Howard Percy Robertson is stored in the archival collection Howard Percy Robertson Papers, Caltech Archives, Pasadena, CA, USA (hereafter HRP). See, especially, box 7, folders 12, 13, and 14.

[6] Jesse L. Greenstein, "Howard Percy Robertson (1903-1961)," *Biographical Memoirs of the National Academy of Sciences*, 51 (1980): 341-365.

[7] See, especially, Jean Eisenstaedt, "La Relativité Générale à l'Étiage: 1925–1955," *Archive for History of Exact Sciences*, 35 (1986):115–185; and Jean Eisenstaedt, "Trajectoires et Impasses de la Solution de Schwarzschild," *Archive for History of Exact Sciences*, 37 (1987):275–357.

[8] Robertson, "Geometry as a Branch" (cit.1), p. 315.

[9] Robertson is referring to Helmholtz's epistemology of geometry exposed in Hermann von Helmholtz, "Über die tatsächlichen Grundlagen der Geometrie," *Nachrichten K. Ges. Wissenschaften zu Göttingen*, 9 (1868): 193-221.

curvature by "measurements made on the surface;" namely, without recourse to the embedding three-dimensional space.[10] Robertson's main aim was to establish the measurement operations that, extending the formulas previously derived for the two-dimensional case, restored the "objective aspect of physical space."[11]

He labeled neo-Kantian the view according to which the geometry of the physical space must necessarily be a congruence geometry. In this perspective, the physicist's problem became to "state clearly those aspects of the physical world which are to correspond to elements of the mathematical system."[12] In other words, Robertson stressed that there exist measurements which allow for a determination of the constant curvature $K$, just as there are measurements made on the surface of the Earth that make us understand that we are not living on an Euclidean plane. The search for clearly defined relationships between axiomatic geometry and measurements is what Robertson called *operational approach* to physical geometry. Once he had outlined his program for congruence geometries, Robertson stressed that the method should be extended to geometries in which the curvature $K$ varies from point to point. In this latter case, Robertson's arguments went, it is possible to establish in any point $P$ the mean curvature of the space at that point as the mean of the various hypersurfaces passing through $P$.

Robertson defined as sound the operational approach above summarized because it gave the possibility to confer a precise value to the curvature at any point. When one follows these procedures, Robertson argued, the choice of the physical geometry becomes a purely empirical problem. In this sense, geometry can be considered as a branch of physics, as the title of Robertson's paper emphasized. Robertson explicitly put this methodology in contrast to Poincare's conventionalism on the relationship between experience and geometry.[13] For him, the criterion of *universality* provided a way to challenge Poincaré's argument. The theory of general relativity theory, he concluded, can successfully serve as a universal physical geometry because the gravitational force acts the same way on all test bodies—a restatement of the principle of equivalence, which Robertson defined as the empirical finding that the observed inertial and gravitational mass of any body are "rigorously proportional for all matter."[14]

## 3   Robertson's place in the philosophical landscape

Robertson's overemphasized reference to the operational approach could be interpreted as an explicit reference to operationalism—a term that was at the time broadly employed by physicists and philosophers alike with particular reference to the views of Percy W. Bridgman.[15] As a doctrine, however, operationalism was

---

[10]Robertson, "Geometry as a Branch" (cit.1), p. 319.
[11]Robertson, "Geometry as a Branch" (cit.1), p. 322.
[12]Robertson, "Geometry as a Branch" (cit.1), p. 322.
[13]For a critique to Robertson's opposition to Poincaré's conventionalism, see Adolf Grünbaum, "Conventionalism in Geometry," in *The Axiomatic Method*, ed. L. Henkin, P. Suppes, and A. Tarski (Amsterdam: North Holland Publishing, 1959), pp. 204-222, esp. 212-213.
[14]Robertson, "Geometry as a Branch" (cit.1), p. 329.
[15]The text where Bridgman first exposed his philosophical thinking is Percy W. Bridgman, *The Logic of Modern Physics* (New York: Macmillan, 1927).

not precisely defined and it was subject to a number of different interpretations. The most common was the inclination to regard Bridgman's operational analysis as a theory of meaning prescribing that the meaning of a concept corresponds to a set of operations. In Robertson's employment of the term, instead, "operational approach" did not refer at all to a prescription to grasp by means of measurement operations otherwise nebulous concepts. Rather, Robertson focuses on the practical need to find measurement methods that could soundly link the elements of mathematical structures to physical phenomena.

More than a commitment to operationalism—as it was usually understood—Robertson's philosophical commitments resembled some of the views Hans Reichenbach had been elaborating since the early 1920s. Robertson's statements that it is possible to choose by means of measurements the appropriate physical geometry between different axiomatic geometries are similar to Reichenbach's argument that once a definition of congruence has been specified "it becomes an empirical question which geometry holds for a physical space."[16] In addition, both Robertson and Reichenbach gave a strong relevance to the notion of universal force. Reichenbach had introduced a definition of universal forces already in 1924 when he argued that it is necessary to establish coordinative definitions in order to choose some metrical indicators of length, where the coordinative definitions had a purely conventional character.[17] A perfectly legitimate choice, Reichenbach argued, is that of rigid infinitesimal measuring rods. In taking into account the distorting forces that modify the length of the infinitesimal rod when it moves from one point of space to another, Reichenbach isolated two kinds of such forces: a) differential forces, which are forces that act in different ways on different materials (such as the deformation due to heat); and b) universal forces, which instead affect all the materials the same way. Reichenbach maintained that employing the rigid measuring rod as the coordinative definition is equivalent to give value zero to the universal forces. Since gravitation has the characteristics of a universal force, Reichenbach recognized that putting the universal gravitational force equal to zero meant that in general relativistic theoretical framework gravity was absorbed by the geometry. Here, Reichenbach's line of reasoning seems to be equivalent to Robertson's argument concerning the role of gravitation as a universal force.

In spite of the various similarities one can uncover between the approach of Robertson and that of Reichenbach, however, it is not possible to interpret Robertson's views as a simplified version of the more detailed philosophical perspective developed by the German proponent of logical empiricism. After an initial attempt to elaborate a neo-Kantian perspective that took into account the success of relativity theories and their implication for physical geometries, Reichenbach came to accept the conventionalist conception of physical geometry as elaborated by Poincaré in his philosophical writings at the beginning of the 20$^{th}$ century.[18]

---

[16] Hans Reichenbach, "Philosophical Significance of Relativity," in *Albert Einstein: Philosopher-Scientist,* ed. Schilpp (cit. 1), pp. 289-311, on p. 197.

[17] Hans Reichenbach, *Axiomatik der relativistischen Raum-Zeit-Lehre* (Braunschweig: Vieweg, 1924).

[18] H. Poincaré, *La Science et L'Hypothése* (Paris: Flammarion, 1902). Michael Friedman argued that Reichenbach deeply changed his perspective after a correspondence exchange with Moritz

Robertson, instead, explicitly challenged the conventional approach to the problem of physical geometry. Indeed, Roberson made no mention of the factorization of the theory of general relativity between its definitional part and its empirical content. Nor can one find in Robertson's account any reference to methodological criteria that could guide the choice between alternative equivalent geometries. In other words, Robertson implicitly dismissed the thesis of the relativity of geometry that became one of the central tenets of Reichenbach's doctrine of space and time after Reichenabach began adopting Poincaré's terminology around 1922.[19]

## 4  The impact of Robertson's scientific work on his methodological stances

Robertson was not a philosopher. He addressed the issue of the relationship between geometry and experience from the perspective of the mathematical physicist who had faced the pressing epistemological problems related to the creation of the new field of relativistic cosmology. As Robertson himself recognized, the feeble contact with empirical data made relativistic cosmology a field particularly dependent on the "general and philosophical predilections of the investigator."[20] The few scientists who worked in this field had to clearly define the methodological criteria to employ in the development of cosmological models as well as in drawing the connections between theories and astronomical observations. As his review article demonstrates, Robertson was actively involved in the program to establish a standard approach to relativistic cosmology as an empirically based discipline in contrast to alternative views of cosmology held by other authoritative scholars.[21]

Robertson's major works on relativistic cosmology covered the period from 1928 to 1936. Although Robertson's methodological approach to relativistic cosmology evolved through this period, some elements maintained a fairly stable position as the fundamental points on which to build what Robertson regarded as an acceptable theory of the universe. Particularly relevant was the role Robertson attributed to the so-called Weyl principle, which from 1929 onward occupied a central position in Robertson's argumentative scheme.[22] In the 1933 review paper, Robertson provided

---

Schlick in the early 1920s. Previously, Reichenbach had exposed a neo-Kantian view that Friedman labeled "relativized a priori," in the book Hans Reichenbach, *Relativitätstheorie und Erkenntnis apriori* (Berlin: Springer, 1920). M. Friedman, "Geometry, convention, and the relativized a priori: Reichenbach, Schlick, and Carnap," in *Reconsidering Logical Positivism* (Cambridge: Cambridge University Press, 1999), pp. 59-70.

[19] For the role of descriptive simplicity in contrast to inductive simplicity in Reichebach's views of the relativity of geometry see H. Reichenbach, *Philosophie der Raum-Zeit Lehre* (Berlin: Walter de Gruyter, 1928), pp. 8-58. See, also, Friedman "Geometry, convention," (cit. 19), esp. p. 64.

[20] Robertson, "Relativistic Cosmology" (cit. 5) p. 62.

[21] In his private correspondence he strongly criticized the approaches of the English astronomers and mathematicians Arthur S. Eddington, James H. Jeans and Edward A. Milne, and explicitly stated that his program was opposed to theirs. See, especially, Robertson to Eric Temple Bell, 15 September 1936, HRP, Box 1, folder 13.

[22] The genesis and evolution of the Weyl principle has been vastly discussed in the historical and philosophical literature. It has been emphasized that the meaning of Weyl's hypothesis, its connection with empirical evidence and even its status as an independent principle changed with time. In view of the interpretative disagreement around the Weyl principle and the different historical reconstructions of its genesis and final integration into the standard model of relativistic

a clear description of what he called "Weyl's coherency assumption" as the necessary hypothesis that allowed for a connection between the exact solutions of Einstein's field equation for the entire universe and the available astronomical data.[23]

Robertson stated that Weyl's assumption according to which in a de Sitter universe "the world lines of all matter belong to a pencil of geodesics which converges toward the past" was equivalent to the utterance that there exists in each region of cosmic space-time a mean motion that represents the actual motion of celestial bodies apart from small and unsystematic deviation.[24] For Robertson, this principle was extrapolated from two distinct astronomical observations: The first empirical finding was that on the large scale astronomical objects seem to be uniformly distributed; the second one was that the relative velocity of such objects in a specific region of space-time is small compared to the velocity of light, and then they might be considered relatively at rest with respect to the main motion of the region under consideration.

The Weyl principle, Robertson's argument went, allowed for a definition of a coordinate framework in which the geodetic lines $x_0 = t$ are chosen in a way that they represent the mean motion of matter in its neighborhood and the spatial hypersurfaces of constant $t$ are orthogonal to the congruence of geodesics so defined. These conditions led to what Robertson defined the "natural" introduction of the cosmic time $t$, which corresponds to the proper time measured by observers who co-move with the mean motion of matter in a certain region.[25] The previous empirically based assumptions could be generalized to an idealized cosmological space-time in which any three-dimensional space-like hypersurfaces of constant cosmic time are homogeneous and isotropic.

In order to persuade the reader that these procedures were sound, Robertson often casted them as "natural." But what was the exact meaning of naturalness in Robertson's epistemology? Which set of ontological commitments and methodological prescriptions were hidden behind this, to say the least, ill-defined notion? Robertson's introduction of the Weyl principle as an extrapolation from astronomical observations might suggest that the term "natural" was a general expression to define those methodological procedures that could be considered as representing a sort of inductive empiricism.

That Robertson was a defender of the use of empiricist methodology in relativis-

---

cosmology, in this paper I will focus only on the way in which Robertson understood and employed the principle. For historical analyses of the Weyl principle, see, John D. North, *The Invented Universe: A History of Modern Cosmology* (Oxford: Clarendon Press, 1965), pp. 74-185; Pierre Kerszberg, "Le Principe de Weyl et l'invention d'une cosmologie non-statique," *Archives for History of Exact Sciences* 35 (1987): 1-89; Sergio Bergia and Lucia Mazzoni, "Genesis and Evolution of Weyl's Reflections on De Sitter's Universe," in *The Expanding Worlds of General Relativity*, ed. Hubert Goenner et al. (Boston: Birkhäuser, 1999), pp. 325-342; and Hubert Goenner, "Weyl's contributions to Cosmology," in *Hermann Weyl's Raum-Zeit-Materie and a general introduction to his scientific work*, ed. Erhard Scholz (Basel: Birkhäuser, 2001), pp. 105-137.

[23] Robertson, "Relativistic Cosmology" (cit. 4) p. 67.
[24] Robertson, "Relativistic Cosmology" (cit. 4) p. 65.
[25] For the suggestion to employ cosmic time as a definition of mind-independent temporal becoming, see Mauro Dorato, *Time and Reality: Spacetime Physics and the Objectivity of Temporal Becoming* (Bologna: CLUEB, 1995), pp. 189-212.

tic cosmology is confirmed by his attitude toward competitive approaches and in particular toward the special relativistic cosmological model elaborated by Edward A. Milne from 1933 onward.[26] Robertson strongly challenged Milne's hypothetico-deductive epistemology according to which cosmologists could elaborate the model they prefer and successively draw the possible connections with observations. The methodological conflict between Robertson and Milne is exemplified by the use they made of the fundamental assumptions on which they based their model. Milne started from the "cosmological principle," which he defined as the assumption that the descriptions of the universe made by two equivalent observers employing their own clocks and associated coordinates coincide. While Milne introduced the principle as an *a priori* axiom without any connection with observations, Robertson considered Milne's cosmological principle to be nothing but a restatement of what he had already defined as the empirically derived Weyl postulate.[27]

Milne argued that from an operational perspective his kinematic model was to be preferred to general relativistic cosmologies because it referred to measuring instruments such as clocks, theodolites, and light signals, and not to unobservable entities such as curved space-time. In 1935-36, Robertson answered to Milne's claims with a three-part article called 'Kinematics and World-Structure.' In it, Robertson offered another derivation of the FLRW line elements, which he had already derived in 1929, and argued that appeal to operational methodology does not consent to take a final decision between alternative cosmological models. He did so by contending that general relativistic cosmology was more complete than Milne's model, for the latter could be interpreted as a special case of the general line element already derived.[28]

Robertson's reasoning in 'Kinematics and World-Structure' had been somewhat misinterpreted by the philosopher of science George Gale, who claimed that Robertson had been converted to Milne's operational methodology and that Robertson's 1935 derivation of the FLRW metric was a consequence of this conversion.[29] While it is true that Robertson began to make explicit reference to the operational methodology in these writings as a direct response to Milne's work, private correspondence shows that Robertson's agenda aimed at demonstrating that the operational approach was consistent with relativistic cosmology. He continued to find unacceptable the lack of any distinction between mathematical theories and physical laws in Milne's approach.[30] A clear distinction between mathematics and physics was

---

[26] Edward A. Milne, *Relativity, Gravitation and World-structure* (Oxford: Clarendon Press, 1935).

[27] Robertson to Eric T. Bell, 15 September 1936, HRP, Box 1, folder 13.

[28] Robertson, "Kinematics and World-Structure" (cit. 3); Robertson, "Kinematics and World-Structure II." *The Astrophysical Journal*, 83 (1936): 187-201; and Robertson, "Kinematics and World-Structure III," *The Astrophysical Journal*, 83 (1936): 257-271.

[29] George Gale and John Urani, "E. A. Milne and the origins of modern cosmology: An essential presence," in *The Attraction of Gravitation: New Studies in the History of General Relativity*, ed. John Earman, Michel Janssen, and John D. Norton (Boston: Birkhäuser, 1993), pp. 390-419; and G. Gale, "Cosmology: Methodological Debates in the 1930s and 1940s," *Stanford Encyclopedia of Philosophy* (Spring 2014 Edition), ed. Edward Zalta URL = http://plato.stanford.edu/archives/spr2014/entries/cosmology-30s/ (retrieved 4 July 2014).

[30] See, e.g., Robertson to Otto Struve, 23 July 1935, HRP, Box 5,Folder 24; Robertson to Edwin

indeed the starting point of what he later labeled "operational approach" in his essay 'Geometry as a Branch of Physics.'

Robertson's focus on extrapolation from observations and the stress on the mathematics physics divide might confirm Gale's view that Robertson championed an empiricist methodology. Gale, however, also stressed that this methodology was coupled to a sort of explanatory realism, which Gale defined in the following way: "if an accepted theory referred to entity $x$, then $x$ was acceptable as a genuinely, physically real object."[31] However, I have not been able to find any clear evidence that Robertson was particularly committed to the reality of the theoretical entities he referred to in his scientific endeavors. Robertson's long-lasting commitment to the theoretical tools of differential geometry and group theory suggests instead that this simple characterization does not completely represent Robertson's views.

In his review of Milne's book *Relativity, Gravitation and World Structure*, Robertson criticized the "cumbrousness and obscureness" of Milne's mathematical exposition, which avoided any employment of group theoretical tools and concepts.[32] This neglect was incomprehensible to Robertson on the grounds that "the theory of groups of automorphisms or motions of a space into itself constitutes the *natural* mathematical tool for the investigation of spaces characterized by a priori symmetry conditions."[33] According to him, the Mach's principle—which he defined in the weaker form as implying that "the metric field is causally determined to within a possible transformation of coordinates by the stress-energy tensor" through the Einstein field equation—justified the extension of group theory to relativistic cosmologist, because the symmetry properties in the material and energetic distribution could be directly interpreted in terms of the line element of space-time.[34]

It is worth noticing that in motivating his commitment to a particular mathematical technique Robertson again employed the term "natural." The question is then whether we can find any common element underlying the use of the notion of naturalness to two seemingly distinct aspects of Robertson's work on relativistic cosmology; namely, the choice of the Weyl principle to ground the natural definition of cosmic time and the use of group theory as the natural theoretical tool for cosmological space-time.

Robertson's expertise in differential geometry made him able to recognize those invariant structures of space-time geometry that might be suitable for group theoretical analysis. It would have been unfeasible to apply group theory to the entire

---

P. Hubble, 15 September 1936, HRP, Box 3, Folder 10; Robertson to Leopold Infeld, 28 September 1940, HRP, Box 3, Folder 15.

[31] Gale, "Methodological Debates," (cit. 29).

[32] Robertson, "Review of Milne's Relativity Gravitation and World-Structure," *Astrophysical Journal*, 83 (1936): 61–66, on p. 65.

[33] *Ibid.*, emphasis mine.

[34] Robertson, "Relativistic Cosmology" (cit. 4) p. 63. For the difficulties associated to the implementation of the Mach's principle in relativistic cosmology, see Michel Janssen, "'No Success Like Failure...':Einstein's Quest for General Relativity, 1907–1920," in *The Cambridge Companion to Einstein*, ed. Michel Janssen and Christoph Lehner (Cambridge: Cambridge University Press, 2014), pp. 167–227; for the different definitions of the Mach's principle, see, Julian Barbour and Herbert Pfister (eds.), *Mach's Principle: From Newton's Bucket to Quantum Gravity*, (Boston: Birkhäuser, 1995).

universe before defining a coordinate system in which the invariant mathematical objects might be clearly defined. Furthermore, Robertson considered necessary to establish a physical meaning for the chosen reference system. The fact that observations fit well within the theoretical needs for the application of group theory was, as I understand it, the reason that led Robertson to consider these lines of reasoning as natural.

It is possible to find an example of the same attitude in Robertson's judgment that the chief achievement of Einstein's gravitational theory was the way in which the theory incorporated the observational equivalence of inertial and gravitational mass. Robertson found it entirely satisfying from an epistemological perspective that within the structure of general relativity "the inertial mass is introduced into the matter-energy tensor and thence *automatically* seeps into the metrical field via the field equation, where it shows up as a gravitational mass."[35] The notion of naturalness one finds in Robertson's writings may well be considered as a translation of this feeling that some theoretical structures might account for observations in an automatic way.

## 5 Concluding remarks

Coming back to the question with which Robertson opened his 1949 essay: Was for Robertson space really curved? No doubt, the explicit response Robertson gave in his essay exhibited some significant similarities with Reichenbach's discourse about the philosophy of space and time. Nevertheless, Robertson explicitly rejected the conventionalist perspective Reichenbach held in his epistemological views of the relativity of geometry from 1922 onward. To better understand Robertson's views on the reality of curved space it is necessary to interpret them in relation to his daily activity. In his work, Robertson consciously defended an empiricist methodology of relativistic cosmology as Gale has correctly recognized. In light of this relation, Robertson's emphasis on operational approach and universality appears as an attempt to translate his methodological prescriptions at a different level that could be suitable for philosophical discussion.

Robertson's methodological perspective does not seem, however, to justify Gale's assertion that Robertson was a naÔf realist about theoretical entities. Neither in his papers nor in his letters, Robertson exposed opinions that could be assimilated to the explanatory realism toward theoretical entities as defined by Gale. Robertson was deeply fascinated by some features of theories such as general relativity according to which observational evidence was accounted automatically by general theoretical structures without adding any further hypothesis. The consonance between observations and pre-existing theoretical tools acted as a persuasive element in the choice between different theories and theoretical approaches. For Robertson, when a theory—like general relativity—showed those kinds of features that he labeled as natural, it was philosophically satisfying, and there was no need to explore other approaches that did not have the same kind of connection with observations. This fascination about the consonance between theoretical structures and

---

[35]Robertson to Subrahmanyan Chandrasekhar, 21 May 1948, HRP, Box 1, Folder 30, emphasis mine.

observations, however, did not become a way through which Robertson exposed a commitment to the reality of the entities described by the theory. Rather, Robertson's discourse always remained anchored at the methodological level without any explicit reference to the truth content of the theory.

# Historical and Philosophical Insights about General Relativity and Space-time from Particle Physics

J. Brian Pitts

ABSTRACT. Historians recently rehabilitated Einstein's "physical strategy" for General Relativity (GR). Independently, particle physicists similarly rederived Einstein's equations for a massless spin 2 field. But why not a light *massive* spin 2, like Neumann and Seeliger did to Newton? Massive gravities are bimetric, supporting conventionalism over geometric empiricism. Nonuniqueness lets field equations explain geometry but not *vice versa*. Massive gravity would have blocked Schlick's critique of Kant's synthetic *a priori*. Finally in 1970 massive spin 2 gravity seemed unstable or empirically falsified. GR was vindicated, but later and on better grounds. However, recently dark energy and theoretical progress have made massive spin 2 gravity potentially viable again.

## 1 Einstein's Physical Strategy Re-Appreciated by GR Historians

Einstein's General Relativity is often thought to owe much to his various principles (equivalence, generalized relativity, general covariance, and Mach's) in contexts of discovery and justification. But a prominent result of the study of Einstein's process of discovery is a new awareness of and appreciation for Einstein's *physical strategy*, which coexisted with his mathematical strategy involving various thought experiments and principles. The physical strategy had as some key ingredients the Newtonian limit, the electromagnetic analogy, coupling of all energy-momentum *including gravity's* as a source for gravity, and energy-momentum conservation as a consequence of the gravitational field equations *alone* [35, 6, 54, 55, 36, 56]. Einstein's mathematical strategy sometimes is seen to be less than compelling [44, 62], leaving space that one might hope to see filled by the physical strategy.

It has even been argued recently, contrary to longstanding views rooted in Einstein's post-discovery claims [22], that he found his field equations using his physical strategy [36]. Just how the physical strategy led to the field equations is still somewhat mysterious, resisting rational reconstruction [56].

## 2 Particle Physicists Effectively Reinvent Physical Strategy

There is, however, an enormous body of relevant but neglected physics literature from the 1920s onward. In the late 1930s progress in particle physics led to Wigner's taxonomy of relativistic wave equations in terms of mass and spin. "Spin" is closely related to tensor rank; hence spin-0 is a scalar field, spin-1 a vector, spin-2 a symmetric tensor. "Mass" pertains to the associated "particles" (quanta) of the field (assuming that one plans to quantize). (The constants $c$ and $\hbar$ are set to 1.) Particle masses are related inversely to the range of the relevant potential, which for a point source takes the form $\frac{1}{r}e^{-mr}$. Hence the *purely classical concepts* involved are merely wave equations (typically second order) that in some cases also have a new fundamental inverse length scale permitting algebraic, not just differentiated, appearance of the potential(s) in the wave equation—basically the Klein-Gordon equation. Despite the facade of quantum terminology—there is no brief equivalent of "massive graviton"—much of particle physics literature is the *systematic exploration of classical field equations covariant under (at least) the Poincaré group* distinctive of Special Relativity—though the larger 15-parameter conformal group or the far more general 'group' of transformations in General Relativity are not excluded. Hence drawing upon particle physics literature is simply what eliminative induction requires for classical field theories.

In this context, Fierz and Pauli found in 1939 that the linearized vacuum Einstein equations are just the equations of a massless spin-2 field [23]. Could Einstein's equations be derived from viewpoints in that neighborhood? Yes: arguments were devised to the effect that, assuming *special* relativity and some standard criteria for viable field theories (especially stability), along with the empirical fact of light bending, Einstein's equations were the unique result—what philosophers call an eliminative induction [37, 29, 22, 71, 47, 16, 68, 4]. The main freedom lay in including or excluding a graviton mass.

If particle physicists effectively reinvented Einstein's physical strategy, how did they get a unique result, in contrast to the residual puzzles found by Renn and Sauer [56]? The biggest difference is a new key ingredient, the elimination of negative energy degrees of freedom, which threaten stability. Eliminating negative energy degrees of freedom nearly fixes the linear part of the theory [68], and fixes it in such a way that the nonlinear part is also fixed almost uniquely. Technical progress in defining energy-momentum tensors also helped. Such derivations bear a close resemblance to Noether's converse Hilbertian assertion [39]—an unrecognized similarity that might have made particle physicists' job easier.

## 3 How Particle Physics Could Have Helped Historians of GR

The main difficulty in seeing the similarity between Einstein's physical strategy and particle physicists' spin-2 derivation of Einstein's equations is the entrenched habits of mutual neglect between communities. If one manages to encounter both literatures, the resemblance is evident. Particle physics derivations subsume Einstein's physical strategy especially as it appears in the little-regarded *Entwurf*, bringing it to successful completion with the correct field equations, using weaker and hence more compelling premises. Thus the *Entwurf* strategy really was viable in principle. In particular, Einstein's appeal to the principle of energy-momentum conservation [21, 40, 6] contains the key ingredient that makes certain particle physics-style derivations of his equations successful [50], namely, that the gravitational field equations alone should entail conservation, without use of the material field equations. Later works derived that key ingredient as a *lemma* from gauge invariance, arguably following from positive energy, arguably following from stability. Einstein's equations follow rigorously from special relativistic classical field theory as the simplest possible local theory of a massless field that bends light and that looks stable by having positive energy [68] (or maybe one can admit only a few closely related rivals); van Nieuwenhuizen overstated the point only slightly in saying that "general relativity follows from special relativity by excluding ghosts" (negative-energy degrees of freedom) [68]. Excluding ghosts nearly fixes the linear approximation. If one does not couple the field to any source, it is physically irrelevant. If a source is introduced, the linearized Bianchi identities lead to inconsistencies unless the source is conserved. The only reasonable candidate is the total stress-energy-momentum, including that of gravity. As a result the initial flat background geometry merges with the gravitational potential, giving an effectively geometric theory, hence with Einstein's nonlinearities [37, 16, 50]. More recently Boulanger and Esole commented that

> it is well appreciated that general relativity is the unique way to consistently deform the Pauli-Fierz action $\int \mathcal{L}_2$ for a free massless spin-2 field under the assumption of locality, Poincaré invariance, preservation of the number of gauge symmetries and the number of derivatives [4].

Familiarity with the particle physics tradition would have shown historians of GR that Einstein's physical strategy was in the vicinity of a compelling argument for his 'correct' field equations. Hence it would not be surprising if his physical strategy played an important role in Einstein's process of discovery and/or justification. Might historians of GR not thus have re-appreciated Einstein's physical strategy decades earlier? Might the apparent tortuous reasoning [56] regarding just how Einstein's physical strategy leads to Einstein's equations have been brought into sharper focus, with valid derivations available to compare with Einstein's trail-blazing efforts? Let $POT$ be the gravitational potential, $GRAV$ a second-order differential operator akin to the Laplacian, and $MASS$ be the total stress-energy-momentum, which generalizes the Newtonian mass density [54]. Whereas the schematic equation $GRAV(POT) = MASS$ is supposedly innocuous, particle physics would also expose the gratuitous exclusion of a mass term, which would require the form $GRAV(POT) + POT = MASS$.

## 4 Massive Gravities?

One might expect that a light massive field of spin-$s$ would approximate a massless spin-$s$ field as closely as desired, by making the mass small enough. Hugo von Seeliger in the 1890s already clearly made a similar point; he wrote (as translated by Norton) that Newton's law was "a purely empirical formula and assuming its exactness would be a new hypothesis supported by nothing." [69, 45] With the intervention of Neumann, which Seeliger accepted, the exponentially decaying point mass potential later seen as characteristic of massive fields was also available in the 1890s. (No clear physical meaning was available yet, however). It is now known that this expectation of a smooth massless limit is true for Newtonian gravity, relativistic spin-0 (Klein-Gordon), spin-1/2 (Dirac), a single spin-1 (de Broglie-Proca massive electromagnetism, classical and quantized), and, in part, a Yang-Mills spin-1 multiplet (classically, but not when quantized) [5]. Hence the idea that gravity might have a finite range due to a non-zero 'graviton mass' was not difficult to conceive. Indeed Einstein reinvented much of the idea in the opening of his 1917 cosmological constant paper [20], intending it as an analog of his cosmological constant. Unfortunately Einstein erred, forgetting the leading zeroth order term [32, 14, 45, 30]. Plausibly, Einstein's mistaken analogy helped to delay conception of doing to GR what Seeliger and Neumann had done to Newton's theory.

Particle physicists would not be much affected by Einstein's mistake, however; Louis de Broglie entertained massive photons from 1922 [11], and the Klein-Gordon equation would soon put the massive scalar field permanently on the map as a toy field theory. Particle physicists got an occasion to think about gravity when a connection between Einstein's theory and the rapidly developing work on relativistic wave equations appeared in the late 1930s [23]. From that time massive gravitons saw sustained, if perhaps not intense, attention until 1970 [64, 48, 12, 19, 47, 14].

One would expect that anything that can be done with a spin-2, can be done more easily with spin-0. Thus the Einstein-Fokker geometric formulation of Nordström's theory (massless spin-0) is a simpler (conformally flat) exercise in Riemannian geometry than Einstein's own theory. There are also many massive scalar gravities [49], and by analogy [47]. The scalar case, though obsolete, is interesting not only because it is easy to understand, but also because massive scalar gravities manifestly make sense as classical field theories. While massive scalar gravity has not been an epistemic possibility since 1919 (the bending of light), it ever remains a metaphysical possibility. Thus the modal lessons about multiple geometries are not hostage to the changing fortunes of massive spin-2 gravity. Massive scalar gravity also shows that (*pace* [38, p. 179] [41]) gravity did not have to burst the bounds of special relativity on account of Nordström's theory having the larger 15-parameter conformal group; massive scalar gravities have just the 10-parameter Poincaré group of symmetries.

## 5 Explanatory Priority of Field Equations over Geometry

In GR, the power of Riemannian geometry to determine the field equations tempts one to think that geometry generically is a good explanation of the field equations. Comparing GR with its massive cousins sheds crucial light on that expectation.

A key fact about massive gravities is the non-uniqueness of the mass term [47], in stark contrast to the uniqueness of the kinetic term (the part that has derivatives of the gravitational potentials), which matches Einstein's theory. The obvious symmetry group for most massive spin-2 gravities is just the Poincaré group of special relativity [47, 14]; the graviton mass term breaks general covariance. If one wishes nonetheless to recover formal general covariance, then a graviton mass term must introduce a background metric *tensor* (as opposed to the numerical matrix $diag(-1, 1, 1, 1)$ or the like), typically (or most simply) flat.

The ability to construct many different field equations from the same geometrical ingredients supports the dynamical or constructive view of space-time theories [7, 8]. The opposing space-time realist view holds that the geometry of space-time instead does the explaining. According to the realist conception of Minkowski spacetime,

> (2) The spatiotemporal interval $s$ between events $(x, y, z, t)$ and $(X, Y, Z, T)$ along a straight [footnote suppressed] line connecting them is a property of the spacetime, independent of the matter it contains, and is given by
>
> $$s^2 = (t - T)^2 - (x - X)^2 - (y - Y)^2 - (z - Z)^2. \qquad (1)$$
>
> When $s^2 > 0$, the interval $s$ corresponds to times elapsed on an ideal clock; when $s^2 < 0$, the interval $s$ corresponds to spatial distances measured by ideal rods (both employed in the standard way). [46]

One might worry that the singular noun "[t]he spatiotemporal interval" is worrisomely ambiguous, as is the adjective "straight." Why can there be only one metric? Resuming:

> (3) Material clocks and rods measure these times and distances because the laws of the matter theories that govern them are adapted to the independent geometry of this spacetime. [46]

But (3) is *false* for massive scalar gravity, in which matter $u$ sees $g_{\mu\nu}$, not the flat metric $\eta_{\mu\nu}$, as is evident by inspection of the matter action $S_{matter}[g_{\mu\nu}, u]$ [37], which lacks $\sqrt{-\eta}$, the volume element of the flat metric. Unlike space-time realism, constructivism makes room for Poincaré-invariant field theories in which rods and clocks do not see the flat geometry, such as massive scalar gravities.

Even if one decides somehow that massive scalar gravities, despite being just Poincaré-invariant, are not theories in Minkowski space-time, thus averting the falsification of space-time realism, it still fails on modal grounds. It simply takes for granted that the world is simpler than we have any right to expect, neglecting a vast array of metaphysical possibilities, some of them physically interesting. Space-time realism, in short, is modally provincial. Norton himself elsewhere decried such narrowness in a different context: one does not want a philosophy of geometry to provide a spurious apparent necessity to a merely contingent conclusion that GR is the best space-time theory [42, pp. 848, 849]. Constructivism, like conventionalism [52, pp. 88, 89] [2, 28, 72], does not assume that there exists a unique geometry; space-time realism, like the late geometric empiricism of Schlick and Eddington, does assume a unique geometry. It is striking that critiques of conventionalism also have usually ignored the possibility of multiple geometries [53, 61, 25, 65, 10, 43].

## 6 Massive Gravity as Unconceived Alternative

The problem of unconceived alternatives or underconsideration [60, 67, 63] can be a serious objection to scientific realism. Massive scalar gravity posed such a problem during the 1910s. Massive spin-2 gravities continued to pose such a problem for philosophers and general relativists at least until 1972, when the unnoticed threat went away. $C.$ 1972 a dilemma appeared: massive spin-2 gravity was either empirically falsified in the pure spin-2 case because of a discontinuous limit of small vs. 0 graviton mass (van Dam-Veltman-Zakharov discontinuity), or it was violently unstable for the spin 2-spin 0 case because the spin-0 has negative energy, permitting spontaneous production of spin-2 and spin-0 gravitons out of nothing. Particle physics gives, but it can also take away. More recently particle physics has given back, reviving the threat to realism about GR due to unconceived alternatives. While underdetermination by approximate but arbitrarily close empirical equivalence has long been clear in electromagnetism, it is now (back) in business for gravitation as well.

For philosophers and physicists interested in space-time prior to 1972, or since 2010, not conceiving of massive gravity means suffering from failure to entertain a rival to GR that is *a priori* plausible (a decently high prior probability $P(T)$ if one is not biased against such theories, and if the smallness of the graviton mass does not seem problematic), has good fit to data (likelihoods $P(E|T)$ approximating those of GR), and, crucially, has significantly different philosophical consequences from GR.

The underdetermination suggested by massive gravities and massive electromagnetism is weaker in four ways than the general thesis often discussed: it is restricted to mathematized sciences, is defeasible rather than algorithmic in generating the rivals, involves a one-parameter family of rivals that work as a team rather than a single rival theory, and is asymmetric: the family (typically) remains viable as long as the massless theory is, but not *vice versa*.

## 7 Schlick's Critique of Kant's Synthetic *A Priori*

The years around 1920 were crucial for a rejection of even a broadly Kantian *a priori* philosophy of geometry, especially due to Moritz Schlick's influence [58, 59, 9, 3, 18], and saw a partial retreat from conventionalism toward geometric empiricism [34, 57, 70]. Schlick argued that GR made even a broadly Kantian philosophy of geometry impossible because the physical truth about the actual world was incompatible with it [58, 59, 57, 9]. Coffa agreed, stuffing half a dozen success terms into two paragraphs in praise of Schlick [9, pp. 196, 197]. That Schlick, brought up as a physicist under Planck, could, in principle, have done to Nordström's and Einstein's theories what Neumann, Seeliger and Einstein had done to Newton's, thus making room for synthetic *a priori* geometry, seems not to have been entertained. Neither was the significance of the 1939 work of Fierz and Pauli [23].

Recognizing massive gravities as unconceived alternatives, one views Schlick's work in a different light. Schlick argued that General Relativity either falsifies or evacuates Kant's synthetic *a priori* [59]. He then quit thinking about space-time, and was assassinated in 1936. But post-1939, the flat background geometry *present in the field equations of massive gravity* would leave a role for Kant's geometrical views even in modern physics after all. (This multi-metric possibility is *not* the old Löze move of retaining flat geometry *via* universal forces! Such entities cannot be independently identified, and turn out to be even more arbitrary than one might have expected due to a new gauge freedom [27, 43]. The observability of the flat metric, indirect though it is, makes the difference [14]. One can ascertain the difference between the two geometries, which is the gravitational potential.) More serious trouble for Kant would arise finally when the van Dam-Veltman-Zakharov discontinuity was discovered. Hence Kant was viable until 1972, not 1920!—and maybe again today.

Massive gravities also bear upon Friedman's claim that the equivalence principle (viewed as identifying gravity and inertia) in GR is constitutively *a priori*, that is, required for this or similar theories to have empirical content [26]. Massive gravities, if the limit of zero graviton mass is smooth as least (true for spin-0, recently arguable for spin-2), have empirical content that closely approximates Nordström's and Einstein's theories, respectively, while the massive spin-0 and (maybe) massive spin-2 sharply distinguish gravity from inertia. The empirical content resides not in principles or in views about geometry, but in partial differential field equations [14, 7].

## 8 Recent Breakthrough in Massive Gravity

In the wake of the seemingly fatal dilemma of 1972, massive gravity was largely dormant until the late 1990s. Then it started to reappear due to the "dark energy" phenomenon indicating that the cosmic expansion is accelerating, casting doubt on the long-distance behavior of GR—the regime where a graviton mass term should be most evident. A viable massive gravity theory must, somehow, achieve a smooth massless limit in order to approximate GR, and be stable (or at least not catastrophically unstable). That such an outcome is possible is now often entertained. Massive gravity is now a "small industry" [33, p. 673] and is worthy of notice by philosophers of science.

Since 2000, Vainshtein's early argument that the van Dam-Veltman-Zakharov discontinuity was an artifact of an approximate rather than exact solution procedure was revived and generalized [66, 15, 1]. Thus pure spin-2 gravity might have a continuous massless limit after all, avoiding empirical falsification. The other problem was that an exact rather than merely approximate treatment of massive gravity shows, apparently, all versions of pure spin-2 gravity at the lowest level of approximation, are actually spin 2-spin 0 theories, hence violently unstable, when treated exactly [5]. This problem was solved by a theoretical breakthrough in late 2010, where it was found how to choose nonlinearities and carefully redefine the fields such that very special pure spin-2 mass terms at the lowest (linear) approximation *remain pure spin-2 when treated exactly* [14, 31].

The answers to deep questions of theory choice and conceptual lessons about space-time theory depend on surprises found in sorting out fine technical details in current physics literature. Thus philosophers should not assume that all the relevant physics has already been worked out long ago and diffused in textbooks.

Lately things have changed rather rapidly, with threats of reversals [17]. Getting the smooth massless limit *via* the Vainshtein mechanism is admittedly "a delicate matter" (as a referee nicely phrased it) [13].

One needs to reexamine all the conceptual innovations of GR that, by analogy to massive electromagnetism, one would expect to fail in massive gravity [14]. Unless they reappear in massive gravity, or massive gravity fails again, then such innovations are optional. Surprisingly many of those innovations *do reappear* if one seeks a consistent notion of causality [51], including gauge freedom, making those the robust and secure conceptual innovations—whether or not massive gravity survives all the intricate questions that have arisen recently. If massive gravity fails, then General Relativity's conceptual innovations are required. If massive gravity remains viable, then General Relativity's conceptual innovations are required only insofar as they also appear in massive gravity. It is striking how the apparent philosophical implications can change with closer and closer investigation.

## BIBLIOGRAPHY

[1] Babichev E., C. Deffayet, and R. Ziour. (2010). "The Recovery of General Relativity in Massive Gravity via the Vainshtein Mechanism", in *Physical Review D* 82:104008. arXiv:1007.4506v1 [gr-qc].

[2] Ben-Menahem Y. (2001). "Convention: Poincaré and Some of His Critics", in *British Journal for the Philosophy of Science*, 52:471–513.

[3] Bitbol M., P. Kerszberg, and J. Petitot, eds. (2009). "Constituting Objectivity: Transcendental Perspectives on Modern Physics". n.p.: Springer.

[4] Boulanger N. and M. Esole. (2002). "A Note on the Uniqueness of $D = 4$, $N = 1$ Supergravity", in *Classical and Quantum Gravity*, 19:2107–2124. gr-qc/0110072v2.

[5] Boulware D. G. and S. Deser. (1972). "Can Gravitation Have a Finite Range?", in *Physical Review D*, 6:3368–3382.

[6] Brading K. (2005). A Note on General Relativity, Energy Conservation, and Noether's Theorems, in *The Universe of General Relativity*, edited by Anne J. Kox and Jean Eisenstaedt, Einstein Studies, volume 11, pp. 125–135. Boston: Birkhäuser.

[7] Brown H. R. (2005). *Physical Relativity: Space-time Structure from a Dynamical Perspective*. New York: Oxford University Press.

[8] Butterfield J. N. (2007). " Reconsidering Relativistic Causality", in *International Studies in the Philosophy of Science*, 21:295–328. arXiv:0708.2189 [quant-ph].

[9] Coffa J. A. (1991). *The Semantic Tradition from Kant to Carnap: To the Vienna Station.* Cambridge: Cambridge University Press. Edited by Linda Wessels.

[10] Coleman R. A. and H. Korté. (1990). " Harmonic Analysis of Directing Fields", in *Journal of Mathematical Physics*, 31:127–130.

[11] de Broglie L. (1922). " Rayonnement noir et quanta de lumière", in *Journal de Physique et la Radium*, 3:422–428.

[12] de Broglie L. (1943). *Théorie Général des Particules a Spin (Method de Fusion)*. Paris: Gauthier-Villars.

[13] de Rham C. (2014). " Massive Gravity", in *Living Reviews in Relativity*, vol. 17. arXiv:1401.4173v2 [hep-th].

[14] de Rham C., G. Gabadadze and A. J. Tolley. (2011). " Resummation of Massive Gravity", in *Physical Review Letters* 106:231101. arXiv:1011.1232v2 [hep-th].

[15] Deffayet C., G. Dvali, G. Gabadadze, and A. I. Vainshtein. (2002). " Nonperturbative Continuity in Graviton Mass versus Perturbative Discontinuity", in *Physical Review D*, 65:044026. hep-th/0106001v2.

[16] Deser S. (1970). " Self-Interaction and Gauge Invariance", in *General Relativity and Gravitation*, 1:9–18. gr-qc/0411023v2.

[17] Deser S. and A. Waldron. (2013). " Acausality of Massive Gravity", in *Physical Review Letters* 110:111101. arXiv:1212.5835.

[18] Domski M., M. Dickson, and M. Friedman, eds. (2010). *Discourse on a New Method: Reinvigorating the Marriage of History and Philosophy of Science.* Chicago: Open Court.

[19] Droz-Vincent P. (1959). " Généralisation des équationes d'Einstein correspondant à l'hypothèse d'une masse non nulle pour la graviton", in *Comptes rendus hebdomadaires des séances de l'Académie des sciences*, 249:2290–2292.

[20] Einstein A. (1923). " Cosmological Considerations on the General Theory of Relativity", in *The Principle of Relativity*, edited by H. A. Lorentz, A. Einstein, H. Minkowski, H. Weyl, A. Sommerfeld, W. Perrett, and G. B. Jeffery. London: Methuen. Dover reprint, New York (1952). Translated from "Kosmologische Betrachtungen zur allgemeinen Relativitätstheorie," *Sitzungsberichte der Königlich Preussichen Akademie der Wissenschaften zu Berlin* (1917) pp. 142-152.

[21] Einstein A. and M. Grossmann. (1996). " Outline of a Generalized Theory of Relativity and of a Theory of Gravitation", in *The Collected Papers of Albert Einstein, Volume 4, The Swiss Years: Writings, 1912-1914, English Translation*, edited by A. Beck and D. Howard, pp. 151–188. Princeton: The Hebrew University of Jerusalem and Princeton University Press. Translated from *Entwurf einer verallgemeinerten Relativitätstheorie und einer Theorie der Gravitation*, Teubner, Leipzig (1913).

[22] Feynman R. P., F. B. Morinigo, W. G. Wagner, B. Hatfield, J. Preskill, and K. S. Thorne. (1995). *Feynman Lectures on Gravitation*. Reading, Mass.: Addison-Wesley. Original by California Institute of Technology (1963).

[23] Fierz M. and W. Pauli. (1939). " On Relativistic Wave Equations for Particles of Arbitrary Spin in an Electromagnetic Field", in *Proceedings of the Royal Society (London) A* 173:211–232.

[24] Freund P. G. O., A. Maheshwari, and E. Schonberg. (1969). " Finite-Range Gravitation", in *Astrophysical Journal* 157:857–867.

[25] Friedman M. (1983). *Foundations of Space-time Theories: Relativistic Physics and Philosophy of Science*. Princeton: Princeton University Press.

[26] Friedman M. (2001). *Dynamics of Reason: The 1999 Kant Lectures at Stanford University*. Stanford: CSLI Publications.

[27] Grishchuk L. P., A. N. Petrov, and A. D. Popova. (1984). " Exact theory of the (Einstein) gravitational field in an arbitrary background space-time", in *Communications in Mathematical Physics* 94:379–396.

[28] Grünbaum A. (1977). " Absolute and Relational Theories of Space and Space-time", jn *Foundations of Space-Time Theories*, Minnesota Studies in the Philosophy of Science, Volume VIII, edited by J. Earman, C. Glymour, and J. Stachel, pp. 303–373. Minneapolis: University of Minnesota.

[29] Gupta S. N. (1954). " Gravitation and Electromagnetism", in *Physical Review* 96:1683–1685.

[30] Harvey A. and E. Schucking. (2000). " Einstein's Mistake and the Cosmological Constant", in *American Journal of Physics*, 68 (8): 723–727.

[31] Hassan S. F. and R. A. Rosen. (2012). " Confirmation of the Secondary Constraint and Absence of Ghost in Massive Gravity and Bimetric Gravity", in *Journal of High Energy Physics*, 1204 (123): 0–16. arXiv:1111.2070 [hep-th].

[32] Heckmann O. (1942). *Theorien der Kosmologie*. Revised. Berlin: Springer. Reprinted 1968.

[33] Hinterbichler K. (2012). " Theoretical Aspects of Massive Gravity", in *Reviews of Modern Physics*, 84:671–710. arXiv:1105.3735v2 [hep-th].

[34] Howard D. (1984). " Realism and Conventionalism in Einstein's Philosophy of Science: The Einstein-Schlick Correspondence", in *Philosophia Naturalis*, 21:618–629.

[35] Janssen M. (2005). " Of Pots and Holes: Einstein's Bumpy Road to General Relativity", in *Annalen der Physik*, 14:S58–S85.

[36] Janssen M. and J. Renn. (2007). " Untying the Knot: How Einstein Found His Way Back to Field Equations Discarded in the Zurich Notebook", in *The Genesis of General Relativity, Volume 2: Einstein's Zurich Notebook: Commentary and Essays*, edited by J. Renn, pp. 839–925. Dordrecht: Springer.

[37] Kraichnan R. H. (1955). " Special-Relativistic Derivation of Generally Covariant Gravitation Theory", in *Physical Review*, 98:1118–1122.

[38] Misner C., K. Thorne, and J. A. Wheeler. (1973). *Gravitation*. New York: Freeman.

[39] Noether E. (1918). " Invariante Variationsprobleme", in *Nachrichten der Königlichen Gesellschaft der Wissenschaften zu Göttingen, Mathematisch-Physikalische Klasse*, pp. 235–257. Translated as "Invariant Variation Problems" by M. A. Tavel, *Transport Theory and Statistical Physics* 1 pp. 183-207 (1971), LaTeXed by Frank Y. Wang, arXiv:physics/0503066 [physics.hist-ph].

[40] Norton J. (1989). " How Einstein Found His Field Equations, 1912-1915", in *Einstein and the History of General Relativity: Based on the Proceedings of the 1986 Osgood Hill Conference*, edited by D. Howard and J. Stachel, Volume 1 of emphEinstein Studies, 101–159. Boston: Birkhäuser.
[41] Norton J. D. (1992). " Einstein, Nordström and the Early Demise of Scalar, Lorentz-covariant Theories of Gravitation", in *Archive for History of Exact Sciences*, 45 (1): 17–94.
[42] Norton J. D.. (1993). " General Covariance and the Foundations of General Relativity: Eight Decades of Dispute", in *Reports on Progress in Physics*, 56:791–858.
[43] Norton J. D.. (1994). " Why Geometry Is Not Conventional: The Verdict of Covariance Principles", in *Semantical Aspects of Spacetime Theories*, edited by U. Majer and H.-J. Schmidt, pp. 159–167. Mannheim: B. I. Wissenschaftsverlag.
[44] Norton J. D.. (1995). " Eliminative Induction as a Method of Discovery: How Einstein Discovered General Relativity", in *The Creation of Ideas in Physics: Studies for a Methodology of Theory Construction*, edited by J. Leplin, Volume 55 of emphThe University of Western Ontario Series in Philosophy of Science, pp. 29–69. Dordrecht: Kluwer Academic.
[45] Norton J. D.. (1999). " The Cosmological Woes of Newtonian Gravitation Theory", in *The Expanding Worlds of General Relativity*, edited by Hubert Goenner, Jürgen Renn, Jim Ritter, and Tilman Sauer, Einstein Studies, volume 7, pp. 271–323. Boston: Birkhäuser.
[46] Norton J. D.. (2008). " Why Constructive Relativity Fails", in *The British Journal for the Philosophy of Science*, 59:821–834.
[47] Ogievetsky V. I. and I. V. Polubarinov. (1965). " Interacting Field of Spin 2 and the Einstein Equations", in *Annals of Physics* 35:167–208.
[48] Petiau G. (1941). " Sur une représentation du corpuscule de spin 2", in *Comptes rendus hebdomadaires des séances de l'Académie des sciences*, 212:47–50.
[49] Pitts J. B. (2011). " Massive Nordström Scalar (Density) Gravities from Universal Coupling", in *General Relativity and Gravitation*, 43:871–895. arXiv:1010.0227v1 [gr-qc].
[50] Pitts J. B. and W. C. Schieve. (2001). " Slightly Bimetric Gravitation", in *General Relativity and Gravitation* 33:1319–1350. gr-qc/0101058v3.
[51] Pitts J. B., W. C. Schieve. (2007). " Universally Coupled Massive Gravity", in *Theoretical and Mathematical Physics*, 151:700–717. gr-qc/0503051v3.
[52] Poincaré H. (1913). " Science and Hypothesis", in *The Foundations of Science*. Lancaster, Pennsylvania: The Science Press. Translated by George Bruce Halsted, reprinted 1946; French original 1902.
[53] Putnam H. (1975). " The Refutation of Conventionalism", in *Mind, Language and Reality: Philosophical Papers, Volume 2*, pp. 153–191. Cambridge: Cambridge University Press.
[54] Renn J. (2005). " Before the Riemann Tensor: The Emergence of Einstein's Double Strategy", in *The Universe of General Relativity*, edited by A. J. Kox and J. Eisenstaedt, Einstein Studies, volume 11, pp. 53–65. Boston: Birkhäuser.
[55] Renn J. and T. Sauer. (1999). " Heuristics and Mathematical Representation in Einstein's Search for a Gravitational Field Equation", in *The Expanding Worlds of General Relativity*, edited by H. Goenner, J. Renn, J. Ritter, and T. Sauer, Volume 7 of *Einstein Studies*, pp. 87–125. Boston: Birkhäuser.
[56] Renn J. and T. Sauer. (2007). " Pathways Out of Classical Physics: Einstein's Double Strategy in his Seach for the Gravitational Field Equations", in *The Genesis of General Relativity, Volume 1: Einstein's Zurich Notebook: Introduction and Source*, edited by J. Renn, pp. 113–312. Dordrecht: Springer.
[57] Ryckman T. (2005). *The Reign of Relativity: Philosophy in Physics 1915-1925*. Oxford: Oxford University Press.
[58] Schlick M. (1920). *Space and Time in Contemporary Physics*. Oxford University. Translated by Henry L. Brose; reprint Dover, New York (1963).
[59] Schlick M. (1921). " Kritische oder empiristische Deutung der neuen Physik?", in *Kant-Studien* 26:96–111. Translated by P. Heath as "Critical or Empiricist Interpretation of Modern Physics?" in H. L. Mulder and B. F. B. van de Velde-Schlick, editors, *Moritz Schlick Philosophical Papers, Volume I (1909-1922)*, pp. 322-334. D. Reidel, Dordrecht (1979).
[60] Sklar L. (1985). " Do Unborn Hypotheses Have Rights?", in *Philosophy and Spacetime Physics*, pp. 148–166. Berkeley: University of California.
[61] Spirtes P. L. (1981). " Conventionalism and the Philosophy of Henri Poincaré". Ph.D. diss., University of Pittsburgh.

[62] Stachel J. (1995). " 'The Manifold of Possibilities': Comments on Norton", in *The Creation of Ideas in Physics: Studies for a Methodology of Theory Construction*, edited by Jarrett Leplin, Volume 55 of *The University of Western Ontario Series in Philosophy of Science*, pp. 71–88. Dordrecht: Kluwer Academic.

[63] Stanford P. K. (2006). *Exceeding Our Grasp: Science, History, and the Problem of Unconceived Alternatives*. New York: Oxford University.

[64] Tonnelat M.-A. (1941). " La seconde quantification dans la théorie du corpuscule de spin 2", in *Comptes rendus hebdomadaires des séances de l'Académie des sciences* 212:430–432.

[65] Torretti R. (1996). *Relativity and Geometry*. New York: Dover. original Pergamon, Oxford, 1983.

[66] Vainshtein A. I. (1972). " To the Problem of Nonvanishing Gravitation Mass", in *Physics Letters B*, 39:393–394.

[67] van Fraassen B. (1989). *Laws and Symmetry*. Oxford: Clarendon Press.

[68] van Nieuwenhuizen P. (1973). " On Ghost-free Tensor Lagrangians and Linearized Gravitation", in *Nuclear Physics B*, 60:478–492.

[69] von Seeliger H. (1895). " Ueber das Newton'sche Gravitationgesetz", in *Astronomische Nachrichten*, 137:129–136. NASA ADS.

[70] Walter S. A. 2010. " Moritz Schlick's Reading of Poincaré's Theory of Relativity", in *Moritz Schlick: Ursprünge und Entwicklungen seines Denkens*, edited by F. O. Engler and M. Iven, Volume 5 of *Schlickiana*, pp. 191–203. Berlin: Parerga.

[71] Weinberg S. (1964). " Derivation of Gauge Invariance and the Equivalence Principle from Lorentz Invariance of the $S$-Matrix", in *Physics Letters*, 9:357–359.

[72] Weinstein S. (1996). " Strange Couplings and Space-Time Structure", in *Philosophy of Science*, 63:S63–S70. Proceedings of the 1996 Biennial Meetings of the Philosophy of Science Association. Part I: Contributed Papers.

# Bohmian Classical Limit in Bounded Regions

Davide Romano

ABSTRACT. Bohmian mechanics is a realistic interpretation of quantum theory. It shares the same ontology of classical mechanics: particles following continuous trajectories in space through time. For this ontological continuity, it seems to be a great candidate for recovering the classical limit of quantum theory. Indeed, in a Bohmian framework, the issue of the classical limit reduces to show how the classical trajectories can emerge from the Bohmian ones, under specific classicality assumptions.

In this paper, we shall focus on a technical problem which arises from the dynamics of a Bohmian system in bounded regions and we suggest that a possible solution is supplied by the action of environmental decoherence. However, we shall show that, in order to implement decoherence in a Bohmian framework, a stronger condition is required (*disjointness of supports*) rather than the usual one (*orthogonality of states*).

## 1 Bohmian mechanics and classical limit

Despite the great success of quantum mechanics, a rigorous and general account of the classical limit has not been reached so far. This means we do not have a clear explanation for the transition from the quantum regime, which describes the short-scale world, to the classical regime, which describes our familiar macroscopic world.

We know that quantum mechanics is a fundamental theory: it applies at every scale[1]. The goal of the classical limit, therefore, is to derive classical mechanics from quantum mechanics, under specific classicality conditions[2].

The problem here is not only mathematical, but also conceptual: in standard quantum mechanics (SQM), the physical state of an N-particle system is described by a state vector, an element of an abstract Hilbert space [3]. Moreover, in SQM the state vector has just a statistical character: for a 1-particle system, the absolute square of the wave function has the meaning of a probability density to find the particle in a definite region if we perform a position measurement on the system. Within this

---

[1] Indeed, it is possible to have macroscopic quantum effects, like superconductivity.

[2] The classicality conditions are the physical conditions that allow for the emergence of a classical regime. For example, in decoherence theory, the classicality condition is the (ubiquitous) entanglement among quantum systems.

[3] If the state vector is expressed in the position basis, then we have the wave function of the system, which is defined over the 3N-dimensional configuration space of the system.

framework, even if we succeeded in deriving the classical equations of motion for a quantum system, should we regard this result as a true classical limit? Probably, we should not. Classical mechanics describes the motion of particles in space, i.e., it describes real paths for the systems (trajectories) and not just 'probability amplitude' paths. How can we derive the former dynamical structure (and ontology) starting from the latter one?[4]

One option is to consider Bohmian mechanics (BM) as the correct interpretation of quantum theory. In BM, a quantum system is described by a wave function together with a configuration of particles, each of them following a continuous trajectory in 3D physical space. Within this framework, both quantum systems and classical systems are composed by matter particles that follow real paths in 3D space[5]. So that the entire issue of the classical limit reduces to the question: *under which conditions do the Bohmian trajectories become Newtonian?*

However, one could object that classical mechanics is just a high level effective theory and that the very concept of 'particle' does not belong to the ontology of the fundamental physical world. In quantum field theory (QFT), for example, the concept of particle might play no role[6]. If we cannot introduce a particle ontology at the level of QFT, then we might not see the necessity of introducing it at the non relativistic quantum level either: a characterization of the theory in terms of the wave function could be enough also for QM. Under this view, the classical limit is obtained by the description of a narrow wave packet following a classical trajectory[7]. This is the standard approach we usually find in SQM textbooks [8], known as *Ehrenfest's theorem*.

However, it is worth noting that some specific QFT models with a particle ontology have been proposed[9], so that the philosophical inquiry about the fundamental ontology of the physical world is still open.

Nevertheless, Ehrenfest's theorem alone cannot provide a proper solution for the quantum to classical transition. First, the wave function of a isolated quantum

---

[4]See, for example, Holland (1993, sect. 6.1) about the conceptual difference between a quantum 'trajectory' and a classical one.

[5]Of course, in BM there is something more: the wave function. Whether the wave function in BM is a real physical entity (i.e., a new physical field) or a nomological entity that only describes how the particles move (the analogy is with the Hamiltonian in classical mechanics) is currently at philosophical debate. Supporters of the first view are, e.g., Holland (1993) and Valentini (1992); supporters of the second view are, e.g., Dürr, Goldstein & Zanghì (2013), Goldstein & Zanghì (2012) and Esfeld *et alii* (2014).

[6]See, e.g., Malament (1996)

[7]We note that, within the SQM framework, this approach seems to miss the conceptual point of the classical limit problem. In SQM, the wave function is not a real entity, but mainly a mathematical tool to extract probabilities of the measurement outcomes. Therefore, a narrow wave packet that follows a classical trajectory simply means that whenever we perform a position measurement on the system, we will obtain a result which is compatible with a classical trajectory. Nonetheless, we cannot extract the picture of a real entity following a classical trajectory from that. In other words, what is problematic is not considering a narrow wave function as a particle, but the statistical interpretation of the wave function as opposed to a real ontological entity (particle) following a trajectory in space.

[8]See, e.g., Merzebacher (1970, ch 4), Shankar (1994, ch. 6), Sakurai (1994, ch. 2). In particular, Shankar sheds also some light on specific limitations of the theorem.

[9]See Dürr *et alii*(2004)

system generally spreads out in a very short time. Moreover, Ballantine shows that Ehrenfest's theorem is neither necessary nor sufficient to obtain a classical dynamical regime for quantum systems[10].

The most convincing approach for the analysis of the quantum to classical transition is actually decoherence theory. So, in order to find out how Newtonian trajectories can emerge from the Bohmian ones, it seems reasonable to check whether and how decoherence theory fits into the Bohmian framework.

The aim of the paper is to focus on a technical problem, which arises in the context of BM in the attempt to derive classical trajectories for a pure state system in bounded regions. The problem follows from the fact that two (or more) Bohmian trajectories of a system cannot cross in the configuration space of the system. So, even if we assume that a macroscopic body, satisfying some specific-classicality conditions (big mass, short wavelength, etc...), starts following at the initial time a classical trajectory, its motion will become highly non classical if, at a later time, different branches of the wave function of the body will be about to cross each other in configuration space.

We argue that a possible solution is offered by the action of environmental decoherence on the system[11]. A relevant point will be clear from the analysis: in order to implement decoherence in the framework of BM, a stronger condition is required (*disjointness of supports*) than the usual one(*orthogonality of states*) for the systems describing the environmental particles that scatter off the (macroscopic) Bohmian system.

In section 2, we will describe the measurement process in BM, focusing on the emergence of the *effective wave function*. In section 3, we will present the problem mentioned above which arises (mainly) in bounded regions. In section 4.1, we will introduce decoherence theory as the crucial ingredient for the quantum to classical transition in every physically realistic situation. In section 4.2, we will show how a simple model of environmental decoherence can solve the problem, thus leading to the emergence of classical trajectories in bounded regions.

## 2 Bohmian mechanics

### 2.1 A short introduction to Bohmian mechanics

Bohmian mechanics is a quantum theory in which the complete physical state of an N-particle system is described by the pair $(Q, \Psi)$, where $Q = (q_1, q_2, \ldots, q_N)$ is the configuration of N particles, each particle $q_k$ ($k = 1, 2, \ldots, N$) living in 3D physical space[12], and $\Psi = \Psi(Q, t)$ is the wave function of the system, which is defined over the 3N-D configuration space of the system. For a non-relativistic spinless N-particle system, the dynamical evolution of the Bohmian system is given by the Schrödinger equation:

$$i\hbar \frac{\partial \Psi(Q,t)}{\partial t} = -\sum_{k=1}^{N} \frac{\hbar^2}{2m_k} \nabla_k^2 \Psi(Q,t) + V\Psi(Q,t)$$

---

[10] See Ballantine (1994), (1996), (1998, sect. 14.1).
[11] This solution has been originally proposed by Allori et alii (2002).
[12] Thus, the configuration $Q$ is defined over the 3N-D configuration space of the system.

which describes the time evolution of the wave function, and the guiding equation:

$$\frac{dq_k}{dt} = \frac{\hbar}{m_k} \text{Im} \frac{\nabla_k \Psi(Q,t)}{\Psi(Q,t)} \; ; \; \text{with } k = 1, 2, \ldots, N$$

which describes the time evolution of each particle position of the total configuration. From the guiding equation, we note the non-local dynamics of the Bohmian particles: the velocity of a single particle ($q_k$) will depend on the position of all the other particles of the total configuration ($Q = (q_1, q_2, \ldots, q_N)$). For obtaining a successful scheme of the quantum to classical transition, we need to explain not only the emergence of classical trajectories but also the passage from a quantum (holistic) non-local dynamics to a classical (separable) non-local dynamics[13].

Bohmian mechanics introduces quantum probabilities as a measure of subjective ignorance on the initial conditions of a system (*epistemic probabilities*): given a system with wave function $\psi$, our maximum knowledge about the actual initial positions of the particles is represented by a statistical distribution of possible configurations, i.e., a *classical ensemble*, according to the absolute square of the wave function:

$$\rho(Q) = |\psi(Q)|^2$$

This is a postulate in BM and it is known as *quantum equilibrium hypothesis*[14]. Moreover, from the Schrödinger equation, it follows that $\rho$ has the property of equivariance:

$$\text{if } \rho(Q,0) = |\psi(Q,0)|^2, \text{ then } \rho(Q,t) = |\psi(Q,t)|^2 \; ; \; \forall t > 0$$

*Quantum equilibrium* and *equivariance* imply that BM provides the same empirical predictions of SQM, once assumed that the result of a measurement is always encoded in a definite position of a pointer[15] and that different positions of a pointer are always represented by (approximately) non-overlapping supports in configurations space[16].

## 2.2 Measurement process in Bohmian mechanics

In this section we analyze a typical measurement process in BM, showing, in particular, how an *effective wave function* of a Bohmian system does emerge. Then, we will show that the condition of disjoint supports for different positions of a pointer

---

[13] In classical mechanics, the potentials which affect the particle motion decay quadratically with the distance, so that we can effectively describe the motion of one particle as autonomous and independent from the motion of a very distant particle (under specific conditions, of course). In BM, instead, the influence of the "quantum potential" on the particle motion does not decay with the distance, so that all the particles belonging to the configuration of a system are holistically related, even if they are located very far each other. See, e.g., Bohm (1987, sect. 3) for a clear explanation about the difference between quantum (Bohmian) and classical non-locality.

[14] The justification of the quantum equilibrium hypothesis is a subtle issue. Two main approaches have been proposed: the typicality approach by Dürr, Goldstein & Zanghì (1992) and the dynamical relaxation approach by Valentini (1991).

[15] We call *pointer* every measurement apparatus that shows a definite outcome after the physical interaction with a quantum system.

[16] We will analyze this condition in more detail in the next section.

is essential for obtaining a clear and definite measurement result.

Let's consider a system $\Psi(x)$, with actual configuration $X$, interacting with an apparatus $\Phi(y)$, with actual configuration $Y$[17]. We suppose that the degrees of freedom of the system and the apparatus are respectively $m$ and $n$, then the support of $\Psi(x)$ is defined over the $m$-dimensional configuration space of the system and the support of $\Phi(y)$ over the $n$-dimensional one of the apparatus[18]. We suppose that the initial state of the system is a superposition of two wave functions:

$$\Psi(x) = \alpha\psi_1(x) + \beta\psi_2(x)$$

with normalization $|\alpha|^2 + |\beta|^2 = 1$.

At the initial time $t = 0$, the system and the apparatus have not interacted yet, so the wave function of the total system (system + apparatus) is factorized:

$$\Psi(x,0)\Phi(y,0) = (\alpha\psi_1(x,0) + \beta\psi_2(x,0))\Phi(y,0)$$

During the time interval $\Delta t = (0,T)$, the system and the apparatus will evolve according to the Schrödinger equation: in a typical measurement interaction, thanks to some coupling term between the two, they will become entangled:

$$\Psi(x,0)\Phi(y,0) \longrightarrow \alpha\Psi_1(x,T)\Phi_1(y,T) + \beta\Psi_2(x,T)\Phi_2(y,T)$$

This is the usual formulation of the measurement problem: the physical state of the total system, after the measurement, represents a coherent superposition of two macroscopically distinct pointer states. In BM, there is a further ingredient that permits to (dis)solve the problem: besides the wave function, every Bohmian system is composed by an actual configuration of particles. So, after the measurement interaction, the macroscopic pointer will show a unique and definite result, the one embodied by the configuration of particles that compose the pointer. In other words, it is the evolution of the particles that finally determines which one of the possible pointer states (described by the evolution of the wave function) has been realized during the measurement process.

We suppose, for example, that $\phi_1$ is the wave function corresponding to the physical state of the pointer "pointing to the left" and $\phi_2$ that of the pointer "pointing to the right": at the time $t = T$, if $Y \in supp(\phi_1)$, then the pointer points to the left, if $Y \in supp(\phi_2)$, then it points to the right. Since the two supports are (macroscopically) disjoint[19], i.e., $supp(\phi_1) \cap supp(\phi_2) \cong \emptyset$ , then the final result is unique and

---

[17]The Bohmian systems are always composed by a wave function and real particles, each of them having a definite position in space. We call *actual configuration* the configuration of particles described by their definite positions in space, and mathematically expressed by $Q = (q_1, q_2, ..., q_N)$.

[18]A *support* of a function is the region of its domain in which it is not zero valued.

[19]It is worth noting that the concept of a perfect *disjointness of supports* is an idealization: the support of a wave function is typically unbounded in configuration space. As a first approximation, we can say that two different supports are disjoint if they have negligible overlap in configuration space. More precisely, we will say that the supports of two different wave functions are (macroscopically) disjoint when their overlap is extremely small in the square norm over any (macroscopic) region.

the superposition disappears[20].

Suppose, for example, that, after the interaction between the system and the apparatus, $Y \in supp(\phi_1)$: in this case, the actual configuration of the particles that compose the apparatus will be so arranged in space to form a physical pointer pointing to the left. Moreover, because of the entanglement[21] between the system and the apparatus during the interaction, the actual configuration of the particles that compose the system will be in the support of $\psi_1$, that is, $X \in supp(\psi_1)$. In this case, we will say that $\psi_1$ is the *effective wave function* (EWF) of the system, i.e., the branch of the total superposition which contains and guides the particles of the system after the interaction, whereas $\psi_2$ is the *empty wave function*, which can be FAPP[22] ignored after the interaction.

Assuming the quantum equilibrium hypothesis and the condition of disjoint supports for any two different pointer states, it is easy to show that the probability distribution of the measurement outcomes is given according to the Born's rule. For example, in the case discussed above, we see that the probability to get the eigenvalue associated to the eigenfunction $\phi_1$ in a measurement is[23]:

$$P(Y(t=T) \in supp(\phi_1)) =$$

$$= \int_{\mathbb{R}^m} d^m x \int_{supp(\phi_1)} d^n y |\alpha \psi_1(x,T)\phi_1(y,T) + \beta \Psi_2(x,T)\phi_2(y,T)|^2 =$$

$$= \int_{\mathbb{R}^m} d^m x \int_{supp(\phi_1)} d^n y |\alpha \psi_1(x,T)\phi_1(y,T)|^2 +$$

$$+ \int_{\mathbb{R}^m} d^m x \int_{supp(\phi_1)} d^n y |\beta \psi_2(x,T)\phi_2(y,T)|^2 +$$

$$+ 2\operatorname{Re} \int_{\mathbb{R}^m} d^m x \int_{supp(\phi_1)} d^n y \; \alpha \beta^* \psi_1(x,T)\psi_2^*(x,T)\phi_1(y,T)\phi_2^*(y,T) \cong$$

$$\cong |\alpha|^2$$

which is in agreement with the Born's rule[24].

In the derivation we have used the quantum equilibrium hypothesis for the first equation and

$$\int_{supp(\phi_1)} d^n y |\phi_2(y,T)|^2 \cong 0$$

$$\int_{supp(\phi_1)} d^n y \; \phi_1(y,T)\phi_2(y,T) \cong 0$$

---

[20]The idea is that, since different macroscopic states of the pointer occupy different regions in 3D physical space, the wave functions describing these states will have disjoint supports in the 3N-D configuration space of the pointer.

[21]During the interaction, the dynamics of the particles of the system is strongly related with that of the particles of the apparatus, so that if $Y \in supp(\phi_{1(2)})$, then $X \in supp(\psi_{1(2)})$.

[22]For All Practical Purposes (acronym introduced by John Bell)

[23]We follow here the derivation presented in Dürr & Teufel (2009, sect. 9.1).

[24]A specular derivation can be done for the other possible outcome of the measurement: in this case we need to integrate in the support of $\phi_2$ and the final probability will be $|\beta|^2$.

because $supp(\phi_1) \cap supp(\phi_2) \cong \emptyset$.

The emergence of the effective wave function of the system $\psi_1(x,T)$ represents a first step in the transition from a holistic regime to a local one[25]: after the measurement, the initial superposition of the total system effectively collapses[26] in just one of the possible branches, which is described by a factorized state between an eigenfunction of the system and one of the apparatus, e.g., $\psi_1(x,T)\phi_1(y,T)$. Hence, the dynamics of the system is now decoupled from that of the apparatus: the further evolution of the particles of the system will be autonomous and independent from that of the particles of the apparatus (because now they belong to distinct and factorized wave functions). Moreover, interference with the empty wave function will result practically impossible, given the condition of disjoint supports for the wave functions of different pointer states.

We might say that the EWF describes a *local* dynamics for the system, since the particle evolution of the sub-system described by $\psi_1$ does not depend on the position of the particles of any external system. Whenever an EWF emerges, the holistic Bohmian non-locality seems, at least temporarily, turned off.

A simple example can help to visualize the situation. Let's consider a typical EPR set up: generally, changing some potentials on one wing of the system, say in the point A, will influence the trajectory of the particle on the other wing, say in the point B[27]. Nevertheless, if, as a consequence of a measurement, an effective wave function emerges (e.g., in the point B), then the trajectory of the particle on the B-side can be influenced only by potentials on its side (i.e., potentials which are connected with B by time-like intervals).

Of course, this is only a first step towards the classical world. The other important step is to show how classical trajectories can emerge starting from the Bohmian ones[28]. In section 3, we will discuss a technical problem arising for the Bohmian classical limit in bounded regions and we will see how decoherence can solve the problem. In section 4, we will briefly introduce decoherence and, finally, we will clarify the mathematical conditions for implementing it in the framework of BM.

## 3 Bohmian classical limit in bounded regions

In this section, we focus on a problem that arises from the dynamics of a Bohmian system in bounded regions[29]. The problem has been originally discussed in Allori

---

[25] With *holistic* I mean the quantum (Bohmian) non-locality, with *local* the classical non-locality. This terminology has been introduced by Esfeld *et alii*(2014)(forthcoming).

[26] In BM, there is never a real collapse of the wave function.

[27] We suppose that the points A and B are space-like separated.

[28] In the following, we will not face the problem of the emergence of classical trajectories in BM. For the interested reader: see, e.g., Rosaler (2014), for a *decoherent histories* approach to the Bohmian classical limit; Appleby (1999) and Sanz, Borondo (2004), for the analysis of specific models where the Bohmian trajectories, implemented in a regime of full decoherence, become classical.

[29] For the sake of clarity, the problem can also arise in unbounded regions: indeed, it is a consequence of a simple mathematical fact, so it is fundamentally independent from the nature (bounded or unbounded) of the space where the system moves in. Nevertheless, since it is more likely to happen in bounded regions than unbounded ones, then it seems more natural to set the

*et alii* (2002, sect. 8). However, for the sake of completeness, we will briefly restate it here.

We consider an infinite potential well of size $L$ in one dimension and a 1-particle Bohmian system in the center of the well. We suppose that the wave function of the system is a linear superposition of two wave packets with opposite momenta. In the classical limit model, the position $x$ of the system will be the center of mass of a macroscopic body whose classical motion we are searching for.

At the initial time $t = 0$, we suppose that the two packets begin to move classically in opposite directions[30]. At the time $t_R$, they (approximately)[31] reach the walls and, for $t > t_R$, they start to converge towards the center. At the time $t_c = 2t_R$ (*first caustic time*), the two wave packets will cross each other in the middle of the well, but, since the Bohmian trajectories of a system cannot cross[32] in the configuration space of the system[33], the two converging trajectories will not cross each other: the trajectory coming from the right-hand side will start to come back to that side after the time $t_c$. In a perfectly symmetric way, the same will happen for the trajectory coming from the left-hand side of the well. So, for example, if the particle is contained, at the beginning, into the wave packet that goes to the right, then it will move in the future only within the right-half part of the well. And this is clearly not a classical behavior[34].

Nevertheless, Allori *et alii* (2002) claim that, in a realistic model, we also need to take into account the interaction with the environment and the problem should vanish. Indeed, an external particle (a neutrino, a photon, an air molecule,...), interacting with the (macroscopic) system before the caustic time $t_c$, will "measure" the actual position of the center of mass of the system, thus eliminating the superposition between the two wave packets of the system. In other words, the interaction between the external particle and the system acts like a position measurement on the system performed by the "environment". Consequently, the environmental interaction will select only one of the two wave packets of the system, which becomes the *effective wave function* of the system.

---

problem in a bounded region.

[30] We suppose to start with classical trajectories for each branch of the wave function, which is equivalent to assume a classical limit in unbounded regions. On this regard, some partial successful result has been achieved so far (I briefly indicate the main approach adopted by the authors for each reference): Allori *et alii* (2002): quantum potential *plus* Ehrenfest's theorem; Holland (1993, ch. 6): quantum potential; Bowman (2005): mixed states *plus* narrow wave packets *plus* decoherence; Sanz & Borondo (2004) and Appleby (1999): decoherence; Rosaler (2014): decoherent histories.

[31] The velocity field in BM is never bi-valued, so the particle arrives very close to the well, but without touching it

[32] Bohmian trajectories cannot cross in configuration space because the guiding equation is a first-order equation, so to each position $x$ corresponds a unique velocity vector $v$.

[33] For a 1-particle system, the configuration space of the system corresponds to the 3D physical space.

[34] Note that this situation is completely different from the case of the "surrealistic trajectories" in BM. In the latter, it is after all not so problematic having odd trajectories, if they finally match with the empirical predictions of QM. In this case, instead, we want to recover the classical dynamics of a macroscopic body, so the empirical predictions to match with are the trajectories of classical mechanics. Thus, every non-classical trajectory of the system cannot match with the empirical result we expect from a classical limit model.

Here the original passage:

> These interactions –even for very small interaction energy– should produce *entanglement* between the center of mass $x$ of the system and the other degrees of freedom $y$, so that their effective role is that of "measuring" the position $X$ and suppressing superpositions of spatially separated wave functions. (Taking these interactions into account is what people nowadays call decoherence [...]). Referring to the above example, the effect of the environment should be to select [...] one of the two packets on a time scale much shorter than the first caustic time $t_c$. (Allori et alii, 2002, sect. 8, p. 12)

The solution proposed by Allori et alii (2002) raises a subtle conceptual issue. As we saw in section 2.2, an EWF emerges in a Bohmian measurement only if the supports of different pointer states are disjoint in configuration space. When the pointer state is a macroscopic state of a classical apparatus, this condition is generally fulfilled. Nevertheless, in the case of the interaction with the environment, the pointer states of the "apparatus" are the *environmental states* of the external particle. Therefore, this solution seems to work only if the supports of different environmental states of the external particle, after the interaction with the macroscopic system, are disjoint in configuration space. So, the question becomes: *is this condition generally satisfied or not*[35]? Indeed, in order to have *effective decoherence*[36] in BM, the condition of disjoint supports for different environmental pointer states has to be satisfied.

It is important to note that this is a stronger condition than the usual one required by decoherence in the standard framework, that is, orthogonality of states.

In the next section, we will analyze a simple but realistic model of decoherence, namely environmental decoherence induced by scattering. The analysis will clarify the difference between the standard condition and the Bohmian one required to have decoherence.

## 4 Decoherence approach to the Bohmian classical limit

### 4.1 A short introduction to decoherence

Decoherence is the local suppression of the phase relations between different states of a quantum system, produced by the entanglement between the system and its environment[37], the latter also described as a quantum system.

We consider a pure state system $|\psi\rangle = \alpha |\psi_1\rangle + \beta |\psi_2\rangle$ and a pure state environmental system $|\xi\rangle$: as long as they do not interact, they remain physically independent and the total wave function is factorized:

$$|\Psi_0\rangle = |\psi\rangle |\xi\rangle = (\alpha |\psi_1\rangle + \beta |\psi_2\rangle) |\xi\rangle$$

---

[35] A related interesting question is: *what happens if the relative environmental states do not have disjoint supports, but they are only (approximately) orthogonal in the Hilbert space of the environment?* At the moment, we have not a rigorous answer to that question.

[36] With *effective decoherence*, we mean a decoherence process, within the framework of BM, which is able to produce an effective wave function for the system.

[37] In general, the environment can be tought either as external or internal degrees of freedom of a (macroscopic) system.

The density operator of the total system can be also factorized into the density operator of the system and that one of the environment:

$$\hat{\rho}^{\Psi_0} = |\Psi_0\rangle \langle \Psi_0| = |\psi\rangle |\xi\rangle \langle \xi| \langle \psi| = \hat{\rho}^\psi \otimes \hat{\rho}^\xi$$

When the system interacts with the environment, the two systems become entangled and they form a new pure state system:

$$|\Psi\rangle = \alpha |\psi_1\rangle |\xi_1\rangle + \beta |\psi_2\rangle |\xi_2\rangle$$

In a realistic physical model, the system will interact (and, then, become entangled) with many environmental states $|\xi_i\rangle$[38] in a very short time. Tracing out the degrees of freedom of the environment, we obtain the reduced density operator of the system. Under the assumption of (approximate) orthogonality of the environmental states, which is essentially the standard condition for decoherence, the reduced density operator formally appears as (approximately) describing a mixture of states:

$$\hat{\rho}^\psi_{red} = Tr_{\xi_i} |\Psi\rangle \langle \Psi| \cong |\alpha|^2 |\psi_1\rangle \langle \psi_1| + |\beta|^2 |\psi_2\rangle \langle \psi_2| \text{ if } \langle \xi_i|\xi_j\rangle \cong \delta_{ij}$$

Nevertheless, it is worth noting that $\hat{\rho}^\psi_{red}$ does not represent a proper mixture of states[39], but an improper mixture, for three main reasons:

1. In SQM, the physical state of a system is mathematically represented by the state vector of the system: in this case, the state vector is assigned only to the global entangled state between the system and the environment, and we cannot assign an individual quantum state to a subsystem ($\psi$) of a larger entangled system ($\Psi$).

2. In SQM, the reduced density operator just describes the statistical distribution of the possible outcomes for an observer who locally performs a measurement on the system. So, it does not carry information about the physical state of the (sub)system *per se*, but only related to the measurements we can perform on it.

3. Decoherence does not select one particular branch of the superposition. All the different branches remain equally real after the action of decoherence: thus, even if the final state of the system looks like a mixture, this is not a proper mixture that can be interpreted in terms of ignorance about the actual state of the system. We might call it an *improper mixture* (see, e.g., Bacciagaluppi (2011, sect. 2.2)).

---

[38] A good approximation for 'many' is the Avogadro number $N_A = 6,022 \times 10^{23}$.
[39] A *proper mixture* of states is an epistemic mixture: the system is in one of the states of the superposition, but we do not know which one of them. An *improper mixture*, instead, is a mathematical expression that looks like a proper mixture, yet it describes an ontological superposition of states (See, e.g., Schlosshauer (2007, sect. 2.4)).

## 4.2 Environmental decoherence induced by scattering

Taking decoherence as realistic background for the classical limit, we firstly introduce the model of environmental decoherence by scattering[40], and, after, we consider if the Bohmian condition of disjoint supports could reasonably fit into the model. As for the mathematical presentation of the model, we will mainly follow Schlosshauer (2007, ch. 3).

We consider a system $S$ that scatters off an external environmental particle, represented by $\xi$. At the initial time $t = 0$, $S$ and $\xi$ are uncorrelated:

$$\hat{\rho}_{S\xi}(0) = \hat{\rho}_S(0) \otimes \hat{\rho}_\xi(0)$$

Representing with $|x\rangle$ the initial state of the center of mass of the system, with $|\chi_i\rangle$ that of the incoming environmental particle and with $\hat{S}$ the scattering operator, we can represent the effect of the scattering of a single environmental particle on the system as follows:

$$|x\rangle |\chi_i\rangle \to \hat{S} |x\rangle |\chi_i\rangle \equiv |x\rangle \hat{S}_x |\chi_i\rangle \equiv |x\rangle |\chi(x)\rangle$$

where $|\chi(x)\rangle$ is the final state of the outgoing environmental particle scattered at $x$ on the system.

From the expression above, we see that if the system is represented by a superposition of different position eigenstates, for example $|x\rangle = \sum_i a_i |x_i\rangle$, then the environmental state and the system state will become entangled: the scattering process is a measurement-like interaction, which establishes correlations between the two systems. The environmental states that scattered off the system can be considered as pointer states which encode information about the position $x$ of the system. The scattering process transforms the initial density operator[41] of the composite system:

$$\hat{\rho}_{S\xi}(0) = \hat{\rho}_S(0) \otimes \hat{\rho}_\xi(0) = \int dx \int dx' \rho_S(x, x', 0) |x\rangle \langle x'| \otimes |\chi_i\rangle \langle \chi_i|$$

into the new density operator:

$$\hat{\rho}_{S\xi} = \int dx \int dx' \rho_S(x, x', 0) |x\rangle \langle x'| \otimes |\chi(x)\rangle \langle \chi(x')|$$

Thus, the reduced density operator of the system after the interaction of a single scattering of an external particle on the system is:

$$\hat{\rho}_S = Tr_\xi \hat{\rho}_{S\xi} = \int dx \int dx' \rho_S(x, x', 0) |x\rangle \langle x'| \langle \chi(x')|\chi(x)\rangle$$

---

[40] The model was originally developed by Joos & Zeh (1985). Recent accounts of the model can be found in Giulini, Joos et alii (2003, ch. 3) and Schlosshauer (2007, ch. 3).

[41] In the following, $\hat{\rho}$ and $\rho$ represents, respectively, the density operator and the density matrix of a system. In general, the density matrix is the density operator expressed in a particular basis, usually in the position basis (like in this case).

Representing the result in the (position basis) density matrix, the evolution of the reduced density matrix of the system under the action of the scattering event can be finally summarized as follows:

$$\rho_S(x, x', 0) \xrightarrow{scattering} \rho_S(x, x', 0) \langle \chi(x') | \chi(x) \rangle$$

This is an important result: in the SQM model of decoherence induced by scattering, the condition for the local suppression of the spatial coherence of the system is given by the orthogonality of the relative environmental states that scattered off the system:

*Standard condition for decoherence*: $\langle \chi(x') | \chi(x) \rangle \cong 0$

In a Bohmian model, this condition is not sufficient to have effective decoherence. Indeed, during the scattering process, the environmental state (the external particle) becomes entangled with the system (a macroscopic body, in the classical limit), thus acting like a pointer that measures the position of the center of mass of the system. Nevertheless, as we saw in section 2.2, a good measurement interaction[42] can be realized in BM only if the wave functions of different states of the pointer have disjoint supports in configuration space. Therefore, for obtaining a local suppression of the spatial coherence of the system, BM requires that the supports of relative environmental states have to be disjoint in configuration space. If $|y\rangle$ indicates a generic position eigenstate of the scattered environmental particle, and $\mathcal{Q}_\xi$ the configuration space of the environment, then the Bohmian condition to have effective decoherence induced by scattering is[43]:

*Bohmian condition for (effective) decoherence*: $\langle \chi(x') | y \rangle \langle y | \chi(x) \rangle \cong 0 \; ; \; \forall y \in \mathcal{Q}_\xi$

or, in terms of the wave function of the scattered environmental particle:

$$supp(\psi_{\chi(x)}(y)) \cap supp(\psi_{\chi(x')}(y)) \cong \emptyset \; ; \; \text{with } supp(\psi_\chi(y)) \in \mathcal{Q}_\xi$$

So, the following question arises: *is the condition of disjoint supports verified in a typical realistic model of environmental decoherence by scattering?*
In the case of a "classic" quantum measurement process[44], we have at least two main reasons to believe that the condition of disjoint supports is fulfilled:

---

[42] That is, a measurement providing a definite outcome.

[43] This result is not new: see, e.g., Rosaler (2014, sect. 5, eq. 20) and references therein. What we are aiming to clarify here is the strong connection between this result and the measurement process in BM as well as its conceptual consequences in the context of the classical limit in BM. Moreover, while Rosaler (2014, sect. 5) assumes that the Bohmian condition for decoherence is always satisfied (Rosaler's justification mainly relies on the high-dimensionality of $\mathcal{Q}_\xi$), we actually don't see any compelling reason for assuming the condition be satisfied for a typical model of environmental decoherence (e.g., in the short-wavelength limit, even a few external particles suffice to produce decoherence, so the high-dimensionality argument of $\mathcal{Q}_\xi$ does not hold in this case). We think, instead, that this issue might deserve a further analysis, even with the help of some quantitative results.

[44] That is, when the pointer states correspond to physical states of a classical apparatus.

1. A classical apparatus is made of an extremely high number of (Bohmian) particles, thereby the configuration space of the apparatus is very high dimensional (proportional to $10^{23}$D). This makes the probability of a significant overlap between the supports of two different macroscopic pointer states very small.
(*high dimensional configuration space*)

2. The wave function of a macroscopic system, like a classical apparatus, is usually very narrow. Moreover, since different macroscopic pointer states occupy different regions in 3D physical space, the wave functions representing these states will be reasonably defined over regions with disjoint supports in configuration space.
(*narrow wave function*)

Nevertheless, the situation changes dramatically when the apparatus is not a macroscopic object, but a microscopic environmental particle, the latter being either a photon, an electron, a neutrino, etc... Indeed, the assumptions mentioned above simply do not apply when the pointer state is a microscopic system:

1. The wave function of a microscopic system is generally not very narrow, and, moreover, it usually spreads out in configuration space in a very short time.
(*wave function spreads out*)

2. In some limiting cases, we can send just few particles that scatter off the system to produce decoherence effects (this is generally true, for example, in the short-wavelength limit[45]). In this case, the configuration space of the environment $\mathcal{Q}_\xi$ is not very high dimensional.
(*low dimensional configuration space*)

Since the traditional arguments[46] for the validity of the condition of disjoint supports do not apply when the measurement apparatus is a microscopic quantum system (like an environmental particle), and *prima facie* we do not have any strong argument for considering the condition satisfied, the question remains open and worth for a future work.

Some final (and more speculative) remarks on the conceptual consequences of the analysis of the conditions for Bohmian decoherence. We note that if the condition of disjoint supports is generally satisfied in a typical model of environmental decoherence, then decoherence fits very well in the framework of BM. Yet, BM could account for the selection of just one trajectory within the branching linear structure

---

[45] See, e.g., Schlosshauer (2007, sect. 3.3.1) and Joos *et alii*(2003, sect. 3.2.1.1).

[46] See, e.g., Dürr & Teufel (2009, sect. 9.1). It is worth noting that in section 9.2 these authors generalize the quantum measurement process, by including the case in which the pointer is a microscopic system. They affirm that is precisely thanks to decoherence processes that an effective wave function is produced «more or less all the time, and more or less everywhere». We agree with them in considering entanglement and decoherence essential for the production of effective wave functions and for the emergence of a (classical non-) local world. Nevertheless, their arguments for the validity of the condition of disjoint supports in the case when the pointer is a microscopic system are pretty qualitative, so they cannot be viewed as a definitive answer on this problem.

produced by the Schrödinger evolution of open quantum systems without the need of a real collapse of the wave function at some stage of the process (SQM) or the introduction of many simultaneous non-detectable existing worlds (Everett, MWI). On the other hand, if the condition of disjoint supports is not generally satisfied in those models, then maybe it would be possible to find some regime in which BM gives different empirical predictions from SQM. Let's consider, for example, a decoherence model in which the condition of orthogonality of states is satisfied, whereas the condition of disjoint supports is not. Under this model, SQM and BM will predict different phenomena: according to SQM, we will obtain decoherence effects; according to BM, we will not. Suppose that we were able to realize an experimental set up that physically implement this model. Performing the experiment, we will hypothetically be able to distinguish whether SQM or BM is true, since the two theories provide different empirical predictions under the same model. Of course, things might be not simple for many reasons. First, we should write a mathematical model in which the condition of orthogonality of states and that one of disjoint supports come apart. Second, the model should be practically implementable into a real physical set up. In any case, what we find interesting is that, if the condition of disjoint supports is really necessary for implementing decoherence in BM, then the possibility is open to find (at least hypothetically) some physical regimes where the Bohmian empirical predictions are different from the SQM ones.

## 5 Conclusion

Decoherence theory is the standard framework to show how classical trajectories and classical states can emerge from the quantum world and it is a crucial ingredient in BM in order to recover the emergence of classical trajectories in bounded regions.

We showed that, in order to implement an *effective decoherence* in BM, i.e., a physical mechanism which gives rise to an effective wave function for a Bohmian system through the interaction with the environment, a condition stronger than the standard orthogonality of states is required: the supports of relative environmental states have to be disjoint in the configuration space of the environment.

Thus, a relevant open issue for recovering the classical limit in BM is to verify whether this condition is satisfied for typical realistic models of environmental decoherence.

## Acknowledgments

A special thanks to Guido Bacciagaluppi and Antonio Vassallo for their helpful comments on earlier drafts of the paper, and for a continuous exchange of ideas about this topic. Their suggestions improved very much my original work.

I am also indebted with the philosophy of physics group in Lausanne: Michael Esfeld, Vincent Lam, Matthias Egg, Andrea Oldofredi and Mario Hubert, for helpful comments and discussions.

This work has been supported by the Swiss National Science Foundation through the research project "*The metaphysics of physics: natural philosophy*".

# BIBLIOGRAPHY

[1] V. Allori, D. Dürr, S. Goldstein, N. Zanghì (2002), *Seven steps towards the classical world*, Journal of Optics B4.
[2] D. M. Appleby (1999), *Bohmian trajectories post-decoherence*, Foundations of Physics 29.
[3] G. Bacciagaluppi (2011), *Measurement and classical regime in quantum mechanics*, in R. Batterman (ed.): *The Oxford Handbook of Philosophy of Physics*, Oxford University Press, 2013.
[4] L. E. Ballentine, *Quantum Mechanics: A Moden Development*, World Scientific, 1998.
[5] L. E. Ballentine, Y. Yang, J. P. Zibin (1994), *Inadequacy of Ehrenfest's theorem to characterize the classical regime*, Physical Review A, vol. 50 (4)
[6] L. E. Ballentine (1996), *The emergence of classical properties from quantum mechanics*, in R. Clifton (ed.): *Perspectives on Quantum Reality*, Kluwer, 1996.
[7] D. Bohm, B. Hiley (1987), *An ontological basis for the quantum theory*, Physics Report, vol. 144 (6)
[8] G. E. Bowman (2005), *On the classical limit in Bohm's theory*, Foundations of Physics, vol. 35 (4)
[9] D. Dürr, S. Goldstein, R. Tumulka, N. Zanghì (2004), *Bohmian mechanics and quantum field theory*, Physical Review Letters 93.
[10] D. Dürr, S. Goldstein, N. Zanghì (1992), *Quantum equilibrium and the origin of absolute uncertainty*, Journal of Statistical Physics 67.
[11] D. Dürr, S. Goldstein, N. Zanghì, *Quantum Physics Without Quantum Philosophy*, Springer, 2013.
[12] D. Dürr, S. Teufel, *Bohmian Mechanics. The Physics and Mathematics of Quantum Theory*, Springer, 2009.
[13] M. Esfeld, D. Lazarovici, M. Hubert, D. Dürr (2014), *The ontology of Bohmian mechanics*, The British Journal for the Philosophy of Science 65.
[14] M. Esfeld, D. Lazarovici, V. Lam, M. Hubert (forthcoming), *The physics and metaphysics of primitive stuff*, forthcoming in The British Journal for the Philosophy of Science.
[15] D. Giulini, E. Joos, C. Kiefer, J. Kupsch, I. Stamatescu, H. Zeh, *Decoherence and the Appearance of a Classical World in Quantum Theory*, Springer, 2003 (II edition).
[16] S. Goldstein, N. Zanghì (2012), *Reality and the role of the wave function in quantum theory*, in D. Albert and A. Ney (eds.): *The Wave Function: Essays in the Metaphysics of Quantum Mechanics*, Oxford University Press, 2012.
[17] E. Joos, H. Zeh (1985), *The emergence of classical properties through interaction with the environment*, Zeitschrift für Physik B – Condensed Matter 59.
[18] P. Holland, *The Quantum Theory of Motion. An Account of the de Broglie-Bohm Causal Interpretation of Quantum Mechanics*, Cambridge University Press, 1993.
[19] D. Malament (1996), *In defense of dogma: why there cannot be a relativistic quantum mechanics of (localizable) particles*, in R. Clifton (ed.): *Perspectives on Quantum Reality*, Kluwer, 1996.
[20] E. Merzebacher, *Quantum Mechanics*, Wiley International, 1970 (II edition).
[21] J. Rosaler (2014), *Is the de Broglie-Bohm theory specially equipped to recover classical behavior?*, in Philosophy of Science Association 24th Biennal Meeting (Chicago, IL).
[22] A. Sanz, F. Borondo (2004), *A Bohmian view on quantum decoherence* (available online on ArXiv: quant-ph/0310096v3). A slightly revised version of the article has been published with the title: *A quantum trajectory description of decoherence* (2007), in European Physical Journal D44 (319).
[23] J. Sakurai, *Modern Quantum Mechanics*, Addison Wesley, 1994 (revised version).
[24] M. Schlosshauer, *Decoherence and the Quantum-to-Classical Transition*, Springer, 2007.
[25] R. Shankar, *Principles of Quantum Mechanics*, Springer, 1994 (II edition).
[26] A. Valentini (1991), *Signal-locality, uncertainty and the sub-quantum H-theorem I*, Physics Letters A, vol. 156 (5).
[27] A. Valentini, *On the Pilot-Wave Theory of Classical, Quantum and Subquantum Physics*, PhD dissertation, SISSA, Trieste, 1992.
[28] D. Wallace, *The Emergent Multiverse. Quantum Theory According to the Everett Interpretation*, Oxford University Press, 2012.
[29] W. H. Zurek (1991), *Decoherence and the transition from quantum to classical*, Physics Today – revised version (2002) in Los Alamos Science 27.
[30] W. H. Zurek, (2003), *Decoherence, einselection, and the quantum origins of the classical*, Review of Modern Physics 75.

# Structural Realism and Algebraic Quantum Field Theory

Emanuele Rossanese

ABSTRACT. The main aim of my paper is to discuss a possible structural interpretation of algebraic quantum field theory (AQFT). I want also to discuss the most serious problem for this interpretation in the context of AQFT, namely the existence of several (unitary) inequivalent representations of the local algebras of observables for the same physical system.

## 1 Introduction

According to a *Received View*, the problems of the particle and field interpretations seem to suggest that one has to look elsewhere in order to find the correct interpretation of algebraic quantum field theory (AQFT). On the one hand, the particle interpretation of AQFT seems to be ruled out by three main arguments. A particle should be a *countable* and *localizable* entity. Moreover, we would like to have an ontology that does not depend on the choice of the frame of reference. However, these three requirements (that is, *countability*, *localizability* and *frame-invariance*) seem to be violated in the context of AQFT. Haag's theorem and the Reeh-Schlieder theorem respectively show that is impossible to define a *unique* total number operator for both free and interacting quantum field systems and that a local number operator is also not definable (see Earman and Fraser 2006 [11] and Reeh and Schlieder 1961 [17]). This undermines the *countability requirement*. Malament's theorem shows that under certain reasonable physical assumptions it is not possible to have a sharp localization of particles in any bounded region of space-time (see Malament 1996 [16]). Finally, the Unruh effect seems to show that the physical content of quantum field systems is observer- or context-dependent (technically speaking is *representation dependent*). In particular, an accelerated observer in a flat space-time would detect a thermal bath of particles exactly when the quantum field in which she moves is in a vacuum state and should be then devoid of particles (see Wald 1994 [20])[1] . Baker (2009) [1] has also showed that the Fock space formalism and the wavefunctional formalism that are naturally considered to respectively ground a particle and a field interpretation are *unitarily equivalent*. This means that the problems of the particle interpretation might undermine also

---

[1] Of course this is only a brief and sketchy presentation of the problems of the particle interpreation of AQFT and it only has the aim to give an idea of what is the Received View. A detailed discussion of these problems is impossible in the context of this paper. The interested reader can see Clifton and Halvorson (2001) [3] and (2002) [4] and Ruetsche (2012) [19].

the field interpretation. For instance, the Unruh effect would show that the physical description of the inertial and accelerated observer would differ also in a field interpretation of AQFT: a minimum level of energy in the first case and an excited state of the field in the second case. Baker then suggests that a structuralist interpretation might be the most correct interpretation of AQFT. This interpretation would in fact emerge *by elimination* from the other interpretative alternatives. In this paper I want to propose a structuralist interpretation of AQFT in order to avoid the above mentioned problems. The main idea is that in the search for the fundamental level, we should look at the level of the mathematical/physical structures that are the basis of the theory. In other words, mathematical/physical structures should be considered as the fundamental ontology of the theory. Moreover, these structures seem to be immune from the problems of the particle and field interpretations briefly described above. In fact, the notion of structure is in a sense weaker than the notions of particle and field, since it does not display all the features that we have mentioned concerning, say, the notion of particle (for instance, the *countability* and *localizability* do not apply to the notion of structure). It is true that also the notion of structure should be clearly defined and has its own problems as we shall see in the second part of the paper. However, if we consider the problems of the other interpretations, I think that a structuralist interpretation of AQFT is the best option avaliable. The paper is structured as follows. In the second section, I will illustrate the basic features of AQFT formalism. In the third section, I will discuss some structuralist interpretations of the theory. Finally, in a conclusive section, I will put forward my personal structuralist interpretation in terms of the superselection formalism and I will discuss some possible objections of this interpretation.

## 2   The Structure of Algebraic Quantum Field Theory

The first thing that we need to mention is that in AQFT the main objects of study are the *algebras of observables* rather than the observables themselves. A crucial concept that we have to define is the important notion of *net of algebras*. The fundamental idea of AQFT is that the physical content of a system described by AQFT is not encoded in an individual algebra of observables but rather in the *mapping* O → **A**(O) from regions O of Minkowski space-time to *algebras of local observables* **A**(O). Such a mapping determines which observables are localized and then take value in O. The physical information is contained in the *net structure* of algebras and not in the individual algebras. Another important notion here is that of *quasilocal algebra* that includes global limits of the local observables as, for example, the total charge observable. The *elements* of an algebra represent, roughly speaking, the physical *operations* that can be performed in a certain space-time region which is associated with that algebra. Given that only finite regions of space-time are considered, we have to work only with local observables and hence with their related *local algebras of observables*. The latter assumption is justified in order to implement the *principle of locality*: *measurements* in a given spatial region must not rely on any *measurement* taken in a different spatial region. According to the Haag-Kastler formulation, the net of local algebras has to satisfy four axioms that impose certain algebraic and physical conditions. These axioms are the following:

(1) *Isotony*: the mapping O → **A**(O) is an inductive system. This means that an observable measurable in the region of space-time O1 is a *fortiori* measurable also in a region of space-time O2 containing O1.

(2) *Microcausality*: if O1 and O2 are space-like separated space-time regions, then $[\mathbf{A}(O1), \mathbf{A}(O2)] = 0$. That is, all observables connected with a space-time region O1 are required to commute with all observables of another algebra which is associated with a space-like separated space-time region O2. This axiom is also called *Einstein causality*.

(3) *Translation covariance*: if **A** is a net of local algebras of observables on an affine space, it is assumed that there exists a faithful and continuous representation $x \to \alpha x$ of the translation group in the group of Aut**A** of automorphisms of **A** and $\alpha x(\mathbf{A}(O)) = \mathbf{A}(O + x)$, for any space-time region O and translation x.

(4) *Spectrum condition*: the support of the spectral measure of the operator associated with a translation is contained in the closed forward light-cone, for all transaltions. This ensure that negative energies cannot occur.

A final point that is important to mention is the appearance of many *(unitary) inequivalent representations* of the same algebra of observables. In the context of non-relativistic quantum mechanics there is a theorem, the Stone-von Neumann uniqueness theorem, which proves that the algebra generated by the canonical commutation relations (*CCRs*) for the position and momentum operators has a representation of these two set of operators in Hilbert space up to unitary equivalence. This means that the specification of the purely algebraic *CCRs* suffices to describe a certain physical system. However, the Stone-von Neumann theorem fails in the context of AQFT, where one has an infinite number of degrees of freedom. The theorem is in fact proved only for system with a finite number of degrees of freedom. In the specific context of AQFT, a *representation* is a *map* that associates every element of an abstract C*-algebra **A** (in which the theory is formulated) with the set of all bounded operators acting on an Hilbert space H. This representation has to be a C*-*homomorphism*, that is, it has to preserve the algebraic structure of the original C*-algebra[2].

The resulting Hilbert space H is then called the *representation space*. We must also consider the fundamental concept of *irreducible representation*: a representation is *irreducible* if the representation space H has no closed invariant subspaces.

---

[2] According to a very general definition, a C*-algebra **A** is a complex algebra of continuous (bounded) linear operators defined on a complex Hilbert space, with the following important proprieties:

(i) **A** is (topologically) closed in the norm topology of operators;

(ii) **A** is closed under the operation of taking adjoints of operators.

An irreducible representation is usually associated with an *elementary system*. One of the first results of AQFT is the acknowledgment of the emergence of many *(unitarily) inequivalent (irreducible) representations* of the same algebra of observables generated by the *CCRs*. This means that for any pair of these (unitary) inequivalent (irreducible) representations of the algebra of observables, there is no unitary operator that can transform one into the other. As we will see in the next section, the existence of (unitarily) inequivalent (irriducilbe) representations is the most serious problem for a structuralist interpretation of AQFT. Before we conclude this section, we must list three definitions that complete this very brief discussion of the basic ideas of AQFT. First, in this specific formalism local observables are defined as self-adjoint elements in local (non-commutative) von Neumann algebras. Second, the state of a physical system is defined as a positive, linear and normalized function that associates elements of the relevant local algebra of observables to real numbers. Finally, we have to introduce the notion of *GNS-representation*. Let $\omega$ be a state on a C*-algebra **A**. Then there exists a Hilbert space H$\omega$, a representation $\pi\omega$ : **A** $\to$ **B**(H$\omega$) of the algebra, and a cyclic vector $|\xi\omega\rangle \in$ H$\omega$, such that for all A $\in$ **A**, the expectation values that the state $\omega$ assigns to the algebraic operator A is duplicated by the expectation value that the vector $|\xi\omega\rangle$ assigns to the Hilbert space operator $\pi$(A). In symbols, $\omega(A) = \langle\xi\omega|\pi\omega(A)|\xi\omega\rangle$ for all A $\in$ **A**. The triple (H$\omega$, $\pi\omega$, $|\xi\omega\rangle$) is a cyclic representation because it contains a cyclic vector and it is called *GNS-representation*. It is unique up to unitarily equivalence. That is, if (H, $\pi$) is a representation of **A** containing a cyclic vector $|\psi\rangle$ such that $\omega(A) = \langle\psi|A|\psi\rangle$, then (H, $\pi$) and (H$\omega$, $\pi\omega$) are unitarily equivalent. A state $\omega$ on a C*-algebra **A** is pure if and only if its GNS-representation is irreducible; if its GNS-representation is reducible, the state is a mixed state.

## 3 Structural Realism and AQFT

The standard basic idea of a structuralist interpretation of a physical theory is that *structures* occupy the most fundamental ontological level of the theory. I will not discuss the debate between different formulations of structural realism. I assume that a general structural interpretation might be formulated and I focus on how this can be done in the specific context of AQFT. Different proposals exist of a structuralist interpretation of AQFT. Haag (1996) [13], for example, notes that the role of fields in AQFT is just a *convenient artefact*. Fields have just the role of *coordinating* the local algebras of observables. Moreover, the basic fields are only linked to the *charge structure* of the local algebra of observables and have no direct connection with some physically observable entity. Haag then claims that the physical content of AQFT is linked to *local operations* that are performed in a certain region of space-time. Accordingly, a field is just a structure that allows to associate an algebra of observable operators to a certain region of space-time. In this sense, a local field is conceived as a *local field operator* that represents only a *physical operation* that is performed in that space-time region. Thus, Haag concludes that the *net structure* of the local algebras of observable operators provides the most fundamental description of what is going on in the context of AQFT. Roberts (2011) [18] provides a more general analysis of the structural content of quantum theories.

He defends a *group structural realism* (GSR) that he defines as follow: The existing entities described by quantum theory are organized into a hierarchy, in which a particular symmetry group occupies the top, most fundamental position (see Roberts 2011 [18], p. 5). According to Roberts, this definition of structural realism allows to have a precise connection to the physical quantities that we actually observe and measure in the lab (see Roberts 2011 [18], p. 5). This last claim seems to be justifies by the *Wigner's legacy*. Wigner's analysis of elementary particles in terms of the classification of the irreducible representations of the Poincaré group is one of the first attempt to define physical objects by means of a symmetry group. Roberts then claims that Wigner's analysis can be considered as one of the first example of GSR. The main idea behind this interpretation of Wigner's analysis is that it has only proved that certain physical quantities (such as *mass*, *spin* and *parity*) can be identified through the *classification* of all the irreducible representations of the Poincaré group. Moreover, Wigner (1939) [21] does not explicitly link these results to the notion of particle. In fact, Wigner only claims that it is possible to correlate the values of physical magnitudes (that is, *quantum numbers*) with certain parameter labelling group representations. Of course, there are not only space-time symmetries. For instance, Roberts mentions and briefly discusses the case of Gell-Mann's and Ne'eman's idea of considering the *SU(3)* as a symmetry group for the strong nuclear force in the context of particle physics[3]. This would in turn allow the definition of a new taxonomy of hadrons classified according to the irreducible representations of the *SU(3)* group. Roberts's GSR is interesting because it stresses the importance of the symmetries in physical theories and in quantum theories in particular. With respect AQFT, GSR might be extremely helpful in pointing out the role of the symmetries that are involved in the *superselection formalism*, as we shall see in the next section of the paper. Lam (2013) [7] proposes an explicit structuralist interpretation of AQFT. According to Lam, the fundamental entities of AQFT are space-time regions instantiating quantum field-theoretic properties. These fundamental entities form the basic structure of quantum field systems. Lam also notes that the Reeh-Schlieder theorem shows that the entanglement is perhaps the central aspect of AQFT and has some features that are absent in the context of the non-relativistic quantum mechanics. The Reeh-Schlieder theorem in fact entails that vacuum is actually entangled across many space-like separated regions and that this result is valid for all bounded energy states; also for non-interacting, free quantum fields. Moreover, Lam takes into exam the specific algebraic structure of the net of local algebras. Local algebras in AQFT are *type III von Neumann algebras*. This feature entails that any global sate is entangled across any diamond or double cone space-time regions and its causal complement. In more formal terms, there is no product state across any type III algebra $\mathbf{A} \subseteq \mathbf{B}(H)$ and its commutant $\mathbf{A'} \subseteq \mathbf{B}(H)$. Thus, a state in a type III algebra is *intrinsically mixed* in the sense that it cannot be represented as a density operator within the *GNS-representation* defined by any pure state. Such intrixically mixed states cannot be then understood as a probability distribution over pure states. In fact, there are no pure states for type-III von Neumann algebras. Since the fundamental entanglement of quantum

---

[3] See Robert (2011) [18] for details.

field systems is exemplified by the type III structure of the algebra, Lam therefore claims that we have to take C*-algebras as the *primitive* and *fundamental* structure of AQFT. Any other structure emerges from the algebraic structure. For instance, one can argue that the topological, differential and metrical space-time structures can be derived from those primitive algebras[4]. However, one can complain that this proposal is too abstract and does not allow to grasp the real physical content of AQFT. Moreover, as we shall see in the next section, if we want to specify a more concrete structure to be the fundamental ontological posit of the theory, this kind of revised proposal may have some problems concerning the role of inequivalent representations.

## 4 Inequivalent Representations and Superselection

French (2012) [12] suggests that AQFT is the most natural ground for a structuralist interpretation. However, French recognizes that there is a problem in this interpretation that concerns the role of the inequivalent representations of the algebra. As said, we may have several (unitary) inequivalent irreducible representations of the same algebra of observables. But if we want to adopt a structuralist interpretation of AQFT, we should be able to choose the representations that have a clear and definite physical meaning. If we are not able to do that, we might have a very abstract algebraic structure that is not *informative* of what is the physical content of the system represented by that algebra. French suggests that the *superselection formalism* might be a possible solution of this problem. This formalism allows to classify all the irreducible representations of the algebra of observables and then sort out all the representations that are not physical. In particular, French suggests to adopt the *DHR superselection theory*. Doplicher, Haag and Roberts (DHR) (1969a [6], 1969b [7], 1971 [8] and 1974 [9]) propose a *superselection criterion* according to which the *physical* representations are those that vary from the vacuum representation only locally. Thank to this criterion, DHR are able to provide an analysis of the superselection formalism in terms of equivalence classes of inequivalent representations corresponding to *charge superselection sectors*. They then prove the following three important results:

(1) It is possible to formulate a *composition law of charges* in terms of the tensor product of group representations;

(2) There is a form of *conjugation* in terms of the complex conjugation represen-

---

[4]Dieks (2001) [5] shows how to recover the space-time structure from the net of local algebras within the context of AQFT. Specifically, he suggests to start from states and operators and distill, Minkowski space-time from them. According to Dieks, one can consider the subalgebras of the total C*-algebra that are partially ordered by a $<$ relation. Then one has several overlapping sets of algebras and some of them can be identified with space-time points. This would create a topological space. The general idea is then to define a group of automorphisms on an index set of the net of algebras, and then interpret this group as the metric-characterising subgroup of the isometry group of the Minkowski space-time. In this sense, then, space and time properties are considered as *ordering parameters* of the fundamental algebraic structure of AQFT. See Dieks (2001) [5], pp. 237-238 for details.

tation;

(3) It is possible to assign a *sign* to each type of charge and this would lead to the well-known *fermionic* and *bosonic statistics*[5].

They conclude their analysis proving that a *GNS-representation* is isomorphic to an object of the *\*-category of localized transportable endomorphisms*. This category allows a product, that is, a tensor product obeying to the *permutation group*. This result seems to provide a natural representation of the permutation group in the symmetric \*-category (see Doplicher and Roberts 1990 [10])[6]. French contends that the DHR analysis proves that all the physical structures that we need in order to grasp the physical content of AQFT naturally arises from the net of algebras and their superselection formalism. I believe that French is on the right track and in the following part of the paper I want to develop a little bit further his idea. DHR propose the following superselection criterion:

## *DHR selection criterion*

The physical representations are the representations which become unitarily equivalent to the vacuum in restriction to the causal complement of any diamond by sufficiently large diameter, $\pi| \mathbf{A}(O') \cong \pi 0 | \mathbf{A}(O')$

This criterion entails that there is a vacuum-like appearances for all the states at very distant regions. In other terms, all the states defined over the algebra of observables are indistinguishable from the vacuum state at space-like distances. The first step is to assume a net of von Neumann algebras $\mathbf{A}(O)$ of local observables defined over the set K of all double cones or diamonds O in Minkowski space-time. It is also assumed that the net of von Neumann algebras $\mathbf{A}(O)$ satisfies the following conditions:

(1) Isotony;

(2) Relativistic covariance under the action of the Poincaré group;

(3) Local commutativity (i.e., locality).

DHR then suggest to apply their criterion to such an algebraic structure in order to select all those states and representations that are (only) locally distinguishable from the vacuum representation. In particular, one has to consider all the GNS-representations $\pi\omega$ corresponding to a certain state $\omega$ to be unitarily equivalent to

---

[5] If we shift to a 2- or 3-dimensional space-time, we need to substitute the permutation group with the braid group, which is a larger group and allow the existence of paraparticles. See Halvorson and Mueger (2007) [14].

[6] It is worth noting that the idea of this category-based structuralist interpretation of AQFT has the same conceptual framework than Roberts's GSR, for both take symmetry as the main criterion to select which physical entities need to be considered as invariants.

the vacuum representation $\omega 0 := \pi 0$ in the space-like distance. It is then possible to define the important notion of *local endomorphisms*. DHR first proves that a GNS-representation $(\pi 0, H0)$ corresponding to $\omega 0$ satisfies Haag's duality: $\pi 0 \ (\mathbf{A}(O'))'$ = $\pi 0 \ (\mathbf{A}(O))''$. This means that the DHR criterion is equivalent to the existence of local endomorphisms $\rho \in End(\mathbf{A})$ such that $\pi \omega = \pi 0 \cdot \rho$ is localized in some $O \in K$ in the sense of $\rho(A) = A$ for all $A \in \mathbf{A}(O')$ and $\rho$ has support in O, that is, supp($\rho$) is contained in O (i.e., supp($\rho$) $\subseteq$ O). It is also possible to define the notion of *transportability* of charges associated with the internal symmetry. For any translation a R4 , there exists $\rho a \in End(\mathbf{A})$ with support in O+a and $\rho \cong \rho a =$ Ad $(ua) \cdot \rho$ with unitary $ua \in \mathbf{A}$. It is then possible to denote $\Delta(O) := \rho \in End(\mathbf{A})$; $\rho$ is transortable and localizable in O as the category of all the transportable endomorphisms. Doplicher and Roberts (1990) [10] goes on to define a *DR-category* in the following way. A DR-category is a C*-tensor category [7] consisting of all objects $\rho \in \Delta = \text{UO}\in k \ \Delta(O)$ and with *morphisms* (or *arrows*) given by *intertwiners* T $\in$ **A** between $\rho, \sigma \in \Delta$ such that $T\rho(\mathbf{A}) = \sigma(\mathbf{A})T$. **T** has a permutation symmetry due to locality, and is closed under direct sums and sub-objects (due to the *property B* following from spectrum condition, locality and weak additivity[8] ). Doplicher and Roberts then proves an important theorem, called *DR reconstruction theorem*. This theorem shows that in spite of being an abstract category $\rho$ of local endomorphisms on the observable algebra **A**, the DR-category **T** determined by the DHR criterion, is isomorphic to the category **RepG** of irreducible group representations with a certain uniquely determined group G to be identified with the global gauge group. Then, the essential result of the DR reconstruction theorem is the following. Given the structure of **T** as a C*-tensor category having the *permutation symmetry*, *direct sums*, *sub-objects* and *conjugates*, it is possible to show that a DR-category assures the existence of a unique internal gauge symmetry group G[9] . Moreover, the DR-category fixes the existence of a unique *field algebra* **F**. One starts from the existence of a field algebra **F** of operators acting on a Hilbert space H and a gauge group G of unitary operators on H[10] . One supposes also that the Hilbert space H contains a vacuum state $\Omega$. It is then possible to consider the observable algebra **A** as the gauge invariants fields (acting on $\Omega$). As said, in the context of AQFT we have an abstract *reducible representation* $\pi$ of **A** on H, since the theory admits several inequivalent irreducible representations of the same algebra of observable. Thanks to the DHR analysis, it is then possible to consider as *physical* all the irreducible subrepresentations of $\pi$ that are superselected via the DHR criterion. In other terms, the physical representations are the superselection sectors that can be reached from the vacuum sector through the action of local (unobservable). These representation are of the form $\pi \omega = \pi 0 \cdot \rho$, where $\rho$ is an endomorphism from the

---

[7]See Doplicher and Roberts (1990) [10] for details.

[8]See Doplicher and Roberts (1990) [10] for details.

[9]The appearance of group structure here is due to the permutation symmetry encoded in **T** coming from the local commutativity in the four dimensional space-time. In the two dimensional case, the permutation symmetry is to be replaced by the *braid group symmetry*, as a consequence of which quantum group symmetry arises instead of the familiar group.

[10]It is possible to consider the gauge group as the representation of a fundamental symmetry group, such as *SU(2)*.

category $\Delta(O)^{11}$. Thus, it is interesting to note that such endomorphisms correspond to the representations that arise by acting on the vacuum representation with (unobservable) fields. This means that there exists a perfect correspondence between the algebraic structure and the physical structure described by AQFT. Moreover, the DR reconstruction theorem proves that the DHR categories are *dual* to compact gauge groups. This means that where there is a compact gauge group, there is also a DHR category and the other way round. In a sense then the analysis of DHR categories and of compact gauge groups is the same. DHR show that it is possible to recover all the properties of quantum fields from the analysis of superselection sectors. As said, they are able to recover the the following structures: (i) properties of quantum number (baryon number, lepton number, and the magnitude of generalized isospin); (ii) composition law and conjugation of charge; (iii) exchange symmetry of identical charges statistics. Interestingly then DHR also proves that charge quantum number structure is in a one-to-one correspondence to the labels of (equivalence class) of irreducible representation of a compact gauge group. Moreover, the composition law is represented by a tensor group of representations belonging to this group. The charge conjugation is represented by the complex conjugate representation. Finally, it is possible to assign a sign to each type of charge and this allows to describe the fermionic or bosonic nature of the particle system. However, DHR criterion cannot account for states with electric charge, since it is possible to determine the electric charge by measuring the total electric flux through an arbitrarily large sphere surrounding a particle, states with an electric charge can be discriminated from the vacuum in the causal complement of any bounded region. Such charges which can be measured at space-like infinity appear typically in gauge field theories, and to fix terminology we will call them gauge-charges. (Buchholz and Fredenhagen 1982 [2], pp. 1-2)[12] This problem is in fact due to the Gauss's law and to the fact that electric charge spread space-like at infinity due to Coulomb's law. For this reason, Buchholz and Fredenhagen (BF) (1982) [2] propose a different criterion where the diamond region O is replaced by an infinitely extended cone around some arbitrary chosen space-like direction they introduce then the notion of *topological charge*. The idea is to consider *almost local algebras* and *almost local operators* in order to have an account of non-localizable charges, as the electric charge. An almost local algebra is the set of all the elements which can be approximated by local observables in a diamond of radius r with an error decreasing in norm faster than any inverse power of r. Then, you can define a space-like plan, called *region C*. Starting with a ball around the origin with radius

---

[11] It is interesting to note that all the representations of **A** that are not physical according to the DHR analysis can be considered as a *surplus structure*.

[12] They continue: The example of the electric charge might suggest that gauge-charges are always connected with long-range forces and the appearance of massless particles. Indeed, this is generally true in Abelian gauge theories. But in non-Abelian gauge theories, such as quantum chromodynamics, the argument fails, and the work on lattice gauge theories indicates that states carrying a gauge-charge may well exist in the absence of massless particles. We shall also see from our general structural analysis that massive particle states might have weaker localization properties than normally assumed. This would admit gauge-charges even in purely massive theories. (Buchholz and Fredenhagen 1982 [2], p. 2) This means that the problems of the DHR analysis are not limited to the treatment of electric charged particles.

r, we draw a straight line from origin to infinity and, around the point at distance r' from the origin on this line, we take a ball with radius r + $\gamma$r' with $\gamma > 0$. C is the union of all these balls for $0 \leq r' \geq \infty$. We also denote by $\mathbf{A}c(C)$ the relative commutant of the algebra $\mathbf{A}(C)$. Now, let O be a diamond space-like separated from the origin a $\in$ R4 an arbitrary point. The region C = a + U$\lambda > 0$ $\lambda \cdot$ O with $0 \leq \lambda \geq \infty$ is called a space-like cone with apex a. It is then possible to define the BF selection criterion.

## *BF selection criterion*

The physical representations are the representations which become unitarily equivalent to the vacuum in restriction to the causal complement of any space-like cone, $\pi| \mathbf{A}(C') \cong \pi 0 | \mathbf{A}(C')$

Therefore, they allow as physically relevant representations all those representations that are unitarily equivalent to the vacuum representation with respect to the causal complement of any space-like cone as defined above. Physically speaking, this is justified by the fact that it is impossible to distinguish the states in the representations $\pi$ and $\pi 0$ by measurements in any region C', because in the region C one can always bring in particles from space-like infinity or remove them without changing the results of measurements in C'. It is then possible to construct a *composition of sectors, charge conjugation* and an *exchange symmetry analysis* also in the context of the BF analysis[13]. In a theory based on a Minkowski space-time, the results of such analysis are equivalent to those of the DHR analysis. What is now the moral of this section? Following French, I think that the superselection formalism provides a solution to the problem of inequivalent representations in the context of a structuralist interpretation of AQFT. In fact, the superselection formalism shows how one can always superselect the physical representations by sorting out all the abstract representations that do not represent certain minimal physical assumptions. Moreover, as said, the superselection formalism allows to derive all the physically interesting features of a quantum field system by considering only the algebraic structure of the theory and some minimal physical assumptions. The main aim of this paper is then to show that French's proposal is correct in order to achieve a clear structuralist interpretation of AQFT. However, my suggestion is to ground this proposal on BF analysis of the superselection sectors rather than on DHR, since the latter does not allow to consider electric charges, while the former seems to provide a formal framework broad enough to consider also those charges. One final remark concerns the fundamenal problem of any structuralist interpretation, namely the definite distinction between *merely formal structures* and *physical structures*. I believe that in the context of AQFT the superselection formalism does the job by identifying the class of physical structures among the set of all possible representations of the algebra of observalbes. We must consider as physical only those representation (that is, structures) that are picked up by the superselection formalism and it is remarkable that such a formalism allows also to derive all the

---

[13] See Buchholz and Fredenhagen (1982) [2] for details.

physical characteristics of a quantum field system by the study of the algebraic properties of the relevant algebraic structure.

## Acknowledgments

I would like to thank Angelo Cei, Mauro Dorato and Matteo Morganti for many helpful comments and discussions on structural realism and the ontology of AQFT. I would also like to thank the audience at the SILFS Triennal Conference 2014, who made very interesting questions and comments that helped me to improve my original work. Finally, I would like to thank an anonymous referee for her/his important suggestions and objections.

## BIBLIOGRAPHY

[1] Baker, D. (2009), Against Field Interpretations of Quantum Field Theory, in *British Journal for the Philosophy of Science*, 60, 585-609.
[2] Buchholz, D., and Fredenhagen, K. (1982), Gauss' Law and the Infraparticle Problem, in *Physics Letter B*, 174, 331-334.
[3] Clifton, R., and Halvorson, H. (2001), Are Rindler Quanta Real? Inequivalent Particle Concepts in Quantum Field Theory, in *British Journal for the Philosophy of Science*, 52 (3), 417-470.
[4] Clifton, R., and Halvorson, H. (2002), No Place for Particles in Relativistic Quantum Theories?, in *Philosophy of Science*, 69, 1-28.
[5] Dieks, D. (2001), Space-Time relationism in Newtonian and Relativistic Physics, in *International Studies in the Philsophy of Science*, 15 (1), 5-17.
[6] Doplicher, S., Haag, R. and Roberts, J.E. (1969a), Fields, Observables and Gauge Transformations. I, in *Communications in Mathematical Physics*, 13, 1-23.
[7] Doplicher, S., Haag, R. and Roberts, J.E. (1969b), Fields, Observables and Gauge Transformations. II, in *Communications in Mathematical Physics*, 15, 173-200.
[8] Doplicher, S., Haag, R. and Roberts, J.E. (1971), Local Observables and Particle Statistics I, in *Communications in Mathematical Physics*, 23, 199-230.
[9] Doplicher, S., Haag, R. and Roberts, J.E. (1974), Local Observables and Particle Statistics II, in *Communications in Mathematical Physics*, 35, 49-85.
[10] Doplicher, S. and Roberts, J.E. (1990), Why There Is a Field Algebra With a Compact Gauge Group Describing the Superselection Structure in Particle Physics, in *Communications in Mathematical Physics*, 131, 51-107.
[11] Earman, J., and Fraser, D. (2006), Haag's Theorem and its Implications for the Foundations of Quantum Field Theory, in *Erkenntnis*, 64, 305-344.
[12] French, S. (2012), Unitary Inequivalence as a Problem for Structural Realism, in *Studies in History and Philosophy of Modern Physics*, 43, 121-136.
[13] Haag, R. (1996), *Local Quantum Physics: Fields, Particles, Algebras*, 2nd edition, Berlin-Heidelberg-New York: Springer.
[14] Halvorson, H. and Mueger, M. (2007), Algebraic Quantum Field Theory, in Butterfield, J. and Earman, J., eds., *Handbook of the Philosophy of Physics, Part A*, Boston: Elsevier.
[15] Lam, V. (2013), The entanglement structure of quantum field systems, in *International Studies in the Philosophy of Science*, 27, 59-72.
[16] Malament, D. (1996), In Defense of Dogma: Why There Cannot Be a Relativistic Quantum Mechanics of (Localizable) Particles, in Clifton, R. (ed.) *Perspectives on Quantum Reality: Non-Relativistic, Relativistic, and Field-Theoretic*, in *The University of Western Ontario Series in Philosophy of Science*, Kluwer Academic Publishers,1-10.
[17] Reeh, H. and Schlieder, S. (1961), Bemerkungen zur Unitaraquivalenz von Lorenzinvarianten Feldern, in *Il Nuovo Cimento*, 22, 1051-1068.
[18] Roberts, B. (2011), Group Structural Realism, in *British Journal for the Philosophy of Science*, 62(1), 47-69.
[19] Ruetsche, L. (2012), *Interpreting Quantum Theories: The Art of the Possible*, Oxford: Oxford University Press.
[20] Wald, R.M. (1994), *Quantum Field Theory in Curved Spacetime and Black Hole Thermodynamics*, Chicago: University of Chicago Press.

[21] Wigner, E.P. (1939), On Unitary Representations of the Inhomogeneous Lorentz Group, in *Annals of Mathematics*, 40 (1), 149-204.

# Symmetries, Symmetry Breaking, Gauge Symmetries

Franco Strocchi

ABSTRACT. The concepts of symmetry, symmetry breaking and gauge symmetries are discussed, their operational meaning being displayed by the observables *and* the (physical) states. For infinitely extended systems the states fall into physically disjoint *phases* characterized by their behavior at infinity or boundary conditions, encoded in the ground state, which provide the cause of symmetry breaking without contradicting Curie Principle. Global gauge symmetries, not seen by the observables, are nevertheless displayed by detectable properties of the states (superselected quantum numbers and parastatistics). Local gauge symmetries are not seen also by the physical states; they appear only in non-positive representations of field algebras. Their role at the Lagrangian level is merely to ensure the validity on the physical states of local Gauss laws, obeyed by the currents which generate the corresponding global gauge symmetries; they are responsible for most distinctive physical properties of gauge quantum field theories. The topological invariants of a local gauge group define superselected quantum numbers, which account for the $\theta$ vacua.

# 1 Introduction

The concepts of symmetries, symmetry breaking and gauge symmetries, at the basis of recent developments in theoretical physics, have given rise to discussions from a philosophical point of view.[1] Critical issues are the meaning of spontaneous symmetry breaking (appearing in conflict with the Principle of Sufficient Reason) and the physical or operational meaning of gauge symmetries.

The aim of this talk is to offer a revisitation of the problems strictly in terms of operational considerations. The starting point (not always emphasized in the literature) is the realization that the description of a physical system involves both the *observables*, identified by the experimental apparatuses used for their measurements, *and* the states, which define the experimental expectations. Since the protocols of preparations of the states may not always be compatible, i.e. obtainable one from the other by physically realizable operations, the states fall into disjoint families, called *phases*, corresponding to incompatible realizations of the system. This is typically the case for infinitely extended systems, where different behaviors or boundary conditions of the states at space infinity identify disjoint phases due to the inevitable *localization* of any realizable operation.

This feature, which generically is not shared by finite dimensional systems, provides the explanation of the phenomenon of spontaneous symmetry breaking, since the boundary conditions at infinity encoded in the ground state represent the cause of the phenomenon in agreement with Curie principle.

The role of the states is also crucial for the physical meaning of gauge symmetries, which have been argued to be non-empirical because they are not seen by the observables. The fact that non-empirical constituents may characterize the theoretical description of subnuclear systems, as displayed by the extraordinary success of the standard model of elementary particle physics, has provoked philosophical discussion on their relevance (see [1]). For the discussion of this issue it is important to distinguish global (GGS) and local gauge symmetries (LGS).

The empirical consequences of the first is displayed by the properties of the states, since invariant polynomials of the gauge generators define elements of the center of the algebra of observables $\mathcal{A}$, whose joint spectrum labels the representations of $\mathcal{A}$ defining *superselected quantum numbers*; another empirical consequence of a global gauge group is the *parastatistics* obeyed by the states. Actually the existence of a gauge group can be inferred from such properties of the states.

At the quantum level, the group of local gauge transformations connected to the identity may be represented non-trivially only in unphysical non-positive representations of the field algebra and therefore they reduce to the identity not only on the observables, but also on the physical states.

From a *technical* point of view, a role of LGS is to identify (through the pointwise invariance under them) the *local* observable subalgebras of auxiliary field algebras (represented in non-positive representations). LGS also provide a useful recipe for writing down Lagrangians which automatically lead to the validity on the physical states of *local Gauss laws* (LGL), satisfied by the currents which generate the cor-

---

[1] An updated and comprehensive account may be found in [1].

responding GGS. Actually, LGL appear as the important physical counterpart of LGS representing the crucial distinctive features of Gauge Quantum Field Theories with respect to ordinary Quantum Field Theory (QFT).

A physical residue of LGS is also provided by their *local* topological invariants, which define elements of the center of the local algebras of observables, the spectrum of which label the inequivalent representations corresponding to the so-called $\theta$ vacua. The occurrence of such local topological invariants explains in particular the breaking of chiral symmetry in Quantum Chromodynamics (QCD), with no corresponding Goldstone bosons.

Finally, since only observables *and* states (*identified* by their expectations of the observables [2] [3]) are needed for a *complete* description of a physical system, and both have a deterministic evolution, the problem of violation of determinism in gauge theories looks rather an artificial issue from a physical and philosophical point of view.

## 2 Symmetries and symmetry breaking

For the clarification of the meaning and consequences of symmetries in physics, from the point of view of general philosophy, a few basic concepts are helpful.

Quite generally, *the description of a physical system* (not necessarily quantum!) is (operationally) given [2] [3] in terms of
1) the **observables**, i.e. the set of measurable quantities of the system, which characterize the system (and generate the so-called *algebra $\mathcal{A}$ of observables*)
2) their **time evolution**
3) the set $\Sigma$ of physical **states** $\omega$ of the system, operationally defined by protocols of preparations and characterized by their expectations of the observables $\{\omega(A), A \in \mathcal{A}\}$.

Operationally, an observable $A$ is identified by the actual experimental apparatus which is used for its measurement, (two apparatuses being *identified* if they yield the same expectations on all the states of the system)

The first relevant point is the *compatible realization of* two different *states*, meaning that they are obtainable one from the other by *physically realizable operations*. This defines a partition of the states into physically disjoint sets, briefly called **phases**, with the physical meaning of describing disjoint realizations of the system, like disjoint thermodynamical phases, disjoint worlds or universes.

For infinitely extended systems, in addition to the condition of *finite energy*, a very strong physical constraint is that the physically realizable operations have inevitably some kind of *localization*, no action at space infinity being physically possible. Thus, for the characterization of the states of a phase $\Gamma$, a crucial role is played by their large distance behavior or by the boundary conditions at space infinity, since they cannot be changed by physically realizable operations. Typically, such a behavior at infinity of the states of a given phase $\Gamma$ is codified by the lowest energy state or ground state $\omega_0 \in \Gamma$, all other states of $\Gamma$ being describable as "localized" modifications of it. Thus, $\omega_0$ identifies $\Gamma$ and defines a corresponding (GNS) representation $\pi_\Gamma(\mathcal{A})$ of the observables in a Hilbert space $\mathcal{H}_\Gamma$, with the cyclic ground state vector $\Psi_0$.[2]

The simplest realization of **symmetries** is *as transformations of the observables commuting with time evolution*, operationally corresponding to the transformations of the experimental apparatuses which identify the observables (e.g. translations, rotations). This is more general than Wigner definition of *symmetries as transformations of the states which leave the transition probabilities invariant* (adapted to the case of the unique Schroedinger phase of atomic systems).

Actually, the disentanglement of symmetry transformations of the observables (briefly **algebraic symmetries**) from those of the states (**Wigner symmetries**), is the crucial revolutionary step at the basis of the concept of spontaneous symmetry breaking, which comes into play when there is more than one phase.

An algebraic symmetry $\beta$ defines also a symmetry of the states of a phase $\Gamma$ (i.e. a Wigner or **unbroken symmetry**) iff it may be represented by unitary operators $U_\beta$ in $\mathcal{H}_\Gamma$.

---

[2] This point is discussed for both classical and quantum systems in [4], [5].

An algebraic symmetry $\beta$ always defines a symmetry of the *whole* set of states $\Sigma$:

$$\omega \to \beta^*\omega \equiv \omega_\beta, \quad \omega_\beta(A) \equiv \omega(\beta^{-1}(A)), \quad \forall A \in \mathcal{A}, \tag{1}$$

but in general $\omega$ and $\omega_\beta$ need not belong to the same phase $\Gamma$, i.e. their preparation may not be compatible, so that the symmetry $\beta$ cannot be experimentally displayed in $\Gamma$ as invariance of transition probabilities, by means of physically compatible operations (**spontaneously broken symmetry**). Thus, the breaking of $\beta$ in $\Gamma$ is characterized by the existence of states $\omega \in \Gamma$ (typically the ground or vacuum state $\omega_0$) such that $\omega_\beta \notin \Gamma$.

The philosophical issue of symmetry breaking, also in connection with Curie principle, has been extensively debated often with misleading or wrong conclusions.

A widespread opinion is that symmetry breaking occurs whenever the ground state is not symmetric, but this is not correct for finite systems, for which (under general conditions) there is only one (pure) phase $\Gamma$, so that both $\omega_0$ and $\omega_{0\,\beta}$ belong to $\Gamma$ and $\beta$ is described by a unitary operator.

Thus, the finite dimensional (mechanical) models, widely used in the literature to illustrate spontaneous symmetry breaking, on the basis of the existence of non-symmetric ground states, are conceptually misleading.[3]

On the other hand, for a *pure phase* of an infinitely extended system, thanks to the uniqueness of the translationally invariant state (implied by the cluster property which characterizes pure phases), the non-invariance of the ground state $\omega_0 \in \Gamma$ *under an internal symmetry $\beta$* (i.e. commuting with space-time translations) implies that $\omega_{0\,\beta}$ cannot belong to $\Gamma$ and $\beta$ is broken in $\Gamma$. *Under these conditions*, the non-invariance of the ground state provides an explanation in agreement with Curie principle, identifying the cause in non-symmetric boundary conditions at infinity encoded in the ground state (see [4] pp. 23, 102). The philosophically deep loss of symmetry requires the existence of disjoint realizations of the system, which is related to its infinite extension.

The existence of an algebraic symmetry reflects on *empirical properties of the states* and may be inferred from them. In fact, an unbroken symmetry implies the validity of Ward identities, which codify the existence of conserved quantities and of selection rules satisfied by the states; for continuous symmetries the conservation laws hold even *locally* by the existence of current continuity equations implied by the first Noether theorem ([5], p.146-7). For a continuous symmetry group $G$ broken in $\Gamma$, even if the generators do not exist as operators in $\mathcal{H}_\Gamma$, the existence of a representation of $G$ at the algebraic level implies **symmetry breaking Ward identities** ([4], Chapter 15), which display corrections given by non-symmetric ground state expectations, called non-symmetric order parameters; an important empirical consequence is the existence of Goldstone bosons, for sufficiently "local" dynamics ([4], Chapters 15-17).

---

[3] The standard models are a particle in a double well or in a mexican hat potential (see also [6] [7]). The example of an elastic bar on top of which a compression force is applied, directed along its axis, exhibits a continuous family of symmetry breaking ground states, but spontaneous symmetry breaking occurs only in the limit of infinite extension of the bar; otherwise, both in the classical as well in the quantum case, there is no obstruction for reaching one ground state from any other.

## 3 Global gauge symmetries

For the debated issue of the empirical meaning of **global gauge symmetries** (GGS) (which by definition act trivially on the observables), a crucial (apparently overlooked) point is that a complete complete description of a physical system involves *both* its algebra of observables *and* the states or representations which describe its possible phases. In fact, *even if* there is no (non-trivial) transformation of the observables corresponding to GGS, GGS are strictly related to the existence of disjoint representations of the observable algebra and their empirical meaning is to provide a classification of them in terms of superselected quantum numbers [8]. This is clearly illustrated by the following examples.

**Example 1.** Consider a free massive fermion field $\psi$ transforming as the fundamental representation of an internal $U(2) = U(1) \otimes SU(2)$ symmetry with the algebra of observables defined by its pointwise invariance under $U(2)$. The existence of the (free) Hamiltonian selects the Fock representation in $\mathcal{H}_F$ for the field algebra $\mathcal{F}$ generated by $\psi$ and this implies the existence of the generator $N$ of $U(1)$ and of the Casimir invariant

$$T^2 \equiv \sum_{\alpha=1}^{3}(Q^\alpha)^2, \quad Q^\alpha \equiv \int d^3x\, \psi^*(\mathbf{x})T^\alpha \psi(\mathbf{x}), \tag{2}$$

with $T^\alpha$, $\alpha = 1,...3$, the representatives of the generators of $SU(2)$. $N$ and $T^2$ are invariant under the gauge group $U(2)$ and as such they (or better their exponentials $U_N(\alpha) = \exp i\alpha N$, $U_T(\beta) = \exp i\beta T^2$, $\alpha, \beta \in \mathbf{R}$) may be taken as elements of the **center $\mathcal{Z}$ of the observable algebra $\mathcal{A}$**. The eigenvalues $n \in \mathbf{N}$ of $N$ and $j(j+1)$ ($j \in \frac{1}{2}\mathbf{N}$) of $T^2$ label the representations of $\mathcal{A}$ in $\mathcal{H}_F$ and the fermion fields $\psi^*$, $\psi$ act as intertwiners between the inequivalent representations of $\mathcal{A}$, by increasing/decreasing the numbers $n$ and $j$.

Had we started by considering only the observable algebra $\mathcal{A}$, we would have found that its representations are labeled by the (superselected) quantum numbers $n$ and $j(j+1)$, corresponding to the spectrum of the central elements $U_N(\alpha)$, $U_T(\beta)$ and that the state vectors of the representations of $\mathcal{A}$ are obtained by applying intertwiners to the $n = 0$, $j = 0$ representation, consisting of the Fock vacuum.

We would then be led to consider a larger (gauge dependent) algebra $\mathcal{F}$ generated by the intertwiners, to interpret $n$ as the spectrum of the generator $N$ of a $U(1)$ group and to infer the existence of an $SU(2)$ group with $j(j+1)$ the eigenvalues of the associated $T^2$. Such a reconstructed $U(2)$ group acts non-trivially on the intertwiners, but trivially on the observables, namely is a global gauge group.

**Example 2.** A familiar physical system displaying the above structure is the quantum system of $N$ identical particles, even if in textbook presentations the relation between the gauge structure and the center of the observables is not emphasized.

The standard treatment introduces the (Weyl algebra $\mathcal{A}_W$ generated by the) canonical variables of $N$ particles and, by the very definition of indistinguishability, the observable algebra $\mathcal{A}$ is characterized by its pointwise invariance under the *non-abelian group $\mathcal{P}$ of permutations*, which is therefore a global gauge group.

As before, its role is that of providing a classification of the inequivalent representations of the observable algebra contained in the unique regular irreducible representation of $\mathcal{A}_W$, (equivalent to the standard Schroedinger representation) in the Hilbert space $\mathcal{H} = L^2(d^{3N}q)$, where $\mathcal{P}$ is unbroken. $\mathcal{H}$ decomposes into irreducible representation of the observable algebra, each being characterized by a Young tableaux, equivalently by the eigenvalues of the characters $\chi_i$, $i = 1,...m$.[9] For our purposes, the relevant point is that the characters are invariant functions of the permutations and, as such, may be considered as elements of the observable algebra, actually elements of its center $\mathcal{Z}$.

Thus, as before, the gauge group $\mathcal{P}$ provides elements of the center of the observables whose joint spectra label the representations of $\mathcal{A}$ defining superselected quantum numbers. Beyond the familiar one-dimensional representations (corresponding to bosons and fermions) there are higher dimensional representations, describing **parastatistics** (i.e. parabosons and parafermions).

Another empirical consequence of a global gauge group is the (*observable*) statistics obeyed by the states, a parastatistics of order $d$ arising as the result of an unbroken (compact) global gauge group acting on ordinary (auxiliary) bosons/fermions fields [10], [11]. In the model of Example 1, an observable consequence of the global gauge group $U(2)$ is that the corresponding particle states are parafermions of order two (meaning that not more than two particles may be in a state). The quarks have the properties of parafermions of order three as a consequence of the color group $SU(3)$ (historically this was one of its motivations).

In conclusion, contrary to the widespread opinion that the gauge symmetries are not empirical, the *global gauge symmetries are displayed by the properties of the states* (**superselected quantum numbers and parastatistics**) and actually can be inferred from them.[4]

It must be stressed that a global gauge symmetry emerges as an empirical property of a system by looking at the *whole set of its different realizations*; in a single factorial representation, the center of the observables is represented by a multiple of the identity and its physical meaning in terms of superselected quantum numbers is somewhat frozen. To reconstruct an operator of the center of $\mathcal{A}$ one must look to its *complete spectrum*, i.e. to *all* factorial representations of $\mathcal{A}$.

A continuous global gauge group becomes particularly hidden in those representations in which the exponentials of localized invariant polynomials of the generators converge to zero when the radius of the localization region goes to infinity. This corresponds to the case in which, in the conventional jargon, the **global gauge group is broken**.

In a representation $\mathcal{H}_\Gamma$ of the field algebra in which the (continuous) gauge group $G$ is broken, briefly called a $G$-broken representation, in contrast with the above examples, the charged fields do no longer intertwine between different representations of the observable algebra; in fact, they are obtainable as weak limits of gauge invariant fields in the Hilbert space $\mathcal{H}_\Gamma$ (*charge bleaching*) [12].

---

[4]The empirical meaning of the invariant functions of the generators of a global gauge group has been pointed out in [5], pp.153-8 and later resumed by Kosso and others; (see also [13], Chapter 7).

**Example 4.** The Bose-Einstein condensation is characterized by the breaking of a global $U(1)$ gauge group (acting on the Bose particle field as the $U(1)$ group of Example 1), as very clearly displayed by the free Bose gas.[5] The $U(1)$ breaking leads to the existence of **Goldstone modes**, the so-called Landau phonons, and the existence of such excitations may in turn indicate the presence of a broken $U(1)$ symmetry.

Finally, the gauge group is also reflected in the *counting of the states*. In $G$-unbroken representations of $\mathcal{A}$, to each irreducible representation of $G$ contained in the field algebra $\mathcal{F}$, there corresponds a single physical state, whereas in the fully broken case to each $d$-dimensional irreducible representation in $\mathcal{F}$, there correspond $d$ different physical states [14] (for a handy account see [5], Part B, Section 2.6).

## 4 Local gauge symmetries

Traditionally, a *local gauge symmetry* group is introduced as an extension of the corresponding global group $G$ by allowing the group parameters to become $C^\infty$ functions of spacetime. It is however better to keep distinct the local gauge group $\mathcal{G}$ parametrized by strictly localized functions (technically of compact support) from the corresponding global one $G$, since the topology of the corresponding Lie algebras is very different and invariance under $\mathcal{G}$ does not imply invariance under $G$ (as displayed by the Dirac-Symanzik electron field, [13], p. 159).

Also from a physical point of view, the two groups are very different, since in *any* (positive) realization (of the system) the group of local gauge transformations connected with the identity is represented trivially, whereas the global gauge group displays its physical meaning through the properties of the states (see the above examples). For example, the $U(1)$ global gauge group is non-trivially represented in Quantum Electrodynamics (QED) by the existence of the charged states, whereas *the local $U(1)$ group reduces to the identity on the physical states* ([13], Section 3.2).

Therefore, the natural question is which is the empirical meaning, if any, of a local gauge symmetry (LGS) $\mathcal{G}$ in QFT. From a technical point of view, pointwise invariance under $\mathcal{G}$ may be used for selecting the *local subalgebra of observables*, from an auxiliary field algebra $\mathcal{F}$, locality (strictly related to causality [11]) not being implied by $G$ invariance (e.g. in QED $\bar\psi(x)\,\psi(y)$ is invariant under $G = U(1)$, but not under $\mathcal{G}$ and is not a *local* observable field).

A deeper insight on the physical counterpart of a LGS is provided by the second Noether theorem, according to which the invariance of the Lagrangian under a group of local gauge transformations $\mathcal{G}$ implies that the currents which generate the corresponding global group $G$ are the divergences of antisymmetric tensors

$$J_\mu^a(x) = \partial^\nu G_{\nu\mu}^a(x) \quad G_{\mu\nu}^a = -G_{\nu\mu}^a. \tag{3}$$

**(local Gauss law ).**

This is a very strong constraint on the physical consequences of $G$ (corresponding to the Maxwell equations in the abelian case). Actually, such a property seems to catch the essential consequence of local gauge symmetry, since $\mathcal{G}$ invariance of the

---
[5] For a simple account see [4], p. 106.

Lagrangian is destroyed by the gauge fixing, whereas the corresponding local Gauss laws (LGL) keep holding on the physical states, independently of the gauge fixing.[6]

Moreover, a LGL implies that $\mathcal{G}$ invariant *local* operators are also $G$ invariant. In the abelian case this implies the **superselection of the electric charge** ([13], Sect.5.3)

Thus, it is tempting to downgrade local gauge symmetry to a merely technical recipe for writing down Lagrangian functions, which automatically lead to LGL for the currents which generate the corresponding global gauge transformations. [7]

The physical relevance of a LGL is that it encodes a general property largely independent of the specific Lagrangian model and in fact, most of the peculiar (welcome) features of Gauge QFT, with respect to standard QFT, may be shown to be direct consequences of the validity of LGL (see [13], Chapter 7):

a) a LGL law implies that *states carrying* a (corresponding) *global gauge charge cannot be localized*; this means that the presence of a charge in the space time region $\mathcal{O}$ can be detected by measuring observables localized in the (spacelike) causal complement $\mathcal{O}'$; this represents a very strong departure from standard QFT, where "charges" in $O$ are not seen by the observables localized in $\mathcal{O}'$;

b) LGL provide direct explanations of the evasion of the Goldstone theorem by global gauge symmetry breaking (Higgs mechanism);

c) particles carrying a gauge charge (like the electron) cannot have a sharp mass (*infraparticle phenomenon*), so that they are *not Wigner particles*;

d) the non-locality of the "charged" fields, required by the Gauss law, opens the possibility of their failure to satisfying the cluster property with the possibility of a linearly raising potential, as displayed by the quark-antiquark interaction, otherwise precluded in standard QFT (where the cluster property follows from locality);

e) a local gauge group may have a non-trivial topology, displayed by components disconnected from the identity, and the corresponding *topological invariants* define elements of the center $\mathcal{Z}$ of the local algebra of observables $\mathcal{A}$; for Yang-Mills theories such elements $\mathcal{T}_n(\mathcal{O})$, localized in $\mathcal{O}$, are labeled by the winding number $n$ and define an abelian group ($\mathcal{T}_n(O)\mathcal{T}_m(O) = \mathcal{T}_{n+m}(O)$); their spectrum $\{e^{i2\pi n\theta}, \theta \in [0,\pi]\}$ labels the factorial representations of the local algebra of observables, the corresponding ground states being the $\theta$-*vacua*. They are unstable under the chiral transformations of the axial $U(1)_A$ and therefore chiral transformations are inevitably broken in *any* factorial representation of $\mathcal{A}$ without Goldstone bosons. Thus, the topology of $\mathcal{G}$ alone provides an explanation of chiral symmetry breaking in QCD, without recourse to the instanton semiclassical approximation ([13], Chap. 8).

In conclusion, LGS are not symmetries of nature in the sense that they reduce to the identity not only on the observables, but also on the states, possibly except for their local topological invariants. From the point of view of general philosophy,

---

[6] A gauge fixing which breaks the global group $G$ involves a symmetry breaking order parameter and it is consistent only if $G$ is broken (see [13], p. 178 and [15]).

[7] The fact that LGL represent the *distinctive physical property* of "local gauge theories" has been discussed and emphasized in [16], [5], p. 146-149, and later rediscovered, without quoting the above references.

they appear in Gauge QFT as merely technical devices to ensure the validity of local Gauss laws (through a mathematical path which uses an invariant Lagrangian *plus* a non-invariant gauge fixing).

By the same reasons, i.e. the realization that the observables and the physical states are the only quantities needed for the complete description of a physical system, the issue of violation of determinism in gauge theories does not deserve physical and philosophical attention, since the observables and the physical states have a deterministic time evolution.

## 5 Additional discussion required by the referee

The aim of the paper is to present logical (mathematically sound) arguments and critical discussion of ideas and proposals which were previously not sufficiently elaborated from a philosophical point of view; in particular the paper aim is to criticize misleading or wrong conclusions drawn from eminent philosophers of physics.

*Empirical meaning of symmetries*

For the discussion of the empirical meaning of symmetries it is important to take into account the basic result of (the first) Noether theorem, by which invariance (of the dynamics) under a continuous one-parameter group of transformations is equivalent to the existence of a conserved quantity; hence, the empirical meaning of a symmetry may be provided by the empirical realizations of the symmetry transformations (e.g. space translations, rotations etc.), *as well as* by the empirical meaning of the associated conserved quantity, which represents the generator of the symmetry. Thus, e.g. the empirical meaning of space translations may be argued by the actual operational realizability of such transformations (in terms of translating observable quantities), as well as by the empirical meaning of the (observable) conserved space momentum. Therefore, it is not appropriate to regard the second manifestation as of *indirect* empirical significance (as stated in [17]), since from an experimental point of view this is by far the more easy way for detecting the existence of a symmetry, as also argued by Morrison [18] : "Conservation laws provide the empirical component or manifestation of symmetries".

The peculiarity of a global gauge symmetry is that it cannot be realized as a group of transformations of the observables (being the identity on them), but nevertheless the associated conserved quantity may have an empirical significance in terms of empirical properties of the states, as it is clearly displayed in Quantum Electrodynamics (QED), where the generator of global gauge transformations describes the electric charge of the states, a very relevant conserved physical property. We therefore essentially adopt the following criterium for empirical significance, stated by Earman [19]: "What is objective or real in the world is described by the behavior of the values of genuine physical magnitudes of the theory", however with the crucial gloss that genuine physical quantities include *both* the observables *and* the states of the given physical systems.

In conclusion, a symmetry has an empirical significance if it is displayed by properties of the observables (e.g. by defining automorphisms of the algebra of observables) or of the physical states (e.g. by providing conserved quantum numbers which classify the states). It follows that global gauge symmetries are empirical, since their generators provide the conserved superselected quantum numbers which label the physical states, but generally local gauge symmetries are not. To my knowledge, the above relevant gloss has been missed in the discussions on the empirical significance of gauge symmetries, even in papers aiming to clarify the philosophical aspects of gauge symmetries [24].

*Empirical meaning of local gauge symmetries*

Practically the whole morning section of the meeting (during which the present paper was presented) was occupied by talks centered on the possible philosophical meaning of local gauge symmetries, dwelling on the philosophical meaning of invariance under local transformations which reduce to the identity on the observables. As argued in Section 4, this looks like a metaphysical issue and, as such, does not deserve scientific attention. The distinction between global and local gauge symmetries is crucial for the discussion of the empirical meaning of gauge symmetries, since only the first have a physical meaning whereas local gauge transformations do not.

To this purpose, I quote the final conclusion by Elena Castellani in her contribution "Symmetry and equivalence" in "Symmetries in Physics" (Ref.1): "Today we believe that global gauge symmetries are unnatural...We now suspect that all fundamental symmetries are local gauge symmetries". In the same book, in the conclusion of his contribution "The interpretation of gauge symmetry" M. Redhead writes "The Gauge Principle is generally regarded as the most fundamental cornerstone of modern theoretical physics. In my view its elucidation is the most pressing problem in current philosophy of physics".

For the discussion of this problem it is crucial to keep distinct the group of gauge transformations which differ from the identity only on compact bounded regions, henceforth called *local*, and the gauge group of *global* (i.e. independent from the point in space time) transformations; englobing both under the name of a local gauge group is, in my opinion, not convenient and likely misleading, because it hides the fact that they have a different status about empirical significance and, moreover, invariance under localized gauge transformations does not imply invariance under the corresponding global ones. Hence, as argued in my paper, the two groups should be taken neatly in separate boxes.

Then, the interesting question is what is the role of local gauge symmetries (equivalently of the Gauge Principle) in the constructions of models of elementary particles and the answer discussed in Section 4 is that they enter only as intermediate steps, doomed to lose any operational and philosophical meaning at the end (except for the related topological invariants, see below). Their merely intermediate role is to lead to the formulation of a dynamics characterized by the validity (on the physical states) of *local Gauss laws* obeyed by the currents which generate the corresponding global gauge symmetries. Such Gauss laws are not spoiled by the inevitable gauge fixing, needed for quantization (the proof of their validity on the physical states is not trivial in general [15], even if it is out of discussion in QED): they are detectable properties of the physical states and, as discussed in Section 4, they provide the physical and philosophical distinctive characterization of gauge quantum field theories.

This pattern is clearly displayed by Quantum Electrodynamics where (one may prove that): 1) the local gauge group reduces to the identity both on the observables as well on the physical states, i.e. does not have any empirical meaning, 2) on the other hand, the local Gauss law (somewhat related to the *intermediate* use of the non-empirical local gauge invariance) has an empirical significance, being one of the Maxwell equations, 3) the global gauge group has an empirical meaning, since its generator is the electric charge, whose corresponding quantum number is superselected.

The recognition that local Gauss laws are the characteristic features of gauge quantum field theories has been argued and stressed in view of quantum theories in [20] [16] [5] and later reproposed, without quoting the above references, by Karatas and Kowalski (1990) [21], Al-Kuwari and Taha (1990) [22], Brading and Brown (2000) [23]. Actually, such papers confine the discussion to the derivation of local Gauss laws from local gauge invariance (second Noether theorem at the *classical level, with no gauge fixing*), missing the crucial fact that at the quantum level local gauge invariance of the Lagrangian has to be broken by the gauge fixing and it is devoid of any empirical (and philosophical) significance, whereas the validity of local Gauss laws keeps being satisfied by the physical states, and it explains the interesting (revolutionary) properties of gauge theories (as explained in Section 4).

In contrast with global gauge symmetries, local gauge symmetries are only useful tricks used in *intermediate* steps (which use an auxiliary unphysical field algebra, initially a Lagrangian which has local gauge invariance, to be next broken by the gauge fixing, a redundant space of vector "states", only a subspace of which describes physical states, on which local gauge symmetries reduce to the identity). The final emerging picture is a description of the physical system characterized by conserved (actually superselected) quantum numbers, provided by the generators of the global gauge symmetry, and by the validity of local Gauss laws (no trace remaining of local gauge invariance).

In my opinion, from a philosophical point of view, one should invest in the meaning of local Gauss laws rather than on local gauge invariance (or on the so-called Gauge Principle).

*Determinism*

The issue of violation of determinism should not even be raised, being discussed with reference to equation of motions for gauge dependent variables which are deprived of objectivity and of reality, the objective description of a physical system involving only (the properties of) observables *and* physical states, whose time evolution is deterministic.

Quite generally, all what is needed for the complete description of a physical system is the determination of the time evolution of its observables and states, but for the solution of the related mathematical problem one may use tricks and auxiliary variables in intermediate steps for which there is no need of a physical (and philosophical) interpretation. Only the final goal and result is relevant and there is a plenty of examples of such a technical strategy in theoretical physics. Thus, in gauge theories it is technically convenient to introduce an auxiliary (gauge dependent) field algebra with well defined dynamics, i.e. such that the (mathematical) Cauchy problem for its time evolution is well posed (existence and uniqueness of solutions). To this purpose one has to introduce a gauge fixing in the Lagrangian, even if it is not necessary to completely fix the gauge; e.g. in QED the Cauchy problem has been proved to be well posed in the Feynman-Gupta-Bleuler gauge, in the temporal gauge, in the Lorentz gauge (all allowing a residual symmetry group of non-constant gauge transformations). The observables are characterized as the functions of such auxiliary fields which are invariant under local gauge symmetry and satisfy locality; this is the (merely) technical role of local gauge symmetry.

In quantum mechanics, once the Hamiltonian $H$ has been defined (as a self-adjoint operator) the time evolution is described by the unitary one-parameter group generated by $H$ and therefore the time evolution is automatically deterministic; thus, for field quantization only those field operator may be introduced which have a deterministic evolution. This is why the quantization of gauge theories requires the introduction of a gauge fixing such that the initial value problem of the (auxiliary) field algebra has a unique solution.

*Infinitely extended systems and SSB*

In order to be (spontaneously) broken, a symmetry, defined as an automorphism/transformation of the observables, must fail to be implementable by unitary operators acting on the states of a physical realization of the system (otherwise one has an unbroken, i.e. Wigner symmetry). This is possible only if there exist disjoint realizations of the system (with the meaning of disjoint phases or worlds) all described by the same algebra of observables with the same time evolution. The physical/empirical meaning of disjointness is that configurations or states of the system belonging to different phases cannot be prepared in the same laboratory, more generally their protocols of preparation are not compatible. In mathematical language this amounts to the impossibility of describing states of different phases by vectors of the same Hilbert space carrying an irreducible or factorial representation of the algebra of observables. SSB in one realization or phase is explained by, and actually equivalent to, the instability of the phase under the symmetry, by the reason that in order to empirically detect the existence of a symmetry one must be able to operationally compare the behavior of each given configuration with that of its transformed one.

For quantum systems described by a finite number of canonical variables (under general regularity conditions, by Stone-von Neumann theorem) there is only one phase and therefore no SSB, even if there are non-symmetric ground states, in contrast with the wrong conclusion drawn from classical finite dimensional models with non-symmetric ground states. This leaves open a possibility for systems described by an infinite number of canonical variables, in particular for infinitely extended systems (which require an infinite number of canonical variables).

Then, the next issue is the existence of disjoint phases for infinitely extended systems; in this case different behaviors or different boundary conditions at space infinity of configurations (or states) of the system imply that their preparations are not compatible, since the inevitable localization of any physically realizable operation (involved in passing from one preparation to another) precludes to change the behavior at infinity. Hence, generically infinitely extended systems exhibit more than one phase, characterized by the boundary conditions at infinity, which are generally encoded in the ground state of the given phase, see Proposition 6.3 of Ref.4) and SSB may occur.

In conclusion, the crucial ingredient for symmetry breaking is the existence of disjoint phases and this occurs for infinitely extended systems (though not exclusively).

*References*

One of the referee request was to comment on a list of papers dealing with overlapping subjects, qualifying the novelties (if any) with respect to them, (a task, which I will reluctantly try).

1) *Brading and Brown [17]*. As in all papers by philosophers of physics, which I know of, the discussion overlooks the important fact that an objective description of a physical system should exclusively be based on (the properties of the) observables *and* states and that the empirical significance of symmetries should be argued in such terms (e.g. automorphisms of the observables and/or conservation laws obeyed by the states, as explained above). The missing clear distinction of global versus local gauge symmetries precludes to immediately reach the conclusion about the empirical significance of the former and the *impossible* empirical significance of the latter. In fact, in that paper local symmetries are identified as those which depend on "arbitrary smooth functions of space and time"; the lack of any localization restriction implies that the so defined group of local symmetries contains the group of global symmetries as a subgroup, since, as every first year student in mathematics knows, the constant functions satisfy the smoothness condition (a tacitly assumed localizability would denote a lack of precision without which mathematics as well as logic do no longer exist).

Had Brading and Brown clearly understood the different status of the two groups and the general argument that local gauge symmetries reduce to the identity both on the observables as well as on the states, they might have reduced their paper to a few lines.

2) *Healy 2010 [25]*. The paper looks as a rather sketchy account of the common (heuristic) wisdom about $\theta$ vacua, ignoring the critical revisitation of such a subject, presented in [26] and later further discussed in Ref. [13]. In my opinion, this is not merely a question of mathematical physics precision, since it is very dangerous and certainly not satisfactory to ground a philosophical discussion on ideas, which may have a useful heuristic value, but have serious problems of mathematical and *logical* consistency.

The winding number $n$ defined in eq. (10), a crucial ingredient of the discussion, requires that $A_i(x)$ are continuous functions and therefore it looses any meaning for relativistic quantum fields, which have been proved to be singular "functions" of space points (technically operator valued tempered distributions). In fact, in order to give a possible meaning to such an equation the standard theoretical physics wisdom is to apply it to regular (euclidean) field configurations in the functional integral formulation (of quantum field theory), the so-called instantons. However, continuity is required and continuous euclidean configurations have zero functional measure (this problem is well known to the eminent theoretical physicists who contributed to this subject, like Coleman, Weinberg etc.). This consistency problem was solved in [26] in a way that has strong philosophical consequences; in fact, no reference is made to the topological structure of the (questionable) semiclassical instanton approximation (of the functional integral) and the proposed solution exclusively exploits the topological invariants of the (non-abelian) local gauge group. It is shown that such topological invariants define elements of the center of the local observable algebra and their spectrum (i.e. the $\theta$ angle) characterizes the $\theta$ vacua. From a general philosophical point of view, the conclusion is that even if the (group of) local gauge transformations connected with the identity reduce to the identity both on the observables as well as on the physical states, the topological invariants which classify the other components disconnected from the identity provide detectable superselected quantum numbers (the $\theta$ angles), which classify the physical states, just as the generators of a global gauge group do. In conclusion, *local gauge symmetries are not empirical except for their topology.*

The first sentence of the paper, with the abstract definition of a symmetry as "an automorphism-transformation that maps the elements of an object onto themselves so as to preserve the structure of that object" is too loose and imprecise. Which elements (observables? states?)? Which structure is preserved? This applies also to the subsequent attempt of formalization (A 1-1 mapping $\phi : S \to S$ of a set of situations...) which uses an undefined (vague) concept ("situations").

The merely intermediate role of local gauge symmetries for the validity of local Gauss laws has been missed.

At the end of Section 3. The last two statements are rather misleading. First, local gauge transformations, as well as the topological invariants provided by them, do not relate configurations associated to different vacua; rather the topological invariants define elements of the center of the observables which label (not relate!) the vacua. The author seems to overlook the crucial difference between the empirical significance of a symmetry displayed by transformations or relations (between observables or states) and the empirical significance displayed by the existence of conservation laws (as argued by Morrison). Similarly, the statement at the end of Section 4, that "a large gauge transformation represents a change from one physical situation to another" is conceptually wrong.

Towards the end of Section 5. The "generator" $\hat{U}$ of a large gauge transformation cannot be defined because the group of large gauge transformation is not continuously connected with the identity. What may be defined, as done in [26], are the elements $T_n$ of the quotient $G/G_0$ of the local gauge group $G$ with the local group $G_0$ of transformations connected with the identity (having zero winding number). Such a quotient is an abelian group, whose elements belong to the center of the local observable algebra and their spectrum (or eigenvalues) are the $\theta$ angles.

The paradox raised at the beginning of Section 6: "a global gauge transformation appears as a special case of a large gauge transformation" is a consequence of the improper choice of not distinguishing global and local gauge transformations (see above discussion).

3) *Struyve 2011 [27]*. The paper is confined to discussing classical field theories, which are known to have serious problems about their physical interpretation, in particular for elementary particles interactions; they may provide some heuristic mathematical information, but they *do not describe nature*, (with the possible exception of classical gravity, which however requires quantum effect for the description of black holes). The most objectionable point is the discussion of SSB in terms of small perturbations around a non-symmetric ground state. As discussed in Ref. 4, in classical field theory, the set of small perturbations around the ground state solution is not stable under time evolution and therefore it looses meaning with the passing of time. The set of "perturbations" of a ground state solution $\phi_0$, which are stable under time evolution are those which define a Hilbert sector or a phase, and are of the form $\phi_0 + \chi$, with $\chi \in H^1$, $\partial_t \chi \in L^2$ (the corresponding theorems are discussed in [4]; neither $\chi$ nor $\dot{\chi}$ remain small!). SSB cannot be identified with the instability under the symmetry of the set of small perturbations ("When considering small perturbations around a particular ground state, the equations of motions will not posses the symmetry of the fundamental equations of motion and one speaks of SSB.", at the beginning of Section 2.2.). The widespread cheap heuristic account/explanation of SSB in terms of small perturbations around a non-symmetric ground state is not (mathematically) correct (as discussed in [4]).

Last but not least, I do not see what the paper significantly add to the gauge invariant account for the Higgs mechanism, in the full quantum case, given by Frohlich-Morchio-Strocchi [14], which does not even appears in the references of Struyve paper.

4) *Smeenk 2006, [28]*. The paper is well written, but most of the general discussion of conceptual problems is not novel and largely taken from [4] [5].

The aim of the paper, stated in the Abstract and in the Introduction ("This article focuses on two problems related to the Higgs mechanism... what is the gauge invariant content of the Higgs phenomenon? and what does it means to break a local gauge symmetry?") is superseded by [14], quoted only at the very end, probably to comply a referee request. The logical and conceptual discussion of the problems of the Higgs mechanism, together with their solutions, already appeared in [5] and in the 2005 edition of [4], which are not even mentioned in the references. E.g. the discussion of SSB in Section 2 heavily relies on [4], in particular for SSB in classical theories, for the exclusion of SSB in finite-dimensional quantum systems by Stone-von Neumann theorem, for the role of the infinite extension for SSB in spin systems. The content of footnote 5 is somewhat misleading, since *both* in Statistical Mechanics (SM) *as well as* in Quantum field theory in order to witness SSB one must consider pure phases, i.e. ground state representations which satisfy the cluster property (this may require a decomposition of the representation obtained in terms of the partition function in SM or of the functional integral in QFT).

In Section 3, the discussion of the Goldstone theorem and the crucial role of locality, usually overlooked in textbook treatments, relies on [4], Chapter 15, especially Section 15.2. The general non-perturbative proof that in local gauges the Goldstone bosons cannot be physical was given in [29], [4], Theorem 19.1, again not even quoted; the evasion of the Goldstone theorem in the Coulomb gauge due to the lack of locality (rather than the lack covariance) is again clearly discussed in the 2005 edition of [4]. The discussion of Elitzur theorem and its consistency with the occurrence of symmetry breaking in several gauges (like e.g. the Coulomb gauge) was clarified in [12] and discussed at length in [5], Part C, Chapter II, 2.5, so that the discussion in Section 5 of Smeenk paper does not seem to add anything new.

# BIBLIOGRAPHY

[1] K. Brading and E. Castellani eds., *Symmetries in Physics: Philosophical Reflections*, Cambridge Univ Press 2003
[2] F. Strocchi, *An Introduction to the Mathematical Structure of Quantum Mechanics*, 2nd ed. World Scientific 2008
[3] F. Strocchi, The physical principles of quantum mechanics. A critical review, Eur. Phys. J. Plus, **127**: 12 (2012)
[4] F. Strocchi, *Symmetry Breaking*, Springer 2005, 2nd ed. 2008
[5] F. Strocchi, *Elements of quantum mechanics for infinite systems*, World Scientific 1985
[6] D.M. Greenberger, Am. Jour. Phys. **46**, 394 (2004)
[7] C. Liu, Philosophy of Science, **70**, 1219 (2003)
[8] S. Doplicher, R. Haag and J.E. Roberts, Comm. Math. Phys. **13**, 1 (1969; **15**, 173 (1969)
[9] P.M.A. Dirac, *The principles of quantum mechanics*, Oxford Univ. Press 1958
[10] K. Drühl, R. Haag and J.E. Roberts, Comm. Math. Phys. **18**, 204 (1970)
[11] R. Haag, *Local Quantum Physics*, Springer 1996
[12] G. Morchio and F. Strocchi, *Infrared problem, Higgs phenomenon and long range interactions*, in *Fundamental Problems of Gauge Field theory*, A.S. Wightman and G. Velo eds., Plenum 1986
[13] F. Strocchi, *An introduction to non-perturbative foundations of quantum field theory*, Oxford Univ. Press 2013
[14] J. Frohlich, G. Morchio and F. Strocchi, Nucl. Phys. **B 190** [FS3] 553 (1981)
[15] G. De Palma and F. Strocchi, Ann. Phys. **336**, 112 (2013)
[16] F. Strocchi, Gauss' law in local quantum field theory, in *Field Theory, Quantization and Statistical Physics*, D. Reidel Publ. 1981
[17] K. Brading and H.J. Brown, British Journal for the Philosophy of Science, **55**, 645 (2004)
[18] M.C. Morrison, Symmetries as Meta-Laws: Structural Metaphysics, in *Laws of Nature: Essays on the Philosophical, Scientific and Historical Dimensions*, Friedel Weinert (ed.) New York: de Gruyter pp.157-88.
[19] J. Earman, Laws, Symmetry and Symmetry Breaking: Invariance Principles and Objectivity, Address to the 2002 meeting og the Philosophy of Science Association, Section 6
[20] F. Strocchi and A.S. Wightman, J. Math. Phys. **15**, 2198 (1974)
[21] D.L. Karatas and K.L. Kowalski, Am. J. Phys. **58**, 123 (1990)
[22] H.A. Al-Kuwari and M.O. Taha, Am. J. Phys. **59**, 363 (1990)
[23] K. Brading and H.J. Brown, Noether's Theorems and Gauge Symmetries, arXiv:hep-th/0009058
[24] S. Friederich, European Journal for Phlosophy of Science, **3**, 157 (2013)
[25] R. Healey, Gauge Symmetry and the Theta-Vacuum, in *EPSA Volume 2:Philosophical Issue in the Sciences*, M. Suarez, M. Dorato and M. Redei eds., Springer 2010, p. 105
[26] G. Morchio and F. Strocchi, Ann. Phys. **324**, 2236 (2009)
[27] W. Struyve, Gauge invariant account of the Higgs mechanism, Studies in History and Philosophy of Modern Physics, **42**, 226 (2011)
[28] C. Smeenk, The elusive Higgs mechanism, Philosophy of Science, **73**, 487 (2006)
[29] F. Strocchi, Comm. Math. Phys. **56**, 57 (1977)
[30] G. Morchio and F. Strocchi, in *Fundamental problems of gauge field theory*, Erice Lectures 1985, Plenum 1986

# A Metaphysical Reflection on the Notion of Background in Modern Spacetime Physics

Antonio Vassallo

ABSTRACT. The paper presents a metaphysical characterization of spatiotemporal backgrounds from a realist perspective. The conceptual analysis is based on a heuristic sketch that encompasses the common formal traits of the major spacetime theories, such as Newtonian mechanics and general relativity. It is shown how this framework can be interpreted in a fully realist fashion, and what is the role of background structures in such a picture. In the end it is argued that, although backgrounds are a source of metaphysical discomfort, still they make a spacetime theory easy to interpret. It is also suggested that this conclusion partially explains why the notion of background independence carries a lot of conceptual difficulties.

**Keywords**: Background structure; spacetime theory; nomic necessity; dynamical sameness; principle of reciprocity; substantive general covariance; background independence.

## 1 Introduction

Tempus absolutum, verum, & mathematicum, in se & natura sua sine relatione ad externum quodvis, æquabiliter fluit [...] Spatium absolutum, natura sua sine relatione ad externum quodvis, semper manet similare & immobile [...]
([11], p. 6)

Newtonian absolute space and time are the epitomes of background structures. Newton's definitions quoted above beautifully express the idea of a background spatiotemporal structure as something whose characteristic properties are insensitive to anything else. Such an idea is indeed straightforward but it is also a source of conceptual discomfort. Starting from the Leibniz/Clarke debate on Newtonian mechanics (NM), and continuing with the aether problem in classical electrodynamics, it became clearer and clearer that the assumption of absolute structures led to differences in the physical description that were not inherent in the phenomena.
These conceptual problems justified a "war" on Newtonian backgrounds that ended victoriously with general relativity (GR), which is quite uncontroversially considered the first spacetime theory that dispenses with background spatiotemporal structures - i.e., it is *background independent*. However, despite the agreement over the fact that GR is a background independent theory, an uncontroversial definition of this

feature is still missing. Having in mind the extremely intuitive characterization of background spatiotemporal structures in NM, we might frown upon this difficulty. The definition of a background independent theory seems straightforward: it is just a theory where no (spatiotemporal) structure bears its properties independently of anything else. Actually, things have proven much more difficult than this, as - for example - the discussion in [9, 15] convincingly shows. The conceptual difficulties in spelling out what background independence exactly amounts to lead not only to interpretational problems for GR (think about the historical debate on the alleged "generalized" principle of relativity initially proposed by Einstein), but also makes it difficult to extend this framework to the quantum regime (see [16], for a technically accessible introduction to the issue of background independence in quantum gravity).

The aim of this short essay is to contribute a reflection on the problem of background independence by revising the metaphysical characterization of spatiotemporal backgrounds under the light of modern spacetime physics. We will start by providing a heuristic sketch that highlights the formal traits that are common, at least, to the major spacetime theories such as NM, special relativity (SR), and GR. We will then discuss a possible way to interpret this unified framework in a straightforward manner, based on some minimal metaphysical commitments that will be assumed as working hypotheses. Finally, we will exploit this conceptual machinery to describe how a background structure would influence the physics of possible worlds where background dependent theories hold. The hope is that, from a metaphysical analysis of possible worlds might come some hint to develop a better physical description of the actual one.

## 2 A Primer on Spacetime Theories

In order to simplify our metaphysical analysis, let us start by providing a simple formal sketch of a spacetime theory that is able to capture, albeit at a heuristic level, the theoretical traits that are common to the most important spacetime theories.[1] For simplicity's sake, we agree that a physical theory can be formalized as a set of relations between mathematical objects, and that each instantiation of such relations - once suitably interpreted - represents a possible state of affairs.

Our main concern, at this stage, is to propose a theoretically ductile picture of spacetime. The first step in this direction is to specify what the building blocks of spacetime are. Again, to keep things simple, we will just say that these primitive elements are called *events*. After a theory is interpreted, then such elements will take a definite physical meaning, such as that of "place-at-a-time", or "physical coincidence". Claiming that spacetime is a set of events $\mathcal{M}$ is for sure general, but rather uninformative, which means that we need to add structure to it. The second step is, then, to equip the set of events with a notion of "surroundings". This can be achieved by defining a new set $M := (\mathcal{M}, \tau)$, which is nothing but our starting set $\mathcal{M}$ together with a family $\tau$ of its subsets satisfying the following requirements:

- The empty set and $\mathcal{M}$ itself belong to $\tau$.

---
[1] The following sketch is based on [7].

- Any union of arbitrarily many elements of $\tau$ is an element of $\tau$.
- Any intersection of finitely many elements of $\tau$ is an element of $\tau$.

$\tau$ is called a *topology* on $\mathcal{M}$, and its elements are called *open sets* in $\mathcal{M}$. A subset $V$ of $M$ is a *neighborhood* for an element $x \in \mathcal{M}$ iff there exists an open set $A \in \tau$ such that $x \in A \subseteq V$. Moreover, we require the elements of $\mathcal{M}$ to be topologically distinguishable and separable, i.e. for any two elements $x$ and $y$ of $\mathcal{M}$, there exists a neighborhood $U$ of $x$ and a neighborhood $V$ of $y$ such that their intersection is the empty set. In this way, we end up with a topological space $M$ with a well-defined criterion for judging whether any two events are numerically distinct or not.

The structure so defined over $M$ is sufficient to introduce a notion of continuity of a function, and this lets us apply a further constraint on the characterization of spacetime, that is, the fact that, locally, it has to appear Euclidean. This constraint is implemented by requiring that for any open set $A$ in $M$ there exist a function $h : A \to \mathbb{R}^n$ that is bijective, continuous and whose inverse is continuous. A function satisfying these conditions is called a *homeomorphism*. Roughly speaking, this condition assures that, for any open set $A$ of $M$, all elements in $A$ can be labelled using a $n$-tuple of real numbers - which usually amounts to saying that $A$ admits a *coordinatization* $\{x^i\}_{i=0,\ldots,n-1}$. Furthermore, we want that, for each two coordinatizations on overlapping neighborhoods, the transition function from one coordinatization to the other - which is entirely defined and acting on $\mathbb{R}^n$ - is differentiable in the ordinary sense. If we have shaped our spacetime judiciously then, in general, to any coordinate transformation $\{x^i\} \to \{y^i\}$ defined in a neighborhood $A$ of $M$ corresponds a map $f : M \to M$ such that, for each point $P$ in $A$, $x^i(f(P)) = y^i(P)$. It can be proven that such a map, also called *intrinsic transformation*, preserves the structure defined so far on $M$. The set of all these structure-preserving transformations is nothing but the group $diff(M)$ of *diffeomorphisms*[2] acting on $M$. The reader not much fond of technicalities can just visualize $diff(M)$ as the group of permutations of elements of $M$ that represent smooth deformations of this space.

So far we have introduced some kind of "canvas" on which an even richer structure - consisting in a variety of geometrical objects - can be defined. The most simple example is that of a (continuous) curve, which is represented by a (continuous) map $\sigma : I \subseteq \mathbb{R} \to M$. In a given coordinate system $\{x^i\}$, the curve acquires the form $x^i = x^i(t), t \in I$. Another possibility is to define a *field-theoretic object* $\Theta$ as a map from $M$ to another space $X$: if $X$ is a space of rank 2 tensors, then $\Theta$ will be a tensor field on $M$ whose components $\Theta_{ij}$ in a coordinate system $\{x^i\}$ will be the elements of a $n \times n$ matrix. These geometrical objects can in general be transformed by the application of a diffeomorphism. For example, if we have a field $\Phi : M \to X$ and we want to apply to this field a transformation $f : M \to M$, this is done by defining such "application" as $f^*\Phi := \Phi \circ f$, which, for all $x \in M$, means that $(f^*\Phi)(x) = \Phi(f(x))$. In case of a map $\gamma : I \to M$, instead, we have $f^*\gamma := f \circ \gamma \Rightarrow (f^*\gamma)(y) = f(\gamma(y))$ for all $y \in \mathcal{Y}$. The fact that there is a (nearly)

---

[2] That is, those mappings from $M$ to itself which are bijective, continuous and differentiable together with their inverses.

one-to-one correspondence between coordinate transition functions and diffeomorphisms allows us to switch from the coordinate language to the intrinsic one without caring for any loss of information.

Among all the geometrical objects definable over $M$, there is a subgroup of them that endow $M$ with more structure than just its topology - indeed, they supply $M$ with a *geometry* properly said. The most important of these objects are the *metric tensor* and the *affine connection*. The former is a rank-2 tensor **g** that is symmetric (i.e. $g_{ij} = g_{ji}$ in all coordinate systems) and non-degenerate (i.e. the determinant $det|g_{ij}|$ of the matrix $|g_{ij}|$ is different from zero in all coordinate systems), and which makes it possible to define the notion of "length" of a curve on $M$. The latter is a derivative operator $\nabla$ (also called *covariant derivative*) that provides a precise meaning to the "change of direction" of a curve on $M$. Hence, for example, a curve that never changes direction is a *straight line* or *affine geodesic* on $M$. Since also **g** permits to define a straight line as the curve of shortest length between two points of $M$, we have also a notion of *metric* geodesic which, in general, does not have to coincide with the affine one. For this reason, the connection is required to be compatible with the metric tensor, i.e. it must always be the case that $\nabla \mathbf{g} = 0$. Once we have a well-defined notion of straight line, we can tell "how much" it corresponds to the usual straight line of Euclidean geometry; this evaluation is made possible by the *Riemann curvature tensor* **Riem[g]**. If the Riemann tensor is identically null all over the manifold, then the geodesics of $M$ are exactly those of Euclidean geometry, and we say that the spacetime is *flat*, otherwise *curved*.

Let us now make some concrete cases. The first example is perhaps the simplest one: the spacetime of special relativity (SR). This theory postulates a spacetime $M$ endowed with the Euclidean topology of $\mathbb{R}^4$, that is, there exists a homeomorphism mapping the entire manifold over $\mathbb{R}^4$. A metric tensor - the *Minkowski* metric $\eta$ - is defined over $M$. As expected, this object takes the form of a $4 \times 4$ matrix in any coordinate system. Moreover, it is always possible to find a coordinate system where $|\eta_{ij}| = diag(-1, 1, 1, 1)$. The Minkowski metric is compatible with a flat connection that basically overlaps with the usual derivative operator of differential calculus: this means that, in SR, the geodesics of $M$ are the usual straight lines of Euclidean geometry.

In NM things are more complicated. We still have that $M$ is globally homeomorphic to $\mathbb{R}^4$, but the geometric structure of the manifold is that of a bunch of Euclidean 3-spaces piled together by a temporal 1-flow - more compactly we write $M = E_3 \times \mathbb{R}$. In order to achieve this structure, we need to postulate a Euclidean 3-metric over each 3-space plus a temporal metric that labels the succession of these spaces. We then fix a flat connection compatible with this building and, finally, we single out a particular class of straight lines that describes the trajectories of bodies at absolute rest. This class of geodesics fixes a notion of "sameness of place through time", while the temporal metric evaluates time intervals in a coordinate-independent manner. In sum, this is the complicated machinery needed to depict an absolute space enduring over absolute time.

Finally, in the case of GR, there is no restriction either on the topology of $M$, or on the metric tensor **g**, or on the affine connection $\nabla$. The only conditions are that

**g** and $\nabla$ are compatible, and that $M$ is *Lorentzian*, which means that it is always possible to find a coordinate system $\{x^i\}$ on a neighborhood $A$ of a point $P \in M$ such that *exactly at that point* **g** reduces to the Minkowski metric.

In technical terms, all the spacetimes described above are instances of a $n$-dimensional (pseudo-)Riemannian manifold. In all cases we had $n = 4$, but in general nothing prevents us from elaborating a theory where the manifold has higher dimensionality. In the Kaluza-Klein approach, for example, a further spatial dimension is added to spacetime, which hence is 5-dimensional.

As we have seen from the above examples, the way we fix all the features of $M$, such as dimensionality, topology, geometry, or even further structures, varies from theory to theory. Some theories fix ab initio just few features, and let the others be dictated by the dynamics, while others presuppose from the outset rigid spatiotemporal structures that are not influenced by the dynamics. Obviously, these possible choices are relevant in determining whether a theory is background independent or not, as it will become clear later.

Now that we have given a formal account of spacetime, we are ready to define a spacetime theory in the following way:

**Definition 1. (Spacetime theory)** A *spacetime theory* $\mathcal{T}$ is a set of mathematical relations $\mathfrak{E}$ involving a set of geometrical objects $\mathcal{O}$ defined over a $n$-dimensional Riemannian manifold $M$:

$$\mathcal{T} = \mathcal{T}(M, \mathcal{O}; \mathfrak{E}).^3 \qquad (1)$$

The power of (1) lies in the fact that this formal unification makes it simpler to spell out the way a spacetime theory is usually interpreted. $M$ plus its additional geometrical structure is taken to be the spacetime properly called; a curve on $M$ describes the motion of a point-like particle (so it is called the *worldline* of that particle), and a generic material field occupying a spacetime region $A$ is represented by a map which assigns to each point in $A$ a tensor (or a vector, or even a scalar). Hence, spacetime is "decorated" by particles' worldlines, which are more or less straight depending on the near presence of material fields, such as the electromagnetic one. If a field is able to bend the worldline of a particle and the particle is able to modify the configuration of a field, then the two are said to be *interacting*. All the possible interactions between physical objects and the resulting motions allowed on $M$ are expressed in terms of relations encoded in $\mathfrak{E}$, which, in a given coordinate system, take the form of differential equations involving the components of the geometrical objects. Here, as a working hypothesis, we will stick to this simple reading, which presupposes a realistic attitude towards the geometric objects of the theory. This means that we will consider all the geometric objects in $\mathcal{O}$ as referring either to real (or at least possible) objects or to properties born by them. Hence, for example, a curve on $M$ will commit us to the (possible) existence of point-like particles mov-

---

[3] Just to be fair, it is not the case that a theory has to be formulated à la (1) in order to be considered a spacetime theory. There are, for example, cases of spacetime theories formulated in Lagragian terms, which cannot be cast in the form (1). However, we do not have to mind this for the present purposes.

ing along that worldline. Since, in general, the objects in $\mathcal{O}$ are field-theoretic in nature, we will be also committed to the existence of fields, which, as we have seen, are further divided into geometric (e.g. metric tensor field) and material (e.g. the electromagnetic field). This "doubly dualistic" metaphysical stance involving mixed particle/field and geometry/matter commitments is of course naive and perfectible. However, the disagreeing reader can just take it as a mere choice of vocabulary, and still follow the conceptual analysis of background structures we are going to perform.

A key motivation to adopt a naive realist attitude towards $\mathcal{O}$ is that, by doing so, we have a more or less clear measure of how much structure a spacetime theory postulates. By claiming this, we accept the line of argument developed in [12], where it is argued that modern physical theories represent objective physical structures in terms of geometric field-theoretic objects. Hence, roughly speaking, the larger $\mathcal{O}$, the more structure is postulated by $\mathcal{T}$.

So far we have agreed to adopt, as a working hypothesis, a naively realistic attitude towards the geometrical objects $\mathcal{O}$ in (1), but this claim by itself is confusing: to what specific theory are we declaring our commitments? The answer is to *all* the theories falling in the scope of definition 1, and this is our second working hypothesis. In order to better spell out this second assumption, we need to introduce another important definition:

**Definition 2.** (Model) A *model* of a spacetime theory $\mathcal{T}$ is a $(k+1)$-tuple $< M, \{O_k\}_{k \in \mathbb{N}} >$ - where $O_i \in \mathcal{O}$ for all $i \leq k$ - that is a solution of $\mathfrak{E}$.

If we think of the space $\mathcal{Q}_\mathcal{T}$ whose points represent each a configuration of *all* the geometrical objects of the theory - which is in fact called *configuration space* of the theory - then $\mathfrak{E}$ selects a subspace $\mathfrak{S}_\mathcal{T} \subset \mathcal{Q}_\mathcal{T}$ comprising all the physically allowed configurations of geometrical objects. This is at the root of the usual distinction between a purely kinematical state of affairs, that is, whatever element of $\mathcal{Q}_\mathcal{T}$, and a physical (or dynamical) state, which belongs to $\mathfrak{S}_\mathcal{T}$.

Definition 2 concerns "total" or "cosmological" models, which means that, in a model $< M, \{O_n\} >$, the geometrical objects are spread throughout the entire manifold $M$. However, it might be the case that a model admits a subclass of "partial" models involving a submanifold $\mathcal{K} \subset M$ and a set of geometrical objects defined on it.

The concept of model is the most important one for interpretational purposes because, from a metaphysical point of view, a model of a theory represents a physically allowed state of affairs. According to our realist attitude, then, a cosmological model of a given theory $\mathcal{T}$ will represent an entire universe where the specific laws of $\mathcal{T}$ hold. In other words, it represents a nomically possible world. By the same token, a submodel of the same theory will be interpreted as a possible local state of affairs in a nomically possible world. In order to make the philosophical analysis easier, we will consider all and only the models of spacetime theories satisfying (1) and we will assume that this set of models represents a cluster of nomically possible situations. Each theory, then, individuates a subset of possible worlds where the particular laws $\mathfrak{E}$ of that theory are at work. Note that this working hypothesis does not restrict us to adopt a particular metaphysical stance neither with respect to possible worlds (they can be mental constructions as well as existent objects), nor with respect to laws of nature ($\mathfrak{E}$ can be either grounded, say, in some genuinely modal feature of the entities inhabiting a possible world, or can be just a description of regularity patterns crafted in that possible world).

The last important definition we need to put forward before digging into metaphysical considerations regards the notion of general covariance:

**Definition 3. (General covariance - Formal version)** A spacetime theory $\mathcal{T}$ is *generally covariant* iff, for all $f \in diff(M)$ and for all $\mathfrak{M} \in \mathfrak{S}_\mathcal{T}$, it is the case that $f(\mathfrak{M}) \in \mathfrak{S}_\mathcal{T}$. $diff(M)$ is the *covariance group* of $\mathcal{T}$.

Here we talk of a "formal" version of general covariance - as opposed to a "substantive" one, which we will encounter later - for the following reason. Since $\mathfrak{E}$ lives on the manifold $M$, i.e., it represents the way the geometrical objects of the theory are related throughout the manifold, and since $diff(M)$ is the group of the structure preserving mappings defined over $M$, then it is trivial to see that, by applying a diffeomorphism to whatever model of the theory, we obtain another model of the theory. Moreover, given that formal general covariance is trivially satisfied by any theory falling in the scope of definition 1, and given that it is possible to formulate extremely different physical theories in the form (1) - just think about the physical abyss that lies between NM and GR -, then it is clear that the notion of general covariance defined above is purely formal and bears no physical import (historically, [10] was the first to acknowledge this fact).

A legitimate question might arise at this point. Given that radically different spacetime theories can be encompassed by the same formal framework, what is it exactly that makes them in fact radically different? To give a precise answer to this question, we need to say something more about the metaphysics of backgrounds.

## 3 A Metaphysical Appraisal of Backgrounds

The notion of background structure we are going to introduce draws from the work of Anderson [1, 2],[4] and is based on the distinction made among the elements of $\mathcal{O}$ between dynamical and non-dynamical objects. Such a distinction will become clearer in a moment. For the time being, let us just say that a background structure $\mathcal{B} \in \mathcal{O}$ is a geometrical object of the theory that is fixed ab initio and, hence, is "persistent" throughout the solution space of the theory.

To inform this notion with physics, consider the special relativistic description of the propagation of a massless scalar field:

$$\Box_\eta \phi = 0, \tag{2}$$

where $\Box_\eta$ is the d'Alembertian operator with components $\eta_{ij} \frac{\partial}{\partial x^i} \frac{\partial}{\partial x^j}$ in some coordinate system.[5] Let us further assume that (2) has two solutions $\phi_1$ and $\phi_2$. According to our metaphysical hypotheses, this means that the SR-cluster admits two possible worlds that are described by the models $<\eta, \phi_1>$ and $<\eta, \phi_2>$.

---
[4] Further refined by Friedman (see [7], in particular chapter II, sections 2 and 3).
[5] The Einstein convention is applied here.

It is obvious to claim that these two worlds share a single feature, namely, the Minkowski metric. The key point is that we can repeat this operation with any two special relativistic worlds, that is, if we inspect the entire space of models of SR, we see that *all* the models of the theory feature $\eta$. From our metaphysical perspective, this translates to the fact that, in all possible worlds belonging to the SR-cluster, there always exists a Minkowski spacetime. Generalizing, we can think of characterizing a background structure by means of its metaphysical necessity or, better, its nomic necessity: a background structure $\mathcal{B}$ of a given spacetime theory $\mathcal{T}$ is an object that such a theory deems necessary, i.e., there are no possible worlds described by $\mathcal{T}$ where $\mathcal{B}$ does not exist.

Along with this first metaphysical feature of background structures comes a clear reason to feel uncomfortable with background dependent theories. A theory that postulates a necessary physical structure is conceptually puzzling, not least because it tells us that there is just *one* physical possibility among many conceivable ones. By the same token, taking a structure as nomically necessary entails that it is physically impossible for it to change although we can conceive of a process in which the structure under scrutiny might in fact change. From an epistemic perspective, we can say that, when a theory accords a nomically necessary status to a spatiotemporal structure $\mathcal{B}$, then it is unable to provide a physically justified answer to the question "why is it $\mathcal{B}$ and not otherwise?". In the case of SR, the theory tells us that the only physically possible spacetime is the Minkowski one, and the only answer this theory can provide to the question "why is it not otherwise?" is "because it is how it is". Some may object that there is nothing really conceptually puzzling here, since it is totally reasonable to expect that the chain of physical justifications provided by a theory stops somewhere - i.e. there always comes a point in which a theory can just answer "because it is how it is". This is fair enough. However, this does not prevent us from putting two claims on the table. The first is: the fewer objects in $\mathcal{O}$ a theory deems nomically necessary, the better. This is because, then, such a theory is likely to exhibit a deeper explanatory structure than other spacetime theories that are more metaphysically "rigid". For example, GR is better than SR with this regard because it explains why and under what circumstances spacetime has a Minkowskian structure. Of course, this claim is not sacrosanct, in the sense that surely some counter-examples can be mounted against it. However, it still is fairly reasonable if applied to the major spacetime theories we have so far. The second claim we want to highlight is: it is not impossible that a theory falling in the scope of definition 1 does not commit us to the nomic necessity of any of the objects in $\mathcal{O}$. Clearly, this second claim does not entail that such a theory admits a bottomless structure of physical justification - although many philosophers would not find anything wrong with that -, but just that the theory fixes ab initio some features other than (full) spatiotemporal structures.

A second important metaphysical feature of spatiotemporal backgrounds comes from the following example. Let us focus on the Newtonian cluster of possible worlds and consider a Newtonian world where there exist a large ship docked on a calm sea. Inside the ship, shut up in the main cabin below decks, there is a man - we can call him Salviati - together with an experimental equipment consisting of jars of flies, fishes in bowls and dripping bottles.[6] Simply speaking, we are dealing with a global model $\mathfrak{M}$, which describes the possible world in its entirety, but we are magnifying just a portion of it, that is, a submodel $\mathfrak{m}$ describing just what happens in the immediate surroundings of the ship. Let us now apply to $\mathfrak{m}$ a transformation $f$ that consists in a rigid spatial translation of the ship. The model $f^*\mathfrak{m}$ will then depict a situation in which the ship is still on a calm sea without wind, but now it is located, say, one meter away from the position it had in $\mathfrak{m}$. In what dynamical aspects does $\mathfrak{m}$ and $f^*\mathfrak{m}$ differ? None: in both cases the ship is at absolute rest and Salviati is unable to spot any difference by looking at the equipment on board. This reasoning can be repeated with rotations. Take $f$ as a 45° rotation of the ship with respect to the original orientation, and again both $\mathfrak{m}$ and $f^*\mathfrak{m}$ will depict a ship at absolute rest, where Salviati's equipment behaves exactly in the same manner as the non-rotated one. We then suspect that the notion of sameness for Newtonian states of affairs is influenced by the underlying background structures. In this case, since Euclidean space is homogeneous and isotropic, the state of absolute rest of the ship is insensitive to where the ship is placed or how it is oriented.

As an acid test, consider another situation where $f^*\mathfrak{m}$ makes Salviati's ship sailing over troubled waters. In this case, it is quite obvious that $\mathfrak{m}$ and $f^*\mathfrak{m}$ depict radically different dynamical situations. The ship in $f^*\mathfrak{m}$ is not in a state of absolute rest (its worldline is not a geodesic at all, let alone a straight line pointing in the privileged "rest direction"), and this has quite disruptive observable consequences: while in $\mathfrak{m}$ Salviati sits down quietly observing his jars of flies, fishes in bowls and dripping bottles, in $f^*\mathfrak{m}$ he[7] is shaking in the main cabin among broken glasses, buzzing flies and asphyxiating fishes.

To sum up, we have individuated another very important metaphysical aspect of backgrounds, namely, that they fix a notion of sameness of dynamical state throughout the cluster of possible worlds of the theories they figure into. From a formal perspective, this means that, if a spacetime theory admits a set of background structures $\{\mathcal{B}_i\}$, then for any two models of the theory related by some transformation $f$, these two models are said to be dynamically indiscernible iff $f^*\mathcal{B}_i = \mathcal{B}_i$, for all $i$, that is, iff $f$ is a transformation (called *isometry*) that leaves all the background structures invariant. We call this set of isometries $iso(\{\mathcal{B}_i\}) \subset diff(M)$ the *symmetry group* of the theory.

This definition of symmetry qualifies as "ontic" in the taxonomy put forward in [5]. The author charges this kind of definition with inferential circularity. In his own words:

> But according to an ontic definition of 'symmetry', in order to check whether a given transformation [$f$] counts as a symmetry of [dynamical] laws, I first need to know which physical features fix the data so that I can check whether [$f$] preserves them. And the problem is that, in many cases, we discover which physical features fix the data by engaging in symmetry-to-reality reasoning!

---

[6] Here, of course, we are referring to the "Gran Naviglio" thought experiment in [8].

[7] Or, if you want, his counterpart, depending on the particular account of possible worlds adopted.

(*Ibid*, p. 28)

Although the above issue is a serious one, worth of extensive philosophical discussion, here we just dodge the charge of inferential circularity by appealing to our naive realist framework. Simply speaking, we do not discover which physical features "fix the data" (in our case, the background structures): we just postulate them ab initio.

At this point, we can go back to the question raised at the end of the previous section, that is, what is it that renders different spacetime theories in fact different? The answer is now crystal clear: the background structures in $\mathcal{O}$. It is in fact thanks to the backgrounds postulated by a theory that we can attribute physical import to a subset of the covariance group $diff(M)$. We have then different theories depending on the subset individuated by the backgrounds. For example, we can say that NM is physically different from SR because the former admits a set of symmetries which form a group called *Galilean*, while the symmetries of the latter belong to the *Poincaré* group.

However, as in the previous case, also this feature of backgrounds may lead to unhappy consequences. To see this let us consider again the docked ship on a calm sea in m, and transform this model in one where the ship is still on a calm sea, but now it is sailing with uniform velocity. Technically, the transformation $f$ involved in this case belongs to the so-called Galilean group. Intuitively speaking, while in m the ship is in a trajectory of absolute rest (straight line pointing in the privileged direction), $f$ just "inclines" the trajectory of an arbitrary angle without "bending" it. We are now in a strange situation: from the global perspective of $\mathfrak{M}$, m and $f^*$m depict different dynamical states - absolute rest vs. motion with uniform absolute velocity, but from Salviati's perspective, there is no empirically observable difference between the two dynamical states! Here, as in the case of nomic necessity, a liberal metaphysician might claim that we should not worry too much and just accept the fact that our theory commits us to the existence of dynamically different yet empirically indistinguishable states of affairs. After all, this is just a metaphysical fact that does not impair in any way the role of physicists. In fact, it is obvious that whatever empirical question regarding the dynamics that Salviati could ask would *always* have an answer, which would be the same irrespective of the fact that the ship is in a state of absolute rest or absolute uniform motion. Again, we concede the point that the existence of dynamically distinct yet empirically indistinguishable states of affairs is not a mortal sin for a theory. But accepting this means accepting that there can be elements of reality that are *totally* opaque to physics! This is a rather embarrassing claim to embrace, especially if we believe that metaphysics must be motivated and informed by science (and physics in particular). At least, it is reasonable to invoke some sort of Occamist norm according to which, among two competing theories with the same empirical consequences, we should prefer the one that commits us to the least structure. Let us try to apply such a norm to NM.

The evidence that the culprit for the above discussed unwanted situation is absolute space is given by the fact that the Galilean group is part of the isometries of all Newtonian background objects except for the class of straight lines that fixes the notion of "sameness of place through time". Fortunately, we can reformulate NM without privileging any set of geodesics and, hence, giving up the commitment to absolute space.[8] In this new framework this particular problem evaporates since now the dynamics of the theory does not distinguish anymore states of rest from states of uniform velocity.

In sum, here lies the second charge against background structures: the more background structures a theory admits, the more it is likely that the theory will consider as dynamically distinct some models that, in fact, admit the very same physical observables.

The last metaphysical feature of a background structure is related to the distinction between dynamical and non-dynamical objects mentioned at the beginning of the section. In short, spatiotemporal backgrounds are non-dynamical objects because they do not enter $\mathfrak{E}$ as elements subjected to the dynamical laws but, rather, they represent the support that renders possible the very formulation of such laws. The problem with the non-dynamicity of background structures is summarized in the following quote:

> [A]n absolute element in a theory indicates a lack of reciprocity; it can influence the physical behavior of the system but cannot, in turn, be influenced by this behavior. This lack of reciprocity seems to be fundamentally unreasonable and unsatisfactory. We may express the converse in what might be called a general principle of reciprocity: Each element of a physical theory is influenced by every other element. In accordance with this principle, a satisfactory theory should have no absolute elements.
> 
> ([1], p. 192)

Anderson effectively summarizes the third peculiarity of backgrounds and the reason why we should feel uneasy about that. However, few comments are in place. First of all, the way Anderson enunciates the principle of reciprocity is too strong and seems to amount to some holistic principle which, most likely, was not the author's intention. Perhaps it would have been better to say that each element of a physical theory *can be* influenced by *some* other element. Secondly, the principle as it stands can be easily challenged on the ground of its vagueness as to how an "element of a physical theory" has to be understood. To see why it is so, we could just consider the Lagrangian formulation of NM. In this framework, the behavior of a mechanical system is fully described by the Lagrange equations: once we fix an appropriate Lagrangian plus initial conditions, we get the full dynamical history of the system in the form of a trajectory in configuration space. In a sense, then, the Lagrangian function is an element of the theory that influences the mechanical system but that is not influenced back, being it a supporting element of the dynamical description. Does it imply that the Lagrangian violates the principle of reciprocity? Here, we are exploiting the vagueness underlying the notion of "element of a physical theory". The Lagrangian is with no doubt an element of the theory, but it would be awkward to interpret it as ontologically on a par with the mechanical system: it is just a descriptive tool that carries dynamical information and, as such, has not to

---
[8]As shown, for example, in [7], chapter III, section 2.

be taken as referring to a concrete object that exists over and above the mechanical system. Evidently, a too broad characterization of an element of the theory led us to a category mistake.

Fortunately, the theoretical framework given by definitions 1 and 2 helps us clarifying the real intentions behind Anderson's quote above. If, in fact, we restrict the scope of the principle of reciprocity to the geometrical objects definable over $M$, we can restate the principle as follows: each element of the set $\mathcal{O}$ must be subjected to the dynamical evolution encoded by $\mathfrak{E}$. This renders the principle of reciprocity less vague and highlights in what sense Anderson characterizes background structures as elements of the theory that violate such a principle. However, we still have the possibility to scupper this characterization. To do so, it is sufficient to reconsider the example of the theory with equation (2). As we have seen, this theory features a background structure, namely the Minkowski metric $\eta$. Now, let us add to (2) a further equation:

$$\mathbf{Riem}[g] = 0. \qquad (3)$$

What have we done here? Leaving aside technical considerations, we have done nothing but "embedding" the fixing condition of the Minkowski metric into $\mathfrak{E}$. Hence, the solution space of this new theory carries absolutely no more physical information than the one associated to (2) alone, and the Minkowski metric is still a background structure satisfying the first two features we have reported. However, now, we have a theory that challenges the utility of the principle of reciprocity as a guide in assessing spacetime theories. In the theory (2)/(3) each element of the set $\mathcal{O}$ is subjected to the dynamical evolution encoded by $\mathfrak{E}$, but still the theory admits a background. This example shows that even the amended version of the principle of reciprocity we have considered is conceptually flawed. Nonetheless, it seems still evident that Anderson's quotation captures a salient feature of backgrounds. Perhaps, we should read this quote in a more straightforward way, and interpret the talk in term of influences as referring to a very concrete notion of physical interaction. In some sense, here we are shifting the problem to what exactly "interacting" amounts to in the modern physical jargon. However, just for the sake of argument, let us assume that an interaction between two elements $\Theta_1$ and $\Theta_2$ of a theory amounts to adding to $\mathfrak{E}$ a coupling relation of the form $F(\Theta_1, \Theta_2, \kappa)$, $\kappa$ being and appropriate coupling parameter. If we reconsider the principle of reciprocity under this light, than it becomes the statement that each field-theoretic object is coupled with some other. The challenge of the theory (2)/(3) is now defused because the background role of the Minkowski metric is restored due to the fact that it does not satisfy this latter version of the principle of reciprocity. Therefore, in the end, we can say that the third metaphysical feature of spatiotemporal background is the one already highlighted by Newton's quotation at the beginning of the paper, namely that they bear their properties without relation to anything else: this feature can be reasonably translated in the language of modern spacetime physics as the fact that they are structures that are not coupled to any material field.

Is this a bad thing, metaphysically speaking? Let us answer with the words of Brown and Lehmkuhl:

If there is a questionable aspect of [the principle of reciprocity], it is less the claim that substances act (how otherwise could their existence be known to us?) than the notion that they are necessarily acted back upon, that action must be reciprocal. If all substances act, they do so in relation to other substances; these other substances therefore cannot be immune from external influences. Now it might seem arbitrary on *a priori* grounds to imagine that the 'sensitivity' of such substances is not universal. That is to say, it might seem arbitrary to suppose that not all substances react to others. But no such abstract qualms can be entirely compelling; Nature must have the last say.
([4], pp. 3, 4)

Otherwise said, pursuing the principle of reciprocity is reasonable but not necessary. To further reflect on this point, let us focus on NM and ask in what sense the absolute backgrounds of this theory influence the motion of bodies. For example, what is it that "forces" an isolated point-particle to move in a straight line? The answer is obviously "nothing", let alone absolute structures: it is just a primitive fact - i.e. not further justifiable via a "why" question - that in every Newtonian world there exists a privileged class of trajectories occupied by bodies in inertial motion. In this sense, absolute structures *define* possible motions but do not push (in an ordinary physical sense) bodies to move that way. Under this light, it does not seems that conceptually hard to withstand a violation of the principle of reciprocity.

## 4 Conclusion: How Easily Can We Dispense with Backgrounds?

In the previous section we have supplied a metaphysical characterization of spatiotemporal backgrounds based on the language of modern spacetime physics. To recap, we have highlighted three features of background structures in a spacetime theory:

1. The theory in which they feature treats them as (nomically) necessary structures.

2. They induce a notion of dynamical sameness among states of affairs throughout the solution space of the theory.

3. Their dynamical influences are not describable as physical interactions.

As we have discussed, with each of this metaphysical traits comes an associated conceptual discomfort. However, we have also highlighted that none of these issues lead to contradictions or physical loopholes. Hence, we are inclined to claim that whether one wants to renounce background structures depends on one's own metaphysical tastes. Otherwise said, one can backup one's commitment to background independence with strong and convincing arguments (and, indeed, many of such arguments can be found in the literature), but she cannot appeal to a requirement of background independence as a physically necessary one.

However, the realist framework we have put forward has made clear that background structures have not only (mild) metaphysical vices, but also metaphysical virtues. The most important among them is the possibility to straightforwardly define the notion of physical symmetry in an ontic manner, without incurring inferential circularity. More generally, once we specify what are the background structures $\{\mathcal{B}_i\}$ of a theory, the interpretation of such a theory becomes a rather smooth business: this is because, once the symmetries of a theory are given, we can identify as referring to real objects or properties those theoretical structures that are invariant under these symmetries. Once again, we stress that this is possible because we assume background structures as postulated ab initio as a matter of ontological fact. In general, in fact, there is no formal criterion that makes an object in $\mathcal{O}$ a background structure, and it can be the case that the very same geometric object can count or not count as background depending on the particular interpretation of the theory chosen ([3], section 3.3, discusses in detail the case of such geometrically ambiguous theories).

So far we have engaged in a conceptual cost-benefit analysis of postulating background structures in our theory. Suppose, now, that we are inclined to buy into the view that a background has more costs than benefits and, hence, we wish to go for background independence. According to our framework, implementing such a requirement amounts - at least - to constructing a theory whose spatiotemporal structures do not satisfy the three conditions listed at the beginning of the section. Here, obviously, we cannot undertake this task, so we will be just content to verify whether GR, which is usually considered the epitome of background independent theory, in fact violates the three metaphysical requirements for background structures.

The dynamical equations of GR have the form $\mathbf{G}[\mathbf{g}] = \kappa \mathbf{T}[\phi, \mathbf{g}]$, where the left-hand side of the relation represents the geometry of spacetime (the so-called Einstein tensor), and the right hand side features the stress-energy tensor, which encodes information regarding the mass-energy distribution over a region of spacetime. We can then say that spacetime in GR is not a background in primis because the theory is about the coupling of the metric field $\mathbf{g}$ with the matter field(s) $\phi$ and, hence, the third requirement above is not met. From the form of the dynamical equations, in the second place, we infer that it is not the case that all the models of the theory feature the same geometric objects and, hence, in the GR-cluster of possible worlds there is no field-theoretic structure that counts as nomically necessary.[9] It seems, then, that also the second condition is not fulfilled. As a matter of fact, as we have hinted at in section 3, there are other features of the models of the theory that bear a physical significance and that show the "persistence" typical of backgrounds. For example, all models of GR feature manifolds of dimensionality 4 and Lorentzian in nature. Hence, although GR does not treat any spatiotemporal structure as nomically necessary, there are some characteristic traits of these structures that are nonetheless preserved throughout the solution space of the theory. Hence, strictly speaking, in GR the spatiotemporal structures do bear at least some properties without relation to anything external.

---

[9]As a matter of fact, some examples might be provided, which challenge this claim (see, e.g., [13]). However, since these examples are not disruptive to our analysis, we can set them aside.

To get rid once and for all of this kind of objections, we can somehow render our distinction between background dependent and independent theories more flexible. Up to now, in fact, we have assumed that, in order to consider a theory background dependent, it is sufficient that it admits at least a background structure. However, this sort of classification might be too coarse or might deliver an unintuitive picture. Consider for example a theory whose equations have two classes of models: one featuring, say, a flat metric, and another featuring a curved one. Clearly, these two metrics would not qualify as backgrounds according to the above characterization, since they are not nomically necessary objects according to the theory. Still, we would feel unconfortable with this conclusion, since such a theory would still be "ontologically rigid". Perhaps, we can establish a well-defined way to count (i) how many physical features in general - not only geometric objects in $\mathcal{O}$ - are deemed nomically necessary by the theory and (ii) how often non-nomically necessary features appear throughout the solution space of the theory. This would imply that the distinction between background dependence and independence would not be so clear-cut, there being different degrees in which they come. If this strategy can be consistently worked out ([3] makes a concrete proposal along these lines), then we would have a measure according to which, say, NM is fully background dependent, while GR is fully background independent modulo minor fixed features.

Finally, let us consider the second requirement and ask, if GR has no background structures, does it still possess a well-behaved notion of dynamical sameness? We face a dilemma here: if we answer no, this would imply that GR is a useless theory incapable of making even the simplest empirical predictions, which is most obviously not the case; if we answer yes, then we have to face a huge controversy. To see why it is so, let us back up our affirmative answer with the following argument:

(P1) The physical symmetries of a spacetime theory are those transformations $f \in diff(M)$ that are isometries for the background structures $\{\mathcal{B}_i\}$;

(P2) GR has no background structures, i.e. $\{\mathcal{B}_i\} = \emptyset$;

Therefore,

(C) In GR, *all* transformations $f \in diff(M)$ are physical symmetries of the theory.

The conclusion of this argument is usually stated as the fact that GR satisfies the requirement of *substantive* general covariance, as opposed to the mere formal version given by definition 3. Note that a similar argument can be mounted, in which (P2) and (C) are switched. In this way, background independence and substantive general covariance would become overlapping concepts. The problem with this line of argument is that it forces us to buy into the view that, trivially, the transformations in $diff(M)$ are all at once isometries of *no* background structure (whatever diffeomorphism applied to nothing does not change anything). But that seems too loose an appeal because the distinction between the whole $diff(M)$ and $iso(\{\mathcal{B}_i\})$ *requires* background structures: if such structures are absent, then we have no means for making the distinction. By the same token, starting from the premise that all diffeomorphisms are physical symmetries of the theory does not provide a firm enough ground to infer that the theory is background independent, since we can always disguise background structures as dynamical objects.

Hence, it seems clear that, in order to define substantive general covariance in a more rigorous way, it is necessary to base the argument for having $diff(M)$ as the set of physical symmetries on an approach different from the one considered in this paper. Earman [6], for example, analyzes substantive general covariance in terms of variational symmetries in the Lagrangian formalism, but this approach does not help with spacetime theories that cannot be rendered in Lagrangian terms.[10] Stachel [17], instead, argues that the problem arises from a wrong way of looking at the structure of spacetime theories. Very simply speaking, Stachel claims that the physically relevant information regarding a spacetime theory is not in general encoded in the manifold $M$, but in a more complex structure, namely, a triple of topological spaces - technically called fiber bundle - $(\mathfrak{X}, M, \mathcal{F})$, with $\mathfrak{X}$ having locally the form $M \times \mathcal{F}$. In this context, the dynamical equations $\mathfrak{E}$ become a set of rules for selecting cross-sections of this fiber bundle.[11] Now, the requirement of substantive general covariance amounts to the fact that all the (geometrical objects referring to) spatiotemporal structures of the theory live on these cross-sections. If some structure still lives on the manifold $M$, then the theory is background dependent. Stachel's approach might prove more effective than that represented by (1) in highlighting the formal differences between spacetime theories - especially with respect to considerations regarding background dependence/independence. However, it does not seem to bring much ontological clarity to the matter. While, in fact, the framework we put forward admits a straightforward interpretation, it is not at all clear how to spell out the way the structure $(\mathfrak{X}, M, \mathcal{F})$ refers to real (or possible) physical structures.

In conclusion, the most important moral we can draw from the analysis developed in this paper is that background structures, albeit showing some metaphysical vices, are nonetheless elements that render the formulation and the interpretation of a spacetime theory sharp and fairly simple. This is why pursuing the requirement of background independence demands a huge conceptual price to be paid.

### Acknowledgements:

I wish to thank an anonymous referee and Davide Romano for helpful comments on an earlier version of this paper. Research contributing to this paper was funded by the Swiss National Science Foundation, Grant no. 105212_149650.

# BIBLIOGRAPHY

[1] Anderson, J. (1964). "Relativity principles and the role of coordinates in physics", in H. Chiu and W. Hoffmann (eds.), *Gravitation and Relativity*, pp. 175–194, New York: W.A. Benjamin, Inc.

[2] Anderson, J. (1967). *Principles of relativity physics*, New York: Academic Press.

---

[10] See [14] for a detailed criticism of Earman's proposal.

[11] Intuitively, if the fiber bundle is a simple vector bundle, then a cross-section of it would be a vector field over $M$.

[3] Belot, G. (2011). "Background-independence", *General Relativity and Gravitation*, vol. 43, pp. 2865–2884. http://arxiv.org/abs/1106.0920.
[4] Brown, H. and D. Lehmkuhl (2015). "Einstein, the reality of space, and the action-reaction principle", forthcoming in Ghose, P. (ed.), *Einstein, Tagore, and the nature of reality*, London: Routledge. http://arxiv.org/abs/1306.4902.
[5] Dasgupta, S. (2015). "Symmetry as an epistemic notion (twice over)", in *British Journal for the Philosophy of Science*, DOI 10.1093/bjps/axu049.
[6] Earman, J. (2006). "Two challenges to the requirement of substantive general covariance", in *Synthese*, vol. 148(2), pp. 443–468.
[7] Friedman, M. (1983). *Foundations of Space-Time Theories. Relativistic Physics and Philosophy of Science*, Princeton: Princeton University Press.
[8] Galilei, G. (1632). *Dialogo sopra i due massimi sistemi del mondo tolemaico e copernicano*, Firenze: Giovanni Battista Landini.
[9] Giulini, D. (2007). "Remarks on the notions of general covariance and background independence", in *Lecture notes in physics*, vol. 721, pp. 105–120. http://arxiv.org/abs/gr-qc/0603087.
[10] Kretschmann, E. (1917). "Über den physikalischen sinn der relativitätspostulate, A. Einsteins neue und seine ursprüngliche Relativitätstheorie", in *Annalen der Physik*, vol. 53, pp 575–614. Italian translation by S. Antoci available at http://fisica.unipv.it/antoci/re/Kretschmann17.pdf.
[11] Newton, I. (1726). *Philosophiae Naturalis Principia Mathematica*, London: The Royal Society of London.
[12] North, J. (2009). "The "structure" of physics: A case study", in *Journal of Philosophy*, vol. 106, pp. 57–88. http://philsci-archive.pitt.edu/4961/.
[13] Pitts, J. (2006). "Absolute objects and counterexamples: Jones-Geroch dust, Torretti constant curvature, tetrad-spinor, and scalar density", in *Studies in History and Philosophy of Modern Physics*, vol. 37(2), pp. 347–351. http://arxiv.org/abs/gr-qc/0506102v4.
[14] Pooley, O. (2010). "Substantive general covariance: Another decade of dispute", in M. Suàrez, M. Dorato, and M. Rèdei (eds.), *EPSA Philosophical Issues in the Sciences: Launch of the European Philosophy of cience Association*, vol. 2, pp. 197–209, Dordrecht: Springer. http://philsci-archive.pitt.edu/9056/1/subgencov.pdf.
[15] Rickles, D. (2008). "Who's afraid of background independence?", in D. Dieks (ed.), *The ontology of spacetime II*, pp. 133–152, Amsterdam: Elsevier B.V. http://philsci-archive.pitt.edu/4223/.
[16] Rozali, M. (2009). "Comments on background independence and gauge redundancies", in *Advanced science letters*, vol. 2(3), pp. 244–250. http://arxiv.org/abs/0809.3962v2.
[17] Stachel, J. (1986). "What a physicist can learn from the discovery of general relativity", in R. Ruffini (ed.), *Proceedings of the Fourth Marcel Grossmann meeting on general relativity*, pp. 1857–1862, Amsterdam: Elsevier B.V.

www.ingramcontent.com/pod-product-compliance
Lightning Source LLC
Chambersburg PA
CBHW051034160426
43193CB00010B/938